Lecture Notes in Computer S<

Edited by G. Goos, J. Hartmanis, and J. v

Springer

Berlin
Heidelberg
New York
Barcelona
Hong Kong
London
Milan
Paris
Tokyo

Bart Preneel (Ed.)

Topics in Cryptology CT-RSA 2002

The Cryptographers Track at the RSA Conference 2002
San Jose, CA, USA, February 18-22, 2002
Proceedings

 Springer

Series Editors

Gerhard Goos, Karlsruhe University, Germany
Juris Hartmanis, Cornell University, NY, USA
Jan van Leeuwen, Utrecht University, The Netherlands

Volume Editor

Bart Preneel
Katholieke Universiteit Leuven, Department of Electrical Engineering
Kasteelpark Arenberg 10, 3001 Leuven-Heverlee, Belgium
E-mail: bart.preneel@esat.kuleuven.ac.be

Cataloging-in-Publication Data applied for

Die Deutsche Bibliothek - CIP-Einheitsaufnahme

Topics in cryptology : the Cryptographers Track at the RSA Conference 2002 ;
proceedings / CT-RSA 2002, San JosØ CA, USA, February 18 - 22, 2002.
Bart Preneel (ed.). - Berlin ; Heidelberg ; New York ; Barcelona ; Hong Kong ;
London ; Milan ; Paris ; Tokyo : Springer, 2002
  (Lecture notes in computer science ; Vol. 2271)
  ISBN 3-540-43224-8

CR Subject Classification (1998): E.3, G.2.1, D.4.6, K.6.5, F.2.1-2, C.2, J.1

ISSN 0302-9743
ISBN 3-540-43224-8 Springer-Verlag Berlin Heidelberg New York

This work is subject to copyright. All rights are reserved, whether the whole or part of the material is
concerned, specifically the rights of translation, reprinting, re-use of illustrations, recitation, broadcasting,
reproduction on microfilms or in any other way, and storage in data banks. Duplication of this publication
or parts thereof is permitted only under the provisions of the German Copyright Law of September 9, 1965,
in its current version, and permission for use must always be obtained from Springer-Verlag. Violations are
liable for prosecution under the German Copyright Law.

Springer-Verlag Berlin Heidelberg New York
a member of BertelsmannSpringer Science+Business Media GmbH

http://www.springer.de

' Springer-Verlag Berlin Heidelberg 2002
Printed in Germany

Typesetting: Camera-ready by author, data conversion by Olgun Computergrafik
Printed on acid-free paper   SPIN: 10846157   06/3142   5 4 3 2 1 0

Preface

This volume continues the tradition established in 2001 of publishing the contributions presented at the Cryptographers' Track (CT-RSA) of the yearly RSA Security Conference in Springer-Verlag's Lecture Notes in Computer Science series.

With 14 parallel tracks and many thousands of participants, the RSA Security Conference is the largest e-security and cryptography conference. In this setting, the Cryptographers' Track presents the latest scientific developments.

The program committee considered 49 papers and selected 20 for presentation. One paper was withdrawn by the authors. The program also included two invited talks by Ron Rivest ("Micropayments Revisited" – joint work with Silvio Micali) and by Victor Shoup ("The Bumpy Road from Cryptographic Theory to Practice").

Each paper was reviewed by at least three program committee members; papers written by program committee members received six reviews. The authors of accepted papers made a substantial effort to take into account the comments in the version submitted to these proceedings. In a limited number of cases, these revisions were checked by members of the program committee.

I would like to thank the 20 members of the program committee who helped to maintain the rigorous scientific standards to which the Cryptographers' Track aims to adhere. They wrote thoughtful reviews and contributed to long discussions; more than 400 Kbyte of comments were accumulated. Many of them attended the program committee meeting, while they could have been enjoying the sunny beaches of Santa Barbara.

I gratefully acknowledge the help of a large number of colleagues who reviewed submissions in their area of expertise: Masayuki Abe, N. Asokan, Tonnes Brekne, Emmanuel Bresson, Eric Brier, Jan Camenisch, Christian Collberg, Don Coppersmith, Jean-Sébastien Coron, Serge Fehr, Marc Fischlin, Matthias Fitzi, Pierre-Alain Fouque, Anwar Hasan, Clemens Holenstein, Kamal Jain, Marc Joye, Darko Kirovski, Lars Knudsen, Neal Koblitz, Anna Lysyanskaya, Lennart Meier, David M'Raihi, Phong Nguyen, Pascal Paillier, Adrian Perrig, David Pointcheval, Tal Rabin, Tomas Sander, Berk Sunar, Michael Szydlo, Christophe Tymen, Frederik Vercauteren, Colin Walter, Andre Weimerskirch, and Susanne Wetzel. I apologize for any inadvertent omissions.

Electronic submissions were made possible by a collection of PHP scripts originally written by Chanathip Namprempre and some perl scripts written by Sam Rebelsky and SIGACT's Electronic Publishing Board. For the review procedure, web-based software was used which I designed for Eurocrypt 2000; the code was written by Wim Moreau and Joris Claessens.

I would like to thank Wim Moreau for helping with the electronic processing of the submissions and final versions, Ari Juels and Burt Kaliski for interfacing with the RSA Security Conference, and Alfred Hofmann and his colleagues at Springer-Verlag for the timely production of this volume.

Finally, I wish to thank all the authors who submitted papers and the authors of accepted papers for the smooth cooperation which enabled us to process these proceedings as a single LaTeX document.

We hope that in the coming years the Cryptographers' Track will continue to be a forum for dialogue between researchers and practitioners in information security.

November 2001   Bart Preneel

RSA Cryptographers' Track 2002

February 18–22, 2002, San Jose, California

The RSA Conference 2002 was organized by RSA Security Inc. and its partner organizations around the world. The Cryptographers' Track was organized by RSA Laboratories (http://www.rsasecurity.com).

Program Chair

Bart Preneel, Katholieke Universiteit Leuven, Belgium

Program Committee

Dan Boneh .... Stanford University, USA
Yvo Desmedt .... Florida State University, USA
Dieter Gollmann .... Microsoft Research, USA
Stuart Haber .... Intertrust, USA
Shai Halevi .... IBM Research, USA
Helena Handschuh .... Gemplus, France
Martin Hirt .... ETH, Switzerland
Markus Jakobssen .... RSA Laboratories, USA
Ari Juels .... RSA Laboratories, USA
Pil Jong Lee .... Postech, Korea
Alfred Menezes .... University of Waterloo, Canada
Kaisa Nyberg .... Nokia, Finland
Tatsuaki Okamoto .... NTT Labs, Japan
Christof Paar .... Worcester Polytechnic Institute, USA
Jean-Jacques Quisquater .... Univ. Cath. de Louvain, Belgium
Jacques Stern .... Ecole Normale Supérieure, France
Michael Wiener .... Canada
Yacov Yacobi .... Microsoft Research, USA
Moti Yung .... Certco, USA
Yuliang Zheng .... Monash University, Australia

Table of Contents

Digital Signatures

Public Key Encryption

Discrete Logarithm

On Hash Function Firewalls
in Signature Schemes

Burton S. Kaliski Jr.

RSA Laboratories
20 Crosby Drive, Bedford, MA 01730, USA
bkaliski@rsasecurity.com

Abstract. The security of many signature schemes depends on the verifier's assurance that the same hash function is applied during signature verification as during signature generation. Several schemes provide this assurance by appending a hash function identifier to the hash value. We show that such "hash function firewalls" do not necessarily prevent an opponent from forging signatures with a weak hash function and we give "weak hash function" attacks on several signature schemes that employ such firewalls. We also describe a new signature forgery attack on PKCS #1 v1.5 signatures, possible even with a strong hash function, based on choosing a new (and suspicious-looking) hash function identifier as part of the attack.

1 Introduction

Nearly all digital signature schemes in practice combine a hash function with other cryptographic operations. In a typical scheme, a signer applies the hash function to the message, then applies a signature primitive to the signer's private key and the hash function output to obtain a signature. The verifier likewise computes a hash value, then applies a verification primitive to the signer's public key, the hash value, and the signature to determine whether the signature is valid. In many schemes, some formatting is applied to the hash value prior to the signature primitive, and in some schemes part or all of the message is included in the input to the signature primitive and can be recovered from the signature.

The security of a signature scheme clearly depends on the security of the hash function. Just as importantly, the security depends on the verifier's assurance that the correct hash function is applied during signature verification. Otherwise, an opponent may be able to take advantage of hash function weaknesses to obtain a forged signature. Suppose that the signer has selected a strong hash function Hash and has signed messages $M_1, \ldots, M_k$, producing signatures $\sigma_1, \ldots, \sigma_k$. Suppose further that the opponent knows a weak hash function WeakHash and can cause the verifier to apply the weak hash function rather than the strong one. If the opponent can find another message M' such that $\mathsf{WeakHash}(M') = \mathsf{Hash}(M_i)$ for some message M_i then the opponent can present M' and σ_i to the verifier, specifying that the weak hash function is to be applied, and they will be accepted.

The verifier's willingness to accept hash functions other than the one the signer has selected is often in the interest of interoperability. A typical verifier may interact with many signers and thus may need to support a large set of hash

B. Preneel (Ed.): CT-RSA 2002, LNCS 2271, pp. 1–16, 2002.
© Springer-Verlag Berlin Heidelberg 2002

functions, even if each individual signer only supports a few hash functions. The verifier may not know which hash functions are acceptable for a given signer, nor whether a weakness has been found in a hash function. (By "weakness," here and elsewhere, we mean the ability for an opponent to invert the hash function efficiently for a significant fraction of hash values.)

Clearly, it is desirable to limit the set of hash functions that a verifier accepts for a given signer. One approach is to permit only a small set of well trusted hash functions in a given domain. The Digital Signature Standard [17], for instance, allows only the SHA-1 hash function [16]; ANSI X9.31 [1] and ANSI X9.62 [2] likewise allow only "ANSI-approved" hash functions. Another approach is to convey the set of permitted hash functions as an attribute of the signer's public key, for instance in a public-key certificate or as part of policy for a certificate domain. In this way, signers can select different hash functions but are not affected by one another's choices.

Each of the methods just mentioned has some drawbacks. Limiting the set of hash functions precludes new hash functions that may be faster or otherwise more attractive; conveying the set as an attribute of a public key may rely on more complex certificate management infrastructure than is conveniently available. Note that identifying the hash function in the message itself is not enough; it is likely as easy for an opponent to control the identifier as any other part of a message when forging a signature.

As a result of these concerns, the hash function is often identified within the formatting that is applied to the hash value. This identifier might be an index to a registry of hash functions, such as IS 10118-3:1998 [10], or it might be based on an OSI object identifier (see [23]), which can be assigned by any organization. The signature primitive is then relied upon to bind the hash function identifier securely to the hash value. In this way the hash function identifier serves as a *firewall* between different hash functions.

A hash function firewall prevents an opponent from causing a verifier to verify a signature with a different hash function than the signer intended. But this protection is only for existing signatures; a separate question is whether an opponent can obtain new signatures that exploit a weak hash function. In this paper, we answer that question, offering the following new contributions:

1. We present new *weak hash function attacks* against the signature schemes in IS 9796-2:1997 [9], IS 14888-2:1999 [11], and IS 9796-3:2000 [12]. Each scheme includes an optional hash function firewall and claims that the firewall offers some level of protection against attack. We show that such claims are incorrect, and that if any registered hash function were found to be weak and a verifier supports it, it may be possible to forge a signature. Significantly, our attacks are possible *without the signer's participation*.

2. We show that the signature schemes in ANSI X9.31 [1] and PKCS #1 v1.5 [23], which also include hash function firewalls, appear to protect against weak hash function attacks. However, in the case of PKCS #1 v1.5 signatures, we show that if the opponent is allowed to register a *new* hash function identifier for a given hash function, it is possible to forge signatures even if the hash function is strong. Moreover, the forged signatures are valid for a wide range of public keys, not just for one signer.

3. We observe that the RSA-PSS construction in IEEE P1363a [8] offers an *implicit* hash function firewall – without an explicit hash function identifier – under reasonable assumptions on the hash function involved (i.e., it is not 'concocted' purely for the purpose of forging a signature), and provided a certain amount of padding is present.

We emphasize that as long as the verifier accepts only strong hash functions for these signature schemes (and, in the case of PKCS #1 v1.5, the verifier is careful about accepting new, suspicious-looking hash function identifiers), the schemes mentioned here remain secure. Moreover, we are not aware of any weaknesses in the normal hash functions employed with these schemes that would be sufficient to enable an attack.

1.1   Related Work

The concept of a hash function firewall appears to have been introduced by Linn in the design of Privacy-Enhanced Mail in 1990 [14]. Referring to the then-contemplated support for the MD2 and MD4 hash functions, Linn wrote,

> "I observe that any accomodation [*sic*] of MD4 in addition to MD2 should incorporate some form of a signed algorithm identifier, perhaps in the high-order [. . .] bits of the input block to the public-key signature operation, to provide *firewalling* for the benefit of users of one algorithm in the event that flaws are found in the other algorithm after placing that other algorithm into use on PEM messages." (emphasis added)

Motivated by Linn's recommendation, the PKCS #1 specification for RSA signatures [23], first published in 1991, adopted a hash function firewall, and a firewall has become standard in most implementations of RSA signatures since that time.

IBM's Transaction Security System [15], also developed in the early 1990s, employed a hash function firewall as well. In TSS, the identifier is called a "hash rule" (see the definition of the Crypto Facility System Signature Record in [15]).

The concept of a hash function firewall was subsequently adopted in several ISO/IEC standards, addressing factoring-based schemes (e.g., RSA) as well as discrete logarithm / elliptic curve-based schemes. The standards also give general guidance on hash function identification. IS 9796-3:2000 [12], for instance, presents four "options for binding signature mechanism and hash-function . . . in order of increasing risk":

- "a) Require a particular hash-function . . . "
- "b) Allow a set of hash-functions and explicitly indicate in every signature the hash-function in use by a hash-function identifier included as part of the signature calculation . . . " (i.e., a firewall)
- "c) Allow a set of hash-functions and explicitly indicate the hash-function in use in the certificate domain parameters."
- "d) Allow a set of hash-functions and indicate the hash-function in use by some other method, e.g., an indication in the message or a bilateral agreement."

The standard states that "[w]ithout such a binding, an adversary may claim the use of a weak hash-function (and not the actual one) and thereby forge the signature."

Most of the literature on signature schemes focuses on the security properties under the assumption that the underlying hash function is strong, and generally assumes there is only one underlying hash function. We are not aware of any formal analysis of the properties of signature schemes that allow multiple hash functions, nor of hash function firewalls in particular.

Additional analysis on several of the signature schemes presented here can be found in a paper by Coron et al. [4]. Their work focuses on the security of the schemes with strong hash functions and does not address weak hash function attacks. Brown and Johnson [3] consider a type of weak-hash function attack in their security analysis of the Pintsov-Vanstone signature scheme [20], but they do not consider hash function firewalls.

1.2 Organization of This Paper

Section 2 defines terminology and notation. In Section 3 we give weak hash function attacks on three signature schemes in ISO standards. Section 4 discusses the firewalls in other signature schemes. Finally, Section 5 draws several conclusions.

An appendix shows why an extension to DSA that includes a hash function firewall, as suggested by at least one author, must be done carefully.

2   Preliminaries

Throughout this paper, we denote a generic instance of a hash function, whether strong or weak, by Hash, and a specific instance of a weak hash function by WeakHash. HashID and WeakHashID denote one-byte identifiers for the corresponding hash functions. In the various ISO/IEC standards, the one-byte identifiers are taken from one of the IS 10118 registries such as IS 10118-3 [10]. However, the specific one-byte identifier values do not affect our results. HashIDLong denotes a variable-length identifier value.

Other notation is as follows:

ℓ_{Hash}	length in bits of the output of Hash
cc	hexadecimal value
0^u (resp. 1^u)	bit string consisting of u '0' (resp. '1') bits
$X.Y$	concatenation of strings X and Y
$X \oplus Y$	bit-wise exclusive-or
$\|X\|$	length in bits of X

In a string expression, $\|X\|$ refers to the 64-bit representation of the length of X, most significant bit first.

We denote conversion between an integer and a bit string as

$$m \Longleftrightarrow \underbrace{M}_{\ell \text{ bits}},$$

where m is an integer, M is the corresponding bit string, and "ℓ bits" indicates the length of the bit string. Conversion is most significant bit first.

A signature scheme consists of a signature operation and a verification operation. In a signature scheme with appendix, the signature operation generates a signature σ given a message and a private key, and the verification operation verifies σ given M and the corresponding public key. In a signature scheme giving message recovery, a message is considered in two parts: a recoverable part M_1 and a non-recoverable part M_2. The signature operation generates a signature σ given M_1, M_2 and the private key; the verification operation verifies the signature σ and recovers M_1 given the M_2 and the public key.

As noted in Section 1, a signature scheme may allow a choice of hash functions. For such a scheme, let $\Sigma_K[\mathsf{Hash}]$ denote the set of all signatures that can be generated by the scheme with hash function Hash under private key K. We say that the scheme has a *hash function firewall* with respect to a set of hash functions $\mathcal{H}$ if for all $\mathsf{Hash}_1, \mathsf{Hash}_2 \in \mathcal{H}$, for all private keys K, $\Sigma_K[\mathsf{Hash}_1]$ and $\Sigma_K[\mathsf{Hash}_2]$ are distinct.

A common way to build a hash function firewall is to append a hash function identifier to the output of the hash function. The resulting value is called a *hash-token* in the ISO/IEC standards. We say that a hash function firewall is *explicit* if it is based on a hash function identifier, and *implicit* otherwise.

A hash function firewall protects signatures generated with a strong hash function from weaknesses in other hash functions. However, it does not necessarily prevent an opponent from exploiting weaknesses in a hash function (i.e., the ability to invert the hash function) to obtain a different signature. We say that a signature scheme has a *weak* hash function firewall if it is vulnerable to such forgery, and we term the attack that obtains such a forgery a *weak hash function attack*.

3   Signature Schemes with Weak Firewalls

We now consider three signature schemes that have hash function firewalls and show that for each one the firewall does not protect against weak hash function attacks.

The attacks all follow a common approach, exploiting a type of existential forgery in terms of the hash-token. Although it is (presumably) difficult to obtain a valid signature on a *predetermined* hash-token for any of the schemes, for all three it is easy to produce a valid (hash-token, signature) pair. If the hash-token identifies a weak hash function, the opponent needs only to invert the weak hash function to forge a signature. The cost of the attack thus consists of the amount of time to produce the (hash-token, signature pair), plus the amount of time to invert the hash function.

In the attacks, part of the input to the hash function is predetermined but the opponent has full control over the rest, which gives considerable flexibility in the choice of message for the forged signature. In the attacks on the schemes giving message recovery, the opponent has no direct control over the recoverable message part, which is effectively random. This may make the forged signatures produced by the attack less plausible, depending on how the recoverable part is interpreted.

3.1   RSA/RW Signatures in IS 9796-2:1997

We first consider the signature scheme giving message recovery in IS 9796-2:1997 [9], which can be based on either the RSA [22] or Rabin-Williams [21] [24] cryptosystem. The scheme has been shown previously to be vulnerable to signature forgery with a strong hash function in certain cases [4]. Our focus here is on vulnerabilities with a weak hash function. We present the RSA version of the scheme, but similar results apply to the Rabin-Williams version.

Let n be a composite modulus and let e be an odd integer relatively prime to $\lambda(n)$ where $e > 1$. We only need to consider the public key, which consists of the pair (n, e). Let $\ell_n = \lfloor \log_2 n \rfloor + 1$ be the length in bits of the modulus n.

We focus on the specification of the scheme in the situation that a hash function identifier is included and that the non-recoverable message part is not empty, as needed by the attack. In this case, the recoverable message part is between $\ell_n - \ell_{\mathsf{Hash}} - 27$ and $\ell_n - \ell_{\mathsf{Hash}} - 20$ bits long.

The scheme employs two predetermined sets of strings for padding and masking (see the standards document for the specific values):

- $\{\mathsf{Pad}_4, \ldots, \mathsf{Pad}_{11}\}$, a prefix-free set where for each u, $\|\mathsf{Pad}_u\| = u$, and where the leftmost bit of each string is 0;
- $\{\mathsf{Mask}_4, \ldots, \mathsf{Mask}_{11}\}$, a set where for each u, $\|\mathsf{Mask}_u\| = \ell_n - \ell_{\mathsf{Hash}} - u - 16$.

An IS 9796-2:1997 signature on a recoverable message part M_1 and a non-recoverable message part M_2, for the conditions just mentioned, is a value s where $0 \le s < n/2$ such that

$$(\pm s^e \bmod n) \Longleftrightarrow \underbrace{\mathsf{Pad}_u.(M_1 \oplus \mathsf{Mask}_u).H.\mathsf{HashID.cc}}_{\ell_n \text{ bits}}$$

for some u such that $4 \le u \le 11$ where $H = \mathsf{Hash}(M_1.M_2)$. (The "$\pm$" indicates that the correspondence must hold for either $s^e \bmod n$ or $(n - s^e) \bmod n$.)

The attack is as follows:

1. Generate s uniformly at random in $\{0, \ldots, (n-1)/2\}$.
2. Let $t = s^e \bmod n$.
3. If $t \equiv 12 \pmod{16}$, then let $m = t$; if $t \equiv n - 12 \pmod{16}$, then let $m = n - t$. Otherwise, go to step 1.
4. Let T be the ℓ_n-bit string corresponding to m.
5. If T is formatted correctly as $T = \mathsf{Pad}_u.M_1'.H.\mathsf{HashID.cc}$ for some u such that $4 \le u < 11$, then parse T to obtain M_1', H and HashID. If not, go to step 1.
6. If $\mathsf{HashID} \ne \mathsf{WeakHashID}$, go to step 1.
7. Let $M_1 = M_1' \oplus \mathsf{Mask}_u$.
8. Find M_2 such that $\mathsf{WeakHash}(M_1.M_2) = H$.
9. Output M_1, M_2 and s.

The probability that step 3 succeeds is about $1/8$. Since each possible value of m in step 3 corresponds to a different and unique value of s and the value of s in step 1 is uniformly distributed in $\{1, \ldots, (n-1)/2\}$, the value of m in step 3 is uniformly distributed in the subset of $\{1, \ldots, n-1\}$ congruent to 12 modulo 16.

For steps 5 and 6 to succeed, the string T in step 5 must begin with one of the eight padding strings and end with WeakHashID.cc. Since for each u, $\|\mathsf{Pad}_u\| = u$ and the leftmost bit of Pad_u is 0, the probability that T begins with Pad_u is at least 2^{-u}. The probability that T begins with one of the eight padding strings is thus at least

$$\sum_{u=4}^{11} 2^{-u} \approx \frac{1}{8}.$$

The probability that T ends with WeakHashID.cc, given that $m \equiv 12 \pmod{16}$, is about 2^{-12}. Thus, the probability that an iteration of steps 1–7 succeeds is at least about $(1/8)(1/8)2^{-12} = 2^{-18}$, so the expected number of iterations is at most about 2^{18}. The cost of the attack is thus about 2^{18} exponentiations, plus the time to invert the weak hash function.

3.2 GQ Signatures in IS 14888-2:1999

We next consider the signature scheme with appendix in IS 14888-2:1999 [11], which is based on the Guillou-Quisquater (GQ) identification scheme [6].

Let n be a composite modulus and let v be an integer relatively prime to $\Phi(n)$ where $v > 1$. The public key is an integer $y \in \{1, \ldots, n-1\}$. In IS 14888-2, as in the original GQ scheme, the public key y is based on the identity of the user, but this property is not relevant to our attack.

An IS 14888-2:1999 signature on a message M is a pair (r, s) where $0 \leq r < n$ and $1 \leq s < n$ such that

$$r = H.\mathsf{HashID}$$

where $H = \mathsf{Hash}(\Pi.M)$ and $\Pi = y^r s^v \bmod n$.

The attack is as follows:

1. Select any $s \in \{1, \ldots, n-1\}$, and any hash value H.
2. Format r as $r = H.\mathsf{WeakHashID}$.
3. Find M such that $\mathsf{WeakHash}(\Pi \| M) = H$.
4. Output M and (r, s).

In contrast to the other attacks in this section that involve multiple iterations, one iteration is sufficient.

In this attack, the opponent has total control over the hash value H. Thus it may be possible to forge signatures if the hash function is weak only for certain hash values H, even if it is still strong for random hash values.

3.3 Nyberg-Rueppel Signatures in IS 9796-3:2000

We finally consider the signature scheme giving message recovery in IS 9796-3:2000 [12], which is based on the Nyberg-Rueppel signature scheme [19]. We present the discrete logarithm version of the scheme, but similar results apply to the elliptic curve version.

Let p and q be odd primes where $q|(p-1)$ and let g be an element of $\mathbf{Z}_p^*$ with multiplicative order q. The public key is an element $y \in \mathbf{Z}_p^*$ where $y = g^x \bmod p$ for some $x \in \{2, \ldots, q-1\}$. Let $\ell_q = \lfloor \log_2 q \rfloor + 1$ be the length in bits of the prime q. In this scheme, the bit length of the hash value, ℓ_{Hash}, must be a multiple of 8.

As in Section 3.1, we focus on the situation that a hash function identifier is included and that the non-recoverable message part is not empty, in which case the recoverable message part is $\lfloor (\ell_q - 1)/8 \rfloor - \ell_{\mathsf{Hash}}/8 - 1$ bytes long. Let $u = ((\ell_q - 1) \bmod 8) + 1$.

An IS 9796-3:2000 signature on a recoverable message part M_1 and a non-recoverable message part M_2, for the conditions just mentioned, is a pair (r, s) where $1 \le r < q$ and $0 \le s < q$ such that

$$((r - \Pi) \bmod q) \Longleftrightarrow \underbrace{0^u.H.\mathsf{HashID}.M_1}_{\ell_q \text{ bits}}$$

where $H = \mathsf{Hash}(\|M_1\|.\|M_2\|.M_1.M_2.\Pi)$ and $\Pi = g^s y^r \bmod p$.

The attack is as follows:

1. Generate $r \in \{1, \ldots, q-1\}$ and $s \in \{0, \ldots, q-1\}$ uniformly at random.
2. Let $\Pi = g^s y^r \bmod p$.
3. Let $m = (r - \Pi) \bmod q$.
4. Let T be the ℓ_q-bit string corresponding to m.
5. If T is formatted correctly as $T = 0^u.H.\mathsf{HashID}.M_1$, then parse T to obtain H, HashID and M_1. If not, go to step 1.
6. If $\mathsf{HashID} \ne \mathsf{WeakHashID}$, go to step 1.
7. Find M_2 such that $\mathsf{WeakHash}(M_1.M_2) = H$.
8. Output M_1, M_2 and (r, s).

For steps 5 and 6 to succeed, the string T in step 5 must begin with 0^u and end with $\mathsf{WeakHashID}$. Assuming that m in step 3 is uniformly distributed in $\{0, \ldots, q-1\}$, the probability that T has the correct format is at least $2^{-(u+8)}$. Thus the expected number of iterations of steps 1–6 at most 2^{u+8}, which is at most 2^{16}.

4   Other Signature Schemes with Firewalls

We now consider additional signature schemes with hash function firewalls and discuss the applicability of weak hash function attacks to each one.

4.1   RSA/RW Signatures in ANSI X9.31

We first consider the signature scheme with appendix in ANSI X9.31 [1], which, like the scheme in IS 9796-2, can be based on either the RSA or the Rabin-Williams cryptosystem. As before, we will focus on the RSA version.

Let the notation be as in Section 3.1, except that only one padding string is defined, Pad_u where $u = \ell_n - \ell_{\mathsf{Hash}} - 16$, a string of bit length u with leftmost bit 0. Let $v = \ell_n - u$. (Actually, ANSI X9.31 employs a different padding string if M is the empty string, but this difference does not affect the result.)

The ANSI X9.31 signature is a value s where $0 \le s < n/2$ such that

$$(\pm s^e \bmod n) \Longleftrightarrow \underbrace{\mathsf{Pad}_u.H.\mathsf{HashID.cc}}_{\ell_n \text{ bits}}$$

where $H = \mathsf{Hash}(M)$.

One approach to a weak hash function attack is to generate s at random until the equation is satisfied. This will take at most about 2^{u+15} iterations. However, with the parameters in ANSI X9.31 – a 160-bit hash value and at least a 1024-bit modulus – u is at least 848, so the attack is infeasible.

Another approach, relevant when e is small (say, $e = 3$ for RSA or $e = 2$ for RW) is to take integer eth roots. Let α be the integer corresponding to the string Pad_u. Considering only the "+" case, the verification equation requires, among other conditions, that

$$\alpha 2^v \le s^e < (\alpha + 1)2^v.$$

To ensure that there is at least one integer s satisfying this inequality, α should be no more than about $c2^{v/(e-1)}$ where $c = e^{(e-1)/e}$. This means that u should be no more than about ℓ_n/e, so again the attack is infeasible for the parameters in ANSI X9.31. Simple extensions of this attack, such as considering the "−" case or adding multiples of n to the target, offer at best a linear improvement in the probability of success. Thus, it appears that the firewall in the ANSI X9.31 signature scheme is strong for typical parameter choices, although we have no proof that this is the case.

4.2 RSA Signatures in PKCS #1 v1.5

We next consider the signature scheme with appendix in PKCS #1 v1.5 [23] (supported in subsequent versions), which is based on the RSA cryptosystem.

Again, let the notation be as in Section 3.1 except that padding strings are defined only for $u \ge 88$ where $u \equiv 0 \pmod{8}$. Let $\mathsf{HashIDLong}$, a byte string, be a variable-length hash function identifier and let $\ell_{\mathsf{HashIDLong}}$ denote its length in bits.

The signature is a value s where $0 \le s < n$ such that

$$s^e \bmod n \Longleftrightarrow \underbrace{\mathsf{Pad}_u.\mathsf{HashIDLong}.H}_{\ell_n \text{ bits}}$$

where $u = \ell_n - \ell_{\mathsf{HashIDLong}} - \ell_{\mathsf{Hash}}$ and $H = \mathsf{Hash}(M)$.

As above, it is not feasible to attack the scheme for typical parameter choices by generating s at random or by taking integer eth roots, since typical $\mathsf{HashIDLong}$ values are relatively short (e.g., 15–19 bytes). Thus, it again appears that the firewall is strong for typical parameter choices and $\ell_{\mathsf{HashIDLong}}$ values, although again we have no proof.

For the purposes of attack, however, there is no reason we must match an *existing* $\ell_{\mathsf{HashIDLong}}$ value. This is true as well in the other attacks, but in those attacks we only had up to 256 choices, all the same length. Here, we have an essentially unlimited number of choices of variable length. Moreover, as discussed further below, except for a short header, a prefix identifying the organization assigning the identifier, and a short trailer, the identifiers can be mostly random.

This leads to the following basic attack, which is a variant of the one in Section 3.1:

1. Generate s uniformly at random in $\{0, \ldots, \lfloor (n-1)^{1/3} \rfloor\}$.
2. Let $m = s^e$. (This is over the integers.)
3. Let T be the ℓ_n-bit string corresponding to m.
4. If T is formatted correctly as $T = \mathsf{Pad}_u.\mathsf{NewHashIDLong}.H$ for some u such that $u \geq 88$ and $u \equiv 0 \pmod 8$, where $\mathsf{NewHashIDLong}$ is a syntactically correct (and potentially new) hash function identifier, then parse T to obtain $\mathsf{NewHashIDLong}$ and H. If not, go to step 1.
5. Find M such that $\mathsf{WeakHash}(M) = H$.
6. Output M, s and $\mathsf{NewHashIDLong}$.

The expected number of iterations of steps 1–4 depends on the probability that T has the correct structure in step 4. Given that $u \geq 88$, the probability will be fairly small. However, the attack can readily be improved by generating s near the eth root of $\alpha 2^v$ where α is the integer corresponding to Pad_u and $v = \ell_n - u$, for some u. As in Section 4.1, u should be no more than about ℓ_n/e. If $e = 3$, this means that $\mathsf{NewHashIDLong}$ will need to be somewhat less than two thirds the length of the modulus. As an additional improvement, one can include the hash function identifier's header and prefix in α. The resulting attack will then require only a small number of iterations.

We can extend this line of attack by restricting attention to values of s such that $s^e \equiv \beta \pmod{2^{\ell_{\mathrm{Hash}}}}$, where β is the integer corresponding to a *predetermined* hash value. If e is odd, this involves an eth root modulo $2^{\ell_{\mathrm{Hash}}}$, which is easy to compute, and we have $s \equiv \beta^{1/e} \pmod{2^{\ell_{\mathrm{Hash}}}}$. Because H is predetermined, an opponent can apply this attack against a *strong* hash function.

Such forgeries are easy to obtain, and we offer an example based on the SHA-256 hash function [18] with a 2048-bit RSA modulus (the particular modulus doesn't matter, only its length) and public exponent $e = 3$. Let

$$M = \text{``abc''}$$
$$H = \mathrm{Hash}(M) = \texttt{ba7816bf 8f01cfea 414140de 5dae2223}$$
$$\texttt{b00361a3 96177a9c b410ff61 f20015ad}$$

In PKCS #1 v1.5, we have

$$\mathsf{Pad}_{88} = \texttt{0001ffff ffffffff ffff00}$$

so we want to find a signature s such that

$$s^3 \Longleftrightarrow \underbrace{\mathsf{Pad}_{88}.\mathsf{NewHashIDLong}.H}_{\ell_n \text{ bits}}$$

for some hash function identifier. The hash function identifier in our example will be $2048 - 88 - 256 = 1704$ bits long, and will have the following form:

$$\mathsf{NewHashIDLong} = \underbrace{\texttt{3001f230 01cd0601 c8}}_{\text{header}}.\mathsf{Prefix}.Z.\underbrace{\texttt{05000420}}_{\text{trailer}}$$

For the example, we will choose the prefix employed by RSA Security Inc. when registering hash functions:

$$\text{Prefix} = \text{2a864886 f70d02},$$

although any other valid prefix would do just as well. The intermediate string Z is restricted only in that the leftmost bit of its rightmost byte should be 0.

Let α and β be the integers corresponding to the high and low ends of the target value:

$\alpha \Longleftrightarrow$ 0001ffff ffffffff ffff0030 01f23001 cd0601c8 2a864886 f70d02

$\beta \Longleftrightarrow$ 05000420 ba7816bf 8f01cfea 414140de 5dae2223 b00361a3
            96177a9c b410ff61 f20015ad

To find the signature, we compute s_0 and s_1 as

$$s_0 \equiv \beta^{1/3} \pmod{2^{288}}$$
$$s_1 = \lceil (\alpha 2^{968}) \rceil$$

and let $s = s_1 2^{288} + s_0$. With high probability, $m = s^3$ will have the correct format. If not, we can replace s_1 by $s_1 + 1$ and try again. In our example, we obtained the following solution:

$s \Longleftrightarrow$ 32cbfd4a 7adc7905 583d74e5 2ca1423a ff0270e9 10024eae
            2460604c c64061aa 40fe53a5 5356533e 51cd254a a365b7e9
            6eb980df 10cb219c 0f9d889d 849b95f8 d1bad59d adb8c65e
            873681eb d71b33e9 1db2bed4 95

The resulting, valid but rather suspicious hash function identifier is

NewHashIDLong = 3001f230 01cd0601 c82a8648 86f70d02 00000000
                00000000 00000000 00000000 00000000 00000030
                4775b412 25df67cf 96248910 8f7b5f9d 077c8bc2
                f3c981ed 0cfe67fa 8d63a5da b8d081c9 e9c3d417
                cf96035d cf7ef23d 08b10670 587dffe4 55edbd8f
                3a666ccc 251ddf71 2b44db81 2fb2061f c233f06d
                cf1ec019 c55e364d 4e9d2070 254bd041 219f3c66
                9cb696f3 d8bf37bc 136b3c45 a9a1a1a3 6e163be5
                cdfa0ae7 c394114b 321ab520 bb25a558 95db62bc
                ecd007c3 e629030e 203bd6dc 13893d30 2bebaa90
                c36d82e9 c0b5168c 4f050004 20

These attacks have the very interesting property that they depend only on the length of the modulus and the public exponent – not on the modulus itself. Thus a forged signature works with *any* public key of appropriate modulus length and public exponent; no particular signer needs to be targeted. However, if the hash function is strong, the attack will yield a signature on only one message for each new hash function identifier.

To effect these "chosen-hash-identifier" attacks, the opponent must propagate NewHashIDLong as a new hash function identifier, for instance in standards documents. Because most identifiers are short and structured, the opponent may need to explain the length and randomness of the value. Also, presumably, the hash function will already be fairly widely implemented. Thus, the opponent will need to explain why a new hash function identifier is needed. One story: "Before this hash function's official identifier was published, we generated a random one to avoid collisions with other identifiers. Some of our software supports the random one. Please use it for interoperability."

4.3   RSA/RW-PSS Signatures in IEEE P1363a

We finally consider the RSA/RW-PSS signature schemes, which are based on a general construction that supports signatures with message recovery and with appendix. The construction is included in the current draft of IEEE P1363a [8]. As usual, we will focus on the RSA version.

Let the notation be as in Section 3.1, and let ℓ_R denote the length of the salt value associated with the scheme (typically, either 0 or ℓ_{Hash}). Let $G(H, \ell)$ denote a mask generation function mapping a hash value H to a bit string of length ℓ.

For convenience, we will consider the signature scheme with appendix as a special case of the general construction where the recoverable message part M_1 is the empty string.

We focus initially on the specification of the scheme in the situation that a hash function identifier is included. In this case, the recoverable message part can be up to $\ell_n - \ell_R - \ell_{\mathsf{Hash}} - 18$ bits long, independent of the length of the non-recoverable message part M_2.

The signature is a value s where $0 \leq s < n$ such that

$$(\pm s^e \bmod n) \Longleftrightarrow \underbrace{0^1.D.H.\mathsf{HashID}.\mathsf{cc}}_{\ell_n \text{ bits}}$$

where

$$D = G(H, \ell_n - \ell_{\mathsf{Hash}} - 17) \oplus (0^u.1^1.M_1.R),$$
$$H = \mathsf{Hash}(\|M_1\|.M_1.\mathsf{Hash}(M_2).R),$$

$u = \ell_n - \ell_R - \|M_1\| - \ell_{\mathsf{Hash}} - 18$, and R is an ℓ_R-bit string[1]. (The "$\pm$" term is optional; a verifier may accept only the "+" case.)

If there is no lower bound on u, this scheme is vulnerable to a weak hash function attack similar to the one in Section 3.1; the expected number of iterations at most about 2^{16}. However, compared to IS 9796-2:1997, the length of the recoverable message part can vary over a wider range when the non-recoverable message is present. Thus, an implementation that is concerned about

---

[1] In the actual scheme, the output of G is processed slightly differently for implementation convenience: an integral multiple of 8 bits is generated, and the leading bits are removed. This does not affect our results.

weak hash function attacks can impose a lower maximum on the length of M_1, say $\|M_1\| \leq \ell_n - \ell_R - \ell_{\mathsf{Hash}} - 82$, so that the minimum value of u is 64 and the expected number of iterations is at least about 2^{80}.

In the signature scheme with appendix, $\|M_1\| = 0$, so u is large for typical parameters, and an attack is infeasible.

The scheme as just described has an explicit hash function firewall. We now argue that even without a hash function identifier, the scheme offers an implicit, probabilistic firewall if the mask generation function depends on the hash function and $\|M_1\|$ is bounded. In this variant, the right side of the signature equation is $0^1.D.H.\mathsf{bc}$, and D is eight bits longer.

Let G_{WeakHash} be a mask generation function based on a predetermined hash function $\mathsf{WeakHash}$. Let s be an existing signature generated with a strong hash function under private key K. We want to show that it is unlikely that $s \in \Sigma_K[\mathsf{WeakHash}]$. Let D be the value computed in the signature equation for s with the strong hash function. If the length of the hash value is the same for the weak hash function as for the strong one, then in order for s to be a valid signature for $\mathsf{WeakHash}$, we need

$$G_{\mathsf{WeakHash}}(H) = D \oplus (0^{u'}.1^1.M_1'.R')$$

for some u', M_1' and R'. Since D is from a signature generated with a strong hash function, it is presumably random. If u is sufficiently larger than ℓ_{Hash}, it is unlikely that $G_{\mathsf{WeakHash}}(H)$ has the correct output, as can be shown by a simple counting argument. This holds even though the hash function is weak, as long as the hash function is selected before the signature. (Similar analysis applies if the lengths of the hash values are different.)

If the hash function is specified after the signature, it is possible that an opponent can 'concoct' the hash function so that $G_{\mathsf{WeakHash}}(H)$ matches the required output, but if we allow opponents to introduce completely new hash functions, many more serious attacks become possible.

5   Conclusions

We have shown that a hash function firewall, while protecting existing signatures, does not necessarily prevent an opponent from producing new signatures with a weak hash function.

Our results are primarily theoretical, since the hash functions typically employed do not have any weaknesses sufficient to enable an attack. Furthermore, the forged signatures produced by our attacks are generally easy to detect (except for the IS 14888-2:1999 scheme), since something will appear suspicious, whether the recoverable message part (too random), the signature (too short), or the hash function identifier (too strange).

We believe that the ISO/IEC standards should be revised. For instance, IS 14888-2 states,

> "The hash-function identifier shall be included in the hash-token *unless* the hash-function is uniquely determined by the signature mechanism or by the domain parameters" (Clause 7.3, emphasis added).

Given these results, the statement should be reversed:

"The hash-function shall be uniquely determined by the signature mechanism or by the domain parameters *even if* the hash-function identifier is included in the hash-function."

Similarly, IS 9796-3's list of options (see Section 1.1) should be reordered, as option b) carries more risk than option c). The revised version of IS 9796-2 [13] provides a more accurate discussion.

The import of our results is a challenge to the false sense of security that may have been associated with hash function firewalls. Hash function firewalls were first proposed in a context where existential forgery of (hash-token, signature) pairs is difficult, due to the formatting of the hash-token. The firewalls were subsequently adopted in other contexts where existential forgery is easy, thus leading to the attack. The lack of any formal model around hash function firewalls has also contributed to the situation.

Hash function identification is one example of the decisions that must be made when interpreting a digital signature. An even more important decision is which signature scheme to apply, and this is not necessarily implied by the type of public key alone. Furthermore, the verifier must decide on how to interpret the message, since different messages may have the same byte string representation and only the byte string is signed by the underlying digital signature scheme. Message interpretation and the choice of signature scheme both especially depend on infrastructure protection, and we believe that the determination of hash function is better left to the infrastructure as well.

Acknowledgements

This research is a result of my work on the IEEE P1363a standards project. I appreciate the support and feedback I've received from IEEE P1363 working group, and from Don Johnson in particular. John Linn and Mike Matyas gave pointers to previous work on hash function firewalls, John Brainard and Bob Silverman helped with the analysis, and Ari Juels gave me ongoing encouragement to write up this result. The paper also benefited from the comments of anonymous referees. Finally, as is my custom, I also thank God (Col. 3:17).

References

1. ANSI. *ANSI X9.31: Digital Signatures Using Reversible Public-Key Cryptography for the Financial Services Industry (rDSA)*, 1998.
2. ANSI. *ANSI X9.62: Public Key Cryptography for the Financial Services Industry: The Elliptic Curve Digital Signature Algorithm*, 1998.
3. D. R. L. Brown and D. B. Johnson. Formal security proofs for a signature scheme with partial message recovery. Technical Report CORR 2000-39, Department of C&O, University of Waterloo, 2000. Available at http://www.cacr.math.uwaterloo.ca/.
4. J.-S. Coron, D. Naccache, and J.P. Stern. On the security of RSA padding. In M. J. Wiener, editor, *Advances in Cryptology – CRYPTO '99 Proceedings*, volume 1666 of *Lecture Notes in Computer Science*, pages 1–18. Springer, 1999.

5. Taher ElGamal. A public key cryptosystem and a signature scheme based on discrete logarithms. *IEEE Transactions on Information Theory*, 31:469–472, 1985.
6. L. C. Guillou and J.-J. Quisquater. A practical zero-knowledge protocol fitted to security microprocessor minimizing both transmission and memory. In C. Günther, editor, *Advances in Cryptology – EUROCRYPT '88 Proceedings*, volume 330 of *Lecture Notes in Computer Science*, pages 123–128. Springer, 1988.
7. IEEE. *IEEE Std 1363-2000: Standard Specifications for Public-Key Cryptography*, 2000.
8. IEEE P1363 Working Group. *IEEE P1363a: Standard Specifications for Public-Key Cryptography: Additional Techniques (draft)*, June 2001. Draft D9. Available from http://grouper.ieee.org/groups/1363.
9. ISO/IEC. *ISO/IEC FCD 9796-2: Security techniques – Digital signature schemes giving message recovery – Part 2: Mechanisms using a hash-function*, 1997.
10. ISO/IEC. *ISO/IEC 10118-3: Security techniques – Hash-functions – Part 3: Dedicated hash-functions*, 1998.
11. ISO/IEC. *ISO/IEC 14888-2: Security techniques – Digital signatures with appendix – Part 2: Identity-based mechanisms*, 1999.
12. ISO/IEC. *ISO/IEC 9796-3: Security techniques – Digital signature schemes giving message recovery – Part 3: Discrete logarithm based mechanisms*, 2000.
13. ISO/IEC. *ISO/IEC FCD 9796-2: Security techniques – Digital signature schemes giving message recovery – Part 2: Integer factorization based mechanisms*, draft, April 28, 2001.
14. J. Linn. RE: re: Interoperability. pem-dev@tis.com message, 15 November 1990. Message-ID ⟨9011151315.AA22619@decpa.pa.dec.com⟩.
15. S. M. Matyas, D. B. Johnson, A. V. Le, R. Prymak, W. C. Martin, W. S. Rohland, and J. D. Wilkins. Public key cryptosystem key management based on control vectors. U.S. Patent No. 5,200,999, 6 April 1993. Filed 27 September 1991.
16. NIST. *FIPS PUB 180-1: Secure Hash Standard*, 1994.
17. NIST. *FIPS PUB 186-2: Digital Signature Standard*, 2000.
18. NIST. *FIPS PUB 180-2 (Draft): Secure Hash Standard*, May 2001.
19. K. Nyberg and R. Rueppel. A new signature scheme based on the DSA giving message recovery. In *First ACM Conference on Computer and Communcations Security*, pages 58–61. ACM Press, 1993.
20. L. Pintsov and S. Vanstone. Postal revenue collection in the digital age. Presented at Fourth International Financial Cryptography Conference, FC '00, February 2000.
21. M. O. Rabin. Digitalized signatures and public-key functions as intractable as factorization. Technical Report MIT/LCS/TR-212, MIT Laboratory for Computer Science, 1979.
22. Ronald L. Rivest, Adi Shamir, and Leonard M. Adleman. A method for obtaining digital signatures and public-key cryptosystems. *Communications of the ACM*, 21:120–126, 1978.
23. RSA Laboratories. *PKCS #1 v1.5: RSA Encryption Standard*, 1993. Available at http://www.rsasecurity.com/rsalabs/pkcs.
24. H. C. Williams. A modification on the RSA public-key encryption procedure. *IEEE Transactions on Information Theory*, 26:726–729, 1980.

A   DSA Signatures with a Hash ID

In the Digital Signature Standard [17], the Digital Signature Algorithm (DSA) is combined only with the Secure Hash Algorithm 1 (SHA-1), so hash function identification is not an issue. Other specifications based on DSA, such as IEEE Std 1363-2000 [7], allow alternate hash functions but do not include a hash function identifier.

In a February 1992 letter to NIST commenting on the then-proposed Digital Signature Standard, this author suggested adding a hash function firewall to DSA, but with only a preliminary security analysis. We show here why such an extension would need to be done carefully. Our results also apply to ECDSA [2].

Let the notation be as in Section 3.3.

We consider an extension to DSA where the hash function identifier is appended to the hash value. A similar analysis could be applied to a scheme where the identifier is prepended.

We assume that the hash values for all the hash functions supported by the signer and verifier are the same length, t bits long, so that the value m is in $\{0, \ldots, 2^{t+8} - 1\}$. To have a firewall among all 2^8 possible identified hash functions, we need $q \geq 2^{t+8}$; otherwise, hash values corresponding to different hash functions might yield the same hash-token modulo q.

In our extension to DSA, a signature on a message M is a pair (r, s) where $0 \leq r < q$ and $1 \leq s < q$ such that

$$r = (g^a y^b \bmod p) \bmod q$$

where $a = ms^{-1} \bmod q$, $b = rs^{-1} \bmod q$, and

$$m \Longleftrightarrow \underbrace{H.\mathsf{HashID}}_{(t+8)\text{ bits}}$$

where $H = \mathsf{Hash}(M)$.

The attack is as follows (this is modeled after the existential forgery presented by ElGamal on the ElGamal signature scheme [5]):

1. Generate $a \in \{0, \ldots, q-1\}$ and $b \in \{1, \ldots, q-1\}$ uniformly at random.
2. Let $r = (y^a g^b \bmod p) \bmod q$.
3. Let $s = rb^{-1} \bmod q$. If $s = 0$, go to step 1.
4. Let $m = as \bmod q$.
5. Let T be the $(t+8)$-bit string corresponding to m. If If $m \geq 2^{t+8}$, go to step 1.
6. Parse T as $T = H.\mathsf{HashID}$.
7. If $\mathsf{HashID} \neq \mathsf{WeakHashID}$, go to step 1.
8. Find M such that $\mathsf{WeakHash}(M) = H$.
9. Output M and (r, s).

Assuming that m in step 5 is uniformly distributed in $\{0, \ldots, q-1\}$, the probability that $m < 2^{t+8}$ in step 5 is $2^{t+8}/q$. The value of m in step 7 is thus uniformly distributed in $\{0, \ldots, 2^{t+8} - 1\}$, so the probability that $\mathsf{HashID} = \mathsf{WeakHashID}$ in step 7 is 2^{-8}. The probability that an iteration of steps 1–7 succeeds is thus $2^t/q$, so the expected number of iterations is $q/2^t$. (We ignore the rare event that $s = 0$ in step 3.)

To avoid the attack, q should be at least about $2^{3t/2}$ so that the number of iterations is about $2^{t/2}$, which is comparable to finding collisions in the hash function. Of course, such a large q will mean more computation for the signer and the verifier.

Observability Analysis
– Detecting When Improved Cryptosystems Fail –

Marc Joye[1], Jean-Jacques Quisquater[2], Sung-Ming Yen[3,*], and Moti Yung[4]

[1] Gemplus Card International, Card Security Group, Gémenos, France
marc.joye@gemplus.com
http://www.geocities.com/MarcJoye/
[2] UCL Crypto Group, Louvain-la-Neuve, Belgium
jjq@dice.ucl.ac.be
http://www.uclcrypto.org/
[3] Dept of Computer Science, National Central University, Taiwan, R.O.C.
yensm@csie.ncu.edu.tw
http://www.csie.ncu.edu.tw/~yensm/
[4] CertCo, New York NY, USA
moti@{certo.com,cs.columbia.edu}
http://www.certco.com/

Abstract. In this paper we show that, paradoxically, what looks like a "universal improvement" or a "straight-forward improvement" which enables better security and better reliability on a theoretical level, may in fact, within certain operational contexts, introduce new exposures and attacks, resulting in a weaker operational cryptosystem. We demonstrate a number of such dangerous "improvements". This implies that careful considerations should be given to the fact that an implemented cryptosystem exists within certain operational environments (which may enable certain types of tampering and other observed information channels via faults, side-channel attacks or behavior of system operators). We use our case studies to draw conclusions about certain investigations required in studying implementations and suggested improvements of cryptosystems; looking at them in the context of their operating environments (combined with their potential adversarial settings). We call these investigations *observability analysis*.

Keywords: Security analysis, observability, cryptanalysis, implementations, side-channel attacks, fault analysis, robustness, cryptosystems.

1 Introduction

The aim of this paper is to highlight that, contrary to the common belief, some popular measures which were suggested to enhance the reliability and security of a basic cryptosystem introduce new attacks, often more insidious (and thus more difficult to identify). These enhancements, *paradoxically*, result in a possibly weaker operational system. The attacks are applicable within a working

---

* Supported in part by the Computer & Communication Research Laboratories, Industrial Technology Research Institute, Republic of China.

B. Preneel (Ed.): CT-RSA 2002, LNCS 2271, pp. 17–29, 2002.
© Springer-Verlag Berlin Heidelberg 2002

environment and an attack model. In particular, we consider adversaries which may inject faults (as was first suggested in [8]), but ones with "limited-feedback channel". Namely, adversaries which receive only a limited feedback from the system (e.g., an indication about the case of fault, which may be observed due to change of behavior of components or parties within the system). The adversaries do not get actual outputs of the decryption device. We also employ such adversaries which introduce faults into ciphertexts (as in [5]), and ones which perform power analysis [21].

Our test case is the RSA system. In fact, RSA is undoubtedly the most widely used and accepted public-key cryptosystem. Owing to this popularity, it is also perhaps the most cryptanalyzed system [7]. Furthermore, many optimizations and improvement measures are known for it. Therefore, exposures resulting from improper use of RSA seem nowadays to be minimal, which contrast with our results. We note that we have chosen the RSA system for concreteness, the same conclusion may remain valid for various other cryptosystems, as well. We believe that our examples are quite basic yet demonstrative and educational.

We note that we do not attempt to claim that we are the first to notice that improvements of an aspect of a system may actually cause problems when the system is analyzed from another angle. In addition, what we call "a paradox" in the sequel, may not always be viewed as a paradox by experienced readers. Yet, we feel that reporting on systematic failures of seemingly universal improvements which, in fact, introduce problems (in certain adversarial settings) is important, regardless of the exact terminology used to describe the phenomena.

Notations. Throughout the paper, the public RSA modulus will be denoted by $n = pq$ for two secret primes p and q; the public encryption key (resp. secret decryption key) will denoted by e (resp. d).

Organization of the Paper. In the next section, we review a simple precaution suggested to avoid the leakage of secrets due to faults in the context of RSA. We then show that this method (or any other method with similar logic behind it, in particular the one suggested for CRT-based RSA) which protects against such leakages, paradoxically, may also be used to recover some secret information. Section 3 illustrates that the optimal asymmetric encryption padding (OAEP) proposed by Bellare and Rogaway, which is certainly one of the best methods currently available to encrypt with RSA (chosen-ciphertext secure in the random oracle model), paradoxically, is susceptible to some attacks (in some settings) which do not apply to the plain RSA encryption (i.e., the RSA primitive). In Section 4, we show that the same conclusion holds for the so-called 'RSA for paranoids', a stronger variant of RSA. Namely, using the stronger RSA version (in conjunction with some provably secure padding) introduces new exposures. Finally, we would like to resolve the "seemingly paradox phenomenon" and indeed we conclude in Section 5 where we explain the underlying issues behind the "paradoxes" where certain exposures which may be observable are in fact a result of an "improved" operational decryption mechanism. (This conclusion is, in fact, independent of and more important than whether the reader will consider our case studies to be really paradoxical or merely "seemingly paradoxical" examples!). We suggest a way to view and analyze possible implementations in light of observable events which may lead to activating possible countermeasures.

2 The Case of Added Reliability:
RSA Enhanced with Fault Analysis Prevention

2.1 How to Implement the RSA?

RSA in Standard Mode. The decryption process in the (plain) RSA goes as follows. Given a ciphertext $c = m^e \bmod n$, one recovers the plaintext as $m = c^d \bmod n$.

Suppose that an error occurs during the computation of $m = c^d \bmod n$; more precisely, suppose that one bit of d, say bit d_j, has flipped (let d' denote the corrupted value of d). It is then very easy to recover the flipped bit and its value [3,8,17]. The decryption process will yield $m' = c^{d'} \bmod n$ instead of m. Let $d = \sum_{i=0}^{k-1} d_i\, 2^i$ denote the binary expansion of d. Since

$$d' = \sum_{\substack{i=0 \\ (i \neq j)}}^{k-1} d_i\, 2^i + \overline{d_j}\, 2^j = d + (\overline{d_j} - d_j)2^j$$

it follows that

$$\frac{(m')^e}{c} \equiv c^{e(\overline{d_j} - d_j)2^j} \equiv \begin{cases} c^{e\,2^j} & (\bmod\ n) \quad \text{if } d_j = 0\,, \\ 1/c^{e\,2^j} & (\bmod\ n) \quad \text{if } d_j = 1\,. \end{cases} \tag{1}$$

Therefore *if an adversary can get access to the value of m'*, she can recover the bit d_j by exhaustively testing whether $(m')^e/c \equiv c^{\pm e\,2^j} \pmod{n}$ for some $0 \leq j \leq k - 1$. This attack readily extends to the case where several bits of secret exponent d have flipped. As noted by Kaliski [19], such an attack is easily defeated by checking whether the decrypted message, m', satisfies $(m')^e \equiv c \pmod{n}$ and outputting the plaintext message $m = m'$ if and only if the comparison is successful.

RSA in CRT Mode. The RSA decryption can be speeded up by a factor of 4 using *Chinese remaindering* (CRT) [25]. From $c = m^e \bmod n$, the corresponding plaintext m is recovered as

$$m = m_p + p[p^{-1}(m_q - m_p) \bmod q] \tag{2}$$

where $m_p = c^{d_p} \bmod p$, $m_q = c^{d_q} \bmod q$, $d_p = d \bmod (p-1)$, and $d_q = d \bmod (q-1)$.

An error within the CRT mode of operation has much more devastating consequences than within the standard mode. Suppose that the computation modulo p is corrupted (we let m'_p denote the corrupted value) while the computation modulo q is not. Then from Eq. (2), the CRT recombination will yield

$m' = m'_p + p[p^{-1}(m_q - m'_p) \bmod q]$ instead of m. Since $m \not\equiv m' \pmod{p}$ and $m \equiv m' \pmod{q}$, it follows that the $\gcd(m'^e - c \pmod{n}, n)$ gives the secret factor q [15], whereas the original fault analysis idea is presented in [8]. Note that this attack is stronger than the previous one: there is no particular assumption on the kind of (induced) errors; the attack is successful as soon as there is an error (*any* error) during the exponentiation modulo one prime factor. Note also that this attack can be used by an adversary to break the system *only if she can get access to the value of m'*.

An elegant method to protect against such kind of failure models was suggested by Shamir at the rump session of *EUROCRYPT '97*, see Adi Shamir's patent [29] and [30] (see also [16]). From a randomly chosen small integer r, the decryption algorithm first computes $m_{rp} = c^{d_{rp}} \bmod rp$ and $m_{rq} = c^{d_{rq}} \bmod rq$ (where $d_{rp} = d \bmod \phi(rp)$ and $d_{rq} = d \bmod \phi(rq)$). Then if $m_{rp} \not\equiv m_{rq} \pmod{r}$, an error has occurred and the system outputs an `error` message[1]; otherwise the computations are supposed correct and the plaintext m is recovered by applying Chinese remaindering on $m_p = m_{rp} \bmod p$ and $m_q = m_{rq} \bmod q$ (see Eq. (2)). The probability that an error is undetected is equal to $1/r$. For example, if r is a 20-bit integer, this probability is already smaller than 10^{-6}. The advantage of Shamir's countermeasure resides in its universality: it is applicable even when the value of e is not available.

2.2   First Paradox

From the above discussion, it appears that the right way to implement the RSA primitive is to use Chinese remaindering (for efficiency) along with Shamir countermeasure (for security in the induced-fault model). However, as we will see, the conclusion is not so straight-forward. The introduction of Shamir's countermeasure, or more generally of any other method for detecting errors, makes the system *irregular*: it now behaves differently depending on whether the computations are error-free or not, namely it outputs the correct decryption or an `error` message, respectively. The system may thus be used as an *oracle* to try to collect some secret information. The next paragraph sketches a possible method to devise such an oracle for many existing implementations of the RSA primitive. See [32] for details and further discussions.

The adversarial setting is the induced-fault model with *limited-feedback* channel. We assume that an adversary is merely an observer who *only knows whether a ciphertext decrypts correctly or not*, and that she has no access to the corresponding plaintext m (or m'). This assumption is much weaker than in the case of an attacker who chooses plaintexts and ciphertexts and such weaker attack is more likely to occur, making the system more vulnerable. We will next outline an attack which in this situation will demonstrate that:

> "A more reliable system (i.e., designed to detect faults) may, in fact, be weaker".

---

[1] Remark that if the decryption algorithm attempts to recompute the answer, this will take a longer time to complete the computation and will be revealed by a timing analysis.

Input: $c, d = (d_{k-1}, \ldots, d_0)_2, n$
Output: $A = c^d \bmod n$

a.1 $A \leftarrow 1;\ B \leftarrow c$
a.2 <u>for</u> i <u>from</u> 0 <u>to</u> $k - 1$
a.3   <u>if</u> $(d_i = 1)$ <u>then</u> $A \leftarrow A\,B \bmod n$
a.4 $B \leftarrow B^2 \bmod n$
a.5 <u>endfor</u>

(a) Square-and-multiply.

Input: $A = (A_{t-1}, \ldots, A_0)_{2^\omega}, B, n$
Output: $R = A\,B \bmod n$

b.1 $R \leftarrow 0$
b.2 <u>for</u> j <u>from</u> $t - 1$ <u>downto</u> 0
b.3 $R \leftarrow (R\,2^\omega + A_j B) \bmod n$
b.4 <u>endfor</u>

(b) Interleaved multiplication.

Fig. 1. RSA exponentiation.

Let us, next, review the attack. RSA exponentiation is usually implemented with the square-and-multiply technique (Fig. 1a) where multiplication and modular reduction are interleaved to fit the word-size $\Omega = 2^\omega$ (Fig. 1b) (e.g., see [20]). Imagine that an adversary wants to guess the value of the ith bit d_i of the decryption exponent d [2]. Suppose that $d_i = 1$, the interleaved multiplication $A\,B \bmod n$ (Line a.3, Fig. 1) is thus performed at iteration i. Suppose furthermore that one or several bits of error are introduced into the more significant positions of register A, or more precisely into some words A_j for $j > j_\tau$ where j_τ represents the current value of counter j (Line b.2, Fig. 1) when the faulty bits are introduced. Since the words containing the errors are no longer required for the next iterations (i.e., for $j = j_\tau, j_\tau - 1, \ldots, 0$, Line b.2, Fig. 1), the computation of $R = A\,B \bmod n$ will be correct. Moreover, since R is restored into register A, $A \leftarrow A\,B \bmod n$ (Line a.3, Fig. 1), the error located in register A will be cleared and the final result $c^d \bmod n$ will be correct, too. Conversely, if the value of bit d_i was 0, then the interleaved multiplication (Line a.3, Fig. 1) is bypassed at iteration i and the errors induced into register A will not be cleared, resulting in an incorrect value for the final result $c^d \bmod n$. Remember that we made the explicit assumption that the adversary knows whether a ciphertext decrypts correctly or not. So, by inducing faulty bits as previously described, she knows the value of bit d_i according to whether the decryption algorithm returns an **error** message or not.

Let us emphasize again that if Shamir's countermeasure (or any other means to detect errors) was not implemented, then the adversary would not be able to guess the correct value of d_i because, due to the limited feedback she is allowed to have, she does not know whether a ciphertext decrypts correctly or not. In this case, the only way for her to recover the value of d_i is to raise the value of $c^d \bmod n$ returned by the decryption algorithm o the e and compare it with the original value of c; if they match then $d_i = 1$, otherwise $d_i = 0$. However, this supposes a much stronger assumption. Namely that (besides inducing faults) the adversary has access to the (raw) decrypted values of many ciphertexts, which, in most cases, is quite unrealistic.

---

[2] Here d has to be understood as the secret exponent involved in the exponentiation. It stands for the decryption key d in standard mode and for $d_p = d \bmod (p - 1)$ (or $d_q = d \bmod (q - 1)$) in CRT mode.

3   The Case of Added Robustness to Stronger Attacks: Optimal Asymmetric Encryption Padding

In this section we will again consider the induced-fault attack model and the same passive attacker who is an observer of the system's reaction.

3.1   How to Encrypt with RSA?

Recall that there is a big difference between the *RSA primitive* also called plain RSA encryption (that is, the modular exponentiation function $f : x \mapsto f(x) = x^e \bmod n$) and an *RSA encryption scheme* (that is, a particular way to use the RSA primitive to encrypt a message). Plain RSA encryption is definitely not a reasonable way to encrypt with RSA: as observed by Goldwasser and Micali [14], an encryption scheme had better be probabilistic. This stems from the fact that a deterministic scheme does not offer, in essence, the desired property of *indistinguishability*. Informally, indistinguishability is defined as the adversary's inability to make the difference between the encryptions of bits '0' and '1', or more generally, given a challenge ciphertext, to learn any information about the corresponding plaintext. This does not imply that the converse is necessarily true: Bleichenbacher [5] has shown that the *(probabilistic)* encryption standard RSA PKCS #1 v1.5 does not achieve indistinguishability and exploited this failure to mount a chosen ciphertext attack on some interactive key establishment protocols (e.g., SSL) constructed from it. Other problems of plain schemes are presented in [9]. The commonly recommended way to encrypt with RSA is the Optimal Asymmetric Encryption Padding (OAEP) by Bellare and Rogaway [4] (which was claimed to achieve chosen-ciphertext security [26])[3]. This method was supported by the RSA standardization process [6] following the Bleichenbacher attack. It was then published as RSA PKCS #1 v2.0, which will be followed by the IEEE and ANSI X9 standards.

A simplified version of OAEP (called basic embedding scheme in [4]) goes as follows. Let $k = \lfloor \log_2(n) \rfloor$. A message $m < 2^{k-k_0}$ is encrypted as

$$c = \left(m \oplus G(r) \,\middle\|\, r \oplus H\big(m \oplus G(r)\big) \right)^e \bmod n \qquad (3)$$

for a (public) "generator" function $G : \{0,1\}^{k_0} \to \{0,1\}^{k-k_0}$, a (public) hash function $H : \{0,1\}^{k-k_0} \to \{0,1\}^{k_0}$ and where r is a random integer uniformly chosen in $\{0,1\}^{k_0}$. To decrypt the ciphertext c, the decryption algorithm computes $x := c^d \bmod n$. Then setting x_0 to the k_0 least significant bits of x and x_1 to the remaining bits of x (i.e., $x = x_1 \| x_0$), it computes $r' = x_0 \oplus H(x_1)$ and returns $x_1 \oplus G(r')$ as the plaintext message corresponding to c.

OAEP differs from the above in that some extra bits are padded: they are used to check the integrity of the message. In particular, OAEP achieves

---

[3] Recently, Shoup [31] has shown that the original proof was enough to claim only security against a non-adaptive adversary [23], and a new proof of security was constructed [12].

plaintext-awareness, that is, informally, the impossibility of producing a cipher-text without the knowledge of the corresponding plaintext, a random oracle property which implies chosen-ciphertext security. Let again $k = \lfloor \log_2(n) \rfloor$. To encrypt a message $m < 2^{k-k_0-k_1}$, choose a random integer r in $\{0,1\}^{k_0}$ and compute

$$c = \left(m\{0\}^{k_1} \oplus G(r) \;\middle\|\; r \oplus H\left(m\{0\}^{k_1} \oplus G(r)\right) \right)^e \bmod n \qquad (4)$$

for $G : \{0,1\}^{k_0} \to \{0,1\}^{k-k_0}$ and $H : \{0,1\}^{k-k_0} \to \{0,1\}^{k_0}$. Then, given c, the decryption algorithm computes $x = c^d \bmod n$ and sets x_0 to the k_0 least significant bits of x and x_1 to the the remaining bits of x (i.e., $x = x_1 \| x_0$). Next, it computes $r' = x_0 \oplus H(x_1)$ and $y = x_1 \oplus G(r')$, and sets y_0 to the k_1 least significant bits of y and y_1 to the remaining bits of y (i.e., $y = y_1 \| y_0$). If $y_0 = \{0\}^{k_1}$ then it returns y_1 as plaintext; otherwise it returns an **error** message.

3.2   Second Paradox

Here too, we see that the decryption algorithm acts as an *oracle*. Hence as in Section 2, we suppose that the RSA exponentiation is carried out with an algorithm such as the one depicted in Fig. 1, then by introducing some errors, an adversary is able to recover the value of the bits d_i of the secret decryption key d according to the decryption is possible or not. It is worth remarking that the "basic embedding scheme" (see Eq. (3)) or even the plain RSA encryption are not susceptible to this attack because they do not check the integrity[4]. In those two cases, the adversary must have access to the decrypted value, encrypt it and finally compare it to the initial ciphertext to guess the value of d_i; with OAEP, the value of d_i is deduced from *the only knowledge that an* **error** *message is returned or not*, the knowledge of the decrypted value is not necessary. We thus have a second paradox:

> *"A more robust implementation (i.e., secure against active attacks) may, in fact, be weaker."*

3.3   Setting-Dependent Robustness in Another Case

The attacks described in this section (and in Section 2 and the subsequent para-doxes) assume some variants of the induced-fault model of [8]. However, the idea that improving a system to increase its robustness in one sense may not suffice for other considerations can be widely applicable (regardless of the specific model). In fact, we can draw exactly this conclusion by considering a model where the decryption is carried out by a device which is really tamper-proof (and no faults are possible). Our attack follows the recent attack by Manger [22] against (some implementations of) RSA PKCS #1 v2.0 combined with the power analy-sis of [21]. RSA PKCS #1 v2.0 [1] is a slight variation of OAEP where the most

---

[4] Note, however, that we do not suggest to use a weaker form of encryption for many other reasons.

significant byte (MSB) of the encoded message, $\widetilde{m}$, being encrypted is forced to 00h to ensure that the resulting padding is always smaller than modulus n.

Current implementations of the decryption operation decrypt c to get $\widetilde{m} = c^d \bmod n$, *check whether the first byte of $\widetilde{m}$ is zero* and if so, check the OAEP integrity of $\widetilde{m}$. If both verifications are successful, the plaintext message m corresponding to $\widetilde{m}$ is output; otherwise, there is an error message. If we can distinguish between an error resulting from a too large message (i.e., MSB($\widetilde{m}$) $\neq$ 00h) and an error resulting from an incorrect decoding, then a chosen ciphertext attack similar in spirit to that of Bleichenbacher [5] can be mounted (see [22] for details).

The trivial solution consists of course in making the two error messages identical, as already recommended in the PKCS #1 v2.0 standard (*cf.* [1, § 7.1.2]). The last PKCS #1 v2.1 draft [2] explicitly requests to made the two kinds of error indistinguishable; in particular, it insists that the execution time of the decryption operation (timing channel) must not reveal whether the first byte is 00h or not. So, it is suggested that, in the case of MSB($\widetilde{m}$) $\neq$ 00h, to proceed to the OAEP integrity check by setting $\widetilde{m}$ to a string of zero octants. However, such an "improved" solution is not satisfactory in all respects, since it is very likely that power analysis (power consumption channel) will reveal information as the power consumption is related to the manipulated data, making it easy to distinguish between the zero string (when MSB($\widetilde{m}$) $\neq$ 00h) and a random-looking string $\widetilde{m}$ (when MSB($\widetilde{m}$) = 00h).

4   The Case of Increased Security Parameter: Unbalanced RSA

We now return to investigate a seemingly paradoxical situation. Actually, this section shows that a seemingly improved variant of RSA is subject to somewhat stronger attacks in the physical sense, since they can be mounted remotely. Namely, we increase the attacker's power to choose ciphertexts, which she can "transmit only remotely" (with or without errors) to the device; at the same time we limit the adversary's device tampering power as before to be an observer of the error messages (and not allowing it access to decrypted values).

4.1   RSA for Paranoids

The security of the RSA system is based on the difficulty of factoring large integers. Therefore, a larger modulus offers further security, at the expense, however, of a larger computational effort. A good compromise is to use an *unbalanced* RSA modulus [28], that is, a modulus $n = pq$ where p, q are primes and $q \gg p$ (e.g., $|p| = 500$ bits and $|q| = 4500$ bits). The best factorization algorithms [24] cannot take advantage of these special moduli and they thus seem as secure as moduli constructed from the product of two 2500-bit primes. Shamir [28] observed that if the plaintext m being encrypted is smaller than p, then, from the corresponding ciphertext $c = m^e \bmod n$, it can be recovered as $m = c^{d_p} \bmod p$

(where $d_p = d \bmod (p-1)$). This follows immediately from Eq. (2) by noting that $m = m \bmod p = m_p$. The resulting system is called *RSA for paranoids*.

4.2 Third Paradox

Is the name 'RSA for paranoids' really justified? If a plaintext m larger than p is encrypted (let $c' = m^e \bmod n$ denote the corresponding ciphertext), then the decryption gives $m' = c'^{d_p} \bmod p = m \bmod p \neq m$. Consequently, $\gcd(m-m', n)$ gives the secret prime p and the system is broken [13]. Note that this can only be turned into an active attack *if the adversary can get access to the value of m'*, which, as in the previous sections, may be considered an unrealistic assumption in many cases.

The usual way to prevent such an attack is to enhance the purely algebraic "plain" scheme and add redundancy and randomness to the plaintext prior to encryption. The redundancy enables one to check the integrity of the plaintext and the randomness serves to avoid many drawbacks inherent to any deterministic encryption algorithm (*cf.* Section 3). We thus assume that the system is implemented using an appropriate embedding of the kind of OAEP[5] (or any other applicable variant thereof); the main point being that the decryption system "internally" checks the *integrity* of the plaintexts (and assures message awareness even against active attacks).

Thus, as before, we now make the weaker assumption that *the adversary only knows whether a ciphertext can be decrypted or not*, but in no way, can she get access to a decrypted value. However, we allow her to choose ciphertexts. The attack demonstrates our principle of observing behavior under an oracle. Technically, it follows the basic properties shown in [13] (see also [18]).

Since the plaintexts being encrypted must be smaller than p, we set $k = \lfloor \log_2(p) \rfloor$ in the OAEP description given by Eq. (4):

$$\mathrm{OAEP}(m) = \underbrace{m\{0\}^{k_1} \oplus G(r)}_{k - k_0 \text{ bits}} \;\big\|\; \underbrace{r \oplus H\big(m\{0\}^{k_1} \oplus G(r)\big)}_{k_0 \text{ bits}}$$

for a message $m \in \{0,1\}^{k-k_0-k_1}$ and a random $r \in \{0,1\}^{k_0}$, where $G : \{0,1\}^{k_0} \to \{0,1\}^{k-k_0}$ and $H : \{0,1\}^{k-k_0} \to \{0,1\}^{k_0}$. The adversary can fix the value of the $(k+1-k_0-k_1)$ most significant bits of $\mathrm{OAEP}(m)$ by an appropriate choice for message m: if she wants that these bits represent a chosen value T, she simply sets $m = T \oplus [G(r)]^{k-k_0-k_1}$ where $[G(r)]^{k-k_0-k_1}$ denotes the $(k - k_0 - k_1)$ most significant bits of $G(r)$. Futhermore, the adversary is not restricted to probe with valid messages; she can choose an m out of the prescribed range (i.e., $m \geq 2^{k-k_0-k_1}$) so that $\mathrm{OAEP}(m)$ will be a $(k+1)$-bit number or more. Consequently, the adversary has a total control on all the bits of $\mathrm{OAEP}(m)$ except the $(k_0 + k_1)$ least significant ones. From this observation, we will now

---

[5] Remark that OAEP, as is, does not apply to 'RSA for paranoids' since its usage is limited to permutations. One has to use, for example, the more general construction of [11].

explain how she can recover the $(k + 1 - k_0 - k_1)$ most significant bits of the secret prime p. The remaining bits of p may be found by appropriate choices for r, exhaustion or more elaborated techniques (e.g., [10]).

As the value of k is public, the adversary knows that p lies in the interval $I_0 = (2^k, 2^{k+1})$. Then she chooses an m so that its OAEP embedding $x_0 := \text{OAEP}(m)$ belongs to I_0 and computes the corresponding ciphertext $c_0 = x_0{}^e \bmod n$. If c_0 can be decrypted[6] then she knows that $x_0 < p$; otherwise (i.e., if an error message is returned) she knows that $x_0 \geq p$. She then reiterates the process with the interval $I_1 = (x_0, 2^{k+1})$ or $I_1 = (2^k, x_0]$, respectively, with a new message each time. And so on... until the $(k + 1 - k_0 - k_1)$ most significant bits of p are disclosed. Note that this attack does not make sense in the standard RSA; that is, when the decryption is computed modulo $n = pq$. So, in this view, the RSA for paranoids with a 5000-bit RSA modulus ($|p| = 500$ bits and $|q| = 4500$ bits) is weaker than the standard RSA using a 1000-bit modulus where $|p| = |q| = 500$ bits. Hence, the paradox:

> "A stronger system with increased length parameters (which possibly make the underlying problem harder) may, in fact, be weaker."

5   Conclusions

What we saw are three improvements that actually in some operational setting induce exposures (vulnerabilities). In each of the three cases, we show that there are implementation environments where:

1. Increasing Reliability may weaken the system;
2. Improving Security and Robustness to attacks (OAEP) may weaken the system;
3. Improving certain security parameters (especially if the growth is not uniform in all parameters) may weaken the system.

The thesis put forth and demonstrated in this paper is that a system exists within certain operational environment which may enable adversaries to tamper and perform other (somewhat limited) observations regarding the system's operations. Paradoxically, what looks like a "universal improvement" or "straightforward improvement" which enables better security and better reliability, may in fact, within these specific operational contexts, introduce new exposures and attacks.

Where do the paradoxes come from? Given the attack, one notices that the improved "more secure and reliable system" takes additional computational measures to improve itself. However, these measures may increase the **observability** of the system w.r.t. certain (possibly newly) introduced "events". Under certain allowed attack scenarios (interaction with the implemented environment) this increased observability in fact makes the systems less smooth in the sense that

---

[6] The attack supposes that the decryption algorithm does not check, implicitly or explicitly, that the decrypted m is $< 2^{k-k_0-k_1}$ (see [18, Section 3]).

the events (reported by the oracle) reveal the inner workings. The adversary in the less smooth operation mode can play with or observe the increased set of possible events. This, in turn, may enable her to break the system. On the other hand, the system without the improvement may operate more smoothly and produce less noticeable differences in reported events, thus not enabling attacks as above. It is expected that as cryptography becomes very strong, future adversaries will concentrate on exploiting observable events and information emitted by the working environment. The basic rule which may resolve the paradoxes can (informally) be stated as follows:

> *"A stronger system whose operation, however, is less smooth than a weaker counterpart (namely where its interaction with the adversarial environment enables enhanced "observability" of events which reflect on its inner workings), may, in fact, be weaker."*

In conclusion, the natural belief that obvious straight-forward security improvements are always good (namely, are universal) is a fallacy. In fact, security measures are *context sensitive*, especially when we consider the full cycle of a cryptographic system (design, implementation, and operation). The interplay of the *basic design* with the *implementation* and *operational environment* and the potential adversaries in each stage should be considered very carefully in the deployment of actual working systems. While the above issue may have been implicitly noticed earlier, our examples demonstrate it on quite modern state-of-the-art constructions. We feel that the notion of "event observability" and its implied exposures and weaknesses have to be taken into consideration when analyzing the security of a working cryptosystem. Our investigation points out that one should conduct what may be called "observability analysis" of the working environment, analyzing events under an expected or reasonable (well modeled) set of exposures. It should then be made sure that the potentially observable events are not made available to the possible adversary (by either technical, physical or operational measures).

References

1. RSA Laboratories. PKCS #1 v2.0: RSA cryptography standard, October 1, 1998. Available at http://www.rsasecurity.com/rsalabs/pkcs/.
2. RSA Laboratories. PKCS #1 v2.1: RSA cryptography standard, Draft 2, January 5, 2001. Available at http://www.rsasecurity.com/rsalabs/pkcs/.
3. F. Bao, R. Deng, Y. Han, A. Jeng, A. D. Narasimhalu, and T.-H. Ngair. Breaking public key cryptosystems on tamper resistant devices in the presence of transient faults. In B. Christianson, B. Crispo, M. Lomas, and M. Roe, eds, *Security Protocols*, vol. 1361 of *Lecture Notes in Computer Science*, pp. 115–124, Springer-Verlag, 1998.
4. Mihir Bellare and Phillip Rogaway. Optimal asymmetric encryption — How to encrypt with RSA. In A. De Santis, ed., *Advances in Cryptology – EUROCRYPT '94*, vol. 950 of *Lecture Notes in Computer Science*, pp. 92–111, Springer-Verlag, 1995.

5. Daniel Bleichenbacher. A chosen ciphertext attack against protocols based on the RSA encryption standard RSA PKCS #1. In H. Krawczyk, ed., *Advances in Cryptology – CRYPTO '98*, vol. 1462 of *Lecture Notes in Computer Science*, pp. 1–12, Springer-Verlag, 1998.
6. Daniel Bleichenbacher, Burt Kaliski, and Jessica Staddon. Recent results on PKCS #1: RSA encryption standard. *RSA Laboratories' Bulletin*, no. 7, June 1998.
7. Dan Boneh. Twenty years of attacks on the RSA cryptosystem. *Notices of the AMS*, 46(2):203–213, 1999.
8. Dan Boneh, Richard A. DeMillo and Richard J. Lipton. On the importance of checking cryptographic protocols for faults. In W. Fumy, ed., *Advances in Cryptology – EUROCRYPT '97*, vol. 1233 of *Lecture Notes in Computer Science*, pp. 37–51, Springer-Verlag, 1997.
9. Dan Boneh, Antoine Joux, and Phong Q. Nguyen. Why Textbook El Gamal and RSA encryption are insecure. In T. Okamoto, ed., *Advances in Cryptology – ASIACRYPT 2000*, vol. 1976 of *Lecture Notes in Computer Science*, pp. 30–43, Springer-Verlag, 2000.
10. Don Coppersmith. Small solutions to polynomial equations, and low exponent RSA vulnerabilities. *Journal of Cryptology*, 10(4):233–260, 1997.
11. Eiichiro Fujisaki and Tatsuaki Okamoto. How to enhance the security of public-key encryption at minimum cost. In H. Imai and Y. Zheng, eds., *Public Key Cryptography*, vol. 1560 of *Lecture Notes in Computer Science*, pp. 53–68, Springer-Verlag, 1999.
12. Eiichiro Fujisaki, Tatsuaki Okamoto, David Pointcheval, and Jacques Stern. RSA–OAEP is secure under the RSA assumption. In J. Kilian, ed., *Advances in Cryptology – CRYPTO 2001*, vol. 2139 of *Lecture Notes in Computer Science*, Springer-Verlag, 2001.
13. Henri Gilbert, Dipankar Gupta, Andrew Odlyzko, and Jean-Jacques Quisquater. Attacks on Shamir's 'RSA for paranoids'. *Information Processing Letters*, 68:197–199, 1998.
14. Shafi Goldwasser and Silvio Micali. Probabilistic encryption. *Journal of Computer and System Sciences*, 28:270–299, 1984.
15. Marc Joye, Arjen K. Lenstra, and Jean-Jacques Quisquater. Chinese remaindering cryptosystems in the presence of faults. *Journal of Cryptology*, 12(4):241-245, 1999.
16. Marc Joye, Pascal Paillier, and Sung-Ming Yen. Secure evaluation of modular functions. In R.J. Hwang and C.K. Wu, eds., *Proc. of the 2001 International Workshop on Cryptology and Network Security (CNS 2001)*, pp. 227–229, Taipei, Taiwan, September 26–28, 2001.
17. Marc Joye, Jean-Jacques Quisquater, Feng Bao, and Robert H. Deng. RSA-type signatures in the presence of transient faults. In M. Darnell, ed., *Cryptography and Coding*, vol. 1355 of *Lecture Notes in Computer Science*, pp. 155–160, Springer-Verlag, 1997.
18. Marc Joye, Jean-Jacques Quisquater, and Moti Yung. On the power of misbehaving adversaries and security analysis of the original EPOC. In D. Naccache, ed., *Topics in Cryptology – CT-RSA 2001*, vol. 2020 of *Lecture Notes in Computer Science*, pp. 208–222, Springer-Verlag, 2001.
19. Burton S. Kaliski Jr. Comments on a new attack on cryptographic devices. RSA Laboratories Technical Note, October 23, 1996.
20. Çetin K. Koç. RSA hardware implementation. Technical Report TR 801, RSA Laboratories, April 1996.

21. Paul Kocher, Joshua Jaffe, and Benjamin Jun. Differential power analysis. In M. Wiener, editor, *Advances in Cryptology – CRYPTO '99*, vol. 1666 of *Lecture Notes in Computer Science*, pp. 388–397, Springer-Verlag, 1999.

22. James Manger. A chosen ciphertext attack on RSA optimal asymmetric encryption padding (OAEP) as standardized in PKCS #1. In J. Kilian, ed., *Advances in Cryptology – CRYPTO 2001*, vol. 2139 of *Lecture Notes in Computer Science*, pp. 230–238, Springer-Verlag, 2001.

23. Moni Naor and Moti Yung. Public-key cryptosystems provably secure against chosen ciphertext attacks. In *Proc. of the 22nd ACM Annual Symposium on the Theory of Computing (STOC '90)*, pp. 427–437, ACM Press, 1990.

24. Andrew Odlyzko. The future of integer factorization. *Cryptobytes*, 1(2):5–12, 1995.

25. Jean-Jacques Quisquater and Chantal Couvreur. Fast decipherment algorithm for RSA public-key cryptosystem. *Electronics Letters*, 18:905–907, 1982.

26. Charles Rackoff and Daniel R. Simon. Non-interactive zero-knowledge proof of knowledge and chosen ciphertext attack. In J. Feigenbaum, ed., *Advances in Cryptology – CRYPTO '91*, vol. 576 of *Lecture Notes in Computer Science*, pp. 433–444, Springer-Verlag, 1992.

27. Ronald L. Rivest, Adi Shamir, and Leonard M. Adleman. A method for obtaining digital signatures and public key cryptosystems. *Communications of the ACM*, 21(2):120–126, 1978.

28. Adi Shamir. RSA for paranoids. *Cryptobytes*, 1(2):1–4, 1995.

29. Adi Shamir. Patent US 5.991.415: Method and apparatus for protecting public key schemes from timing and fault attacks, 12 May 1997.

30. Adi Shamir. How to check modular exponentiation. Presented at the rump session of *EUROCRYPT '97*, Konstanz, Germany, 11–15th May 1997.

31. Victor Shoup. OAEP reconsidered. In J. Kilian, ed., *Advances in Cryptology – CRYPTO 2001*, vol. 2139 of *Lecture Notes in Computer Science*, Springer-Verlag, 2001.

32. Sung-Ming Yen and Marc Joye. Checking before output may not be enough against fault-based cryptanalysis. *IEEE Transactions on Computers*, 49(9):967–970, 2000.

Precise Bounds
for Montgomery Modular Multiplication
and Some Potentially Insecure RSA Moduli

Colin D. Walter

Comodo Research Laboratory
10 Hey Street, Bradford, BD7 1DQ, UK
www.comodo.net

Computation Department, UMIST
PO Box 88, Sackville Street, Manchester, M60 1QD, UK
www.co.umist.ac.uk

Abstract. An optimal upper bound for the number of iterations and precise bounds for the output are established for the version of Montgomery Modular Multiplication from which conditional statements have been eliminated. The removal of such statements is done to avoid timing attacks on embedded cryptosystems but it can mean greater execution time. Unfortunately, this inefficiency is close to its maximal for standard RSA key lengths such as 512 or 1024 bits. Certain such keys are then potentially subject to attack using differential power analysis. These keys are identified, but they are rare and the danger is minimal. The improved bounds, however, lead to consequent savings in hardware.

Key words: Cryptography, RSA cryptosystem, Montgomery modular multiplication, differential power analysis, DPA.

1 Introduction

Modular multiplication forms the basis of RSA encryption [11], El-Gamal encryption [5], and Diffie-Hellman key exchange [2] as well as many other cryptographic functions. One algorithm which is particularly efficient for such multiplication is that of Peter Montgomery [9]. Since the algorithm makes modular adjustments unrelated to the magnitude of the accumulating partial product, precise bounds on its output are of interest in order to determine how much space is required for operands and also how many iterations of its main loop need to be performed.

If M is the modulus in question, it is easy to show that the main loop of Montgomery's algorithm will generate an output less than $2M$ if the inputs satisfy that same condition and enough iterations are done [3]. Thus exponentiation can be performed more simply by omitting the final conditional subtraction of M from each multiplication until the last one. However, with standard choices of RSA key length, such as 512 or 1024 bits, $2M$ usually exceeds the word length by just 1 bit. Hence, for efficiency reasons, the conditional subtractions are usually included.

But, recent studies of embedded cryptographic systems have revealed the importance of eliminating conditional statements from code because they may

B. Preneel (Ed.): CT-RSA 2002, LNCS 2271, pp. 30–39, 2002.
© Springer-Verlag Berlin Heidelberg 2002

enable timing attacks to be mounted [7]. For example, the conditional subtraction of M occurs with different probability during an exponentiation depending on whether the operation being performed is a square or a multiplication [13]. This can lead to the recovery of the secret key. So, in [12] it was shown that *all* conditional statements, even the very last one, can be removed from exponentiation code using Montgomery modular multiplication if sufficiently many iterations are performed. Hachez and Quisquater [6] subsequently showed that the sufficient conditions given in [12] can lead to an excessive number of iterations. In fact, the problem only arises close to word boundaries.

The main purpose of this paper is to improve further on the efficiency issues which arise from [6] and [12]. Specifically, we establish the precise conditions for avoiding unnecessary iterations, and for selecting optimal hardware for certain common modulus lengths. Unfortunately, inefficiency turns out to be close to its maximum for standard RSA key lengths such as 512 or 1024 bits. For such key lengths the normal $2M$ bound yields a topmost "overflow" digit of either 0 or 1. Because of the different power used during multiplication by zero and non-zero words, this could lead to an attack on the cryptosystem using *differential power analysis* (DPA) [8], [14]. A secondary aim of this paper is therefore to identify the moduli for which this is a risk and assess the danger. The very fortunate conclusion is that almost all standard keys are safe from this form of DPA attack, and for those which are vulnerable the risk is extremely small. As a by-product, we show that hardware savings are possible for all standard RSA moduli by ensuring that the output has the same number of digits as the modulus.

2   Montgomery Modular Multiplication (MMM)

For the notation, suppose the crypto-system is supported by a multiplier that performs k-bit $\times$ k-bit multiplications. Let $r = 2^k$. The $i{+}1$st digit of A in base r will be written $A[i]$ and so the value of A is $A = \sum_{i=0}^{n-1} A[i]r^i$ if it has n digits. Then the appropriate form of the MMM algorithm is as below. It excludes the usual final conditional statement:

```
If P >= M then P <- P-M
```

which would be applied when the inputs A and B are less than M.

MONTGOMERY'S MODULAR MULTIPLICATION ALGORITHM (MMM)

> { Pre-conditions: A has n digits base r, each in the range 0 to $r{-}1$,
>   and M is prime to r. }
> ```
> P <- 0 ;
> For i <- 0 to n-1 do
> Begin
>       q <- (-M⁻¹(P+A[i]B)) mod r ;
>       P <- (P + A[i]B + qM) div r ;
>       { Invariant: 0 ≤ P < M + B }
> End
> ```
> { Post-conditions: $Pr^n \equiv A{\times}B \bmod M$, $ABr^{-n} \leq P < M{+}ABr^{-n}$ }

Here M^{-1} denotes the multiplicative inverse modulo r. The computation of q is done with the multiplier using the lowest digits of M, P and B, and provides a result in the range 0 to $r-1$. The choice ensures exact divisibility by r in the second line of the loop. So the MMM procedure computes $A \times B$ mod M with some shift, which is readily seen to be r^n because there are n iterations. Clearly $P < M+B$ is a loop invariant because it holds initially and, assuming it holds at the end of one iteration, it must hold at the end of the next iteration because

$$(P+A[i]B+qM) \text{ div } r \quad < \quad ((M+B)+(r-1)B+(r-1)M)/r \quad = \quad M+B$$

In particular, it must also hold on termination.

To achieve an output bound of $2M$ when A and B are bounded above by $2M$, it may be necessary to perform extra iterations, i.e. to increase n. Thus, if n is big enough that $A[n-1] \leq (r-2)/2$ then, with working as above, the bound on the final output becomes $M+B/2$, which is at most $2M$. But, for standard RSA key lengths, many of the inputs A can be expected to have the same number of bits as M, and hence the top non-zero digit of A is often likely to equal or exceed $r/2$. It therefore seems that an additional iteration with a zero digit of A is necessary to achieve the usual output bound of $2M$.

However, a more precise output bound is claimed in the above code for MMM. Thus, let $\alpha_i = \sum_{j=0}^{i-1} A[j]r^{j-i}$. Then, by induction, $\alpha_{i+1} = (\alpha_i+A[i])/r < 1$ for all i. We will now prove that the loop invariant

$$\alpha_{i+1}B \quad \leq \quad P \quad < \quad M+\alpha_{i+1}B$$

holds at the end of the iteration involving $A[i]$. Let P_{i+1} denote the value of P at this point and let q_i be the value of q set during the iteration. Prior to the loop commencement, the property $\alpha_0 B = 0 = P_0 < M = M+\alpha_0 B$ holds. This starts the proof by induction. For the inductive step,

$$\begin{aligned} \alpha_{i+1}B &= (\alpha_i B + A[i]B)/r \\ &\leq (P_i + A[i]B + q_i M)/r \\ &= (P_i + A[i]B + q_i M) \text{ div } r \\ &= P_{i+1} \end{aligned}$$

and

$$\begin{aligned} P_{i+1} &= (P_i + A[i]B + q_i M) \text{ div } r \\ &= (P_i + A[i]B + q_i M)/r \\ &< ((M+\alpha_i B) + A[i]B + (r-1)M)/r \\ &= M + \alpha_{i+1}B \end{aligned}$$

At the end of the last iteration, $\alpha_n = Ar^{-n}$, and so the loop invariant provides:

Theorem 1. $ABr^{-n} \leq P < M+ABr^{-n}$ is a post-condition of MMM.

An alternative derivation of this can be obtained via the partial quotient formed from the successive values of the digit q. Using the same notation, define $\gamma_i = \sum_{j=0}^{i-1} q_j r^{j-i}$. Then it is easy to see that the invariant

$$P_i = \alpha_i \times B + \gamma_i \times M$$

holds at the start of the loop iteration [4]. As in the case of α_i, γ_i satisifes $0 \le \gamma_i < 1$ for all i. Therefore taking $i = n$ for the invariant at the end of the final iteration yields Theorem 1[1].

The theorem improves on $0 \le P < M+B$ because $Ar^{-n} < 1$, and it is substantially better if A has zero or non-maximal top digits. It is also the best possible bound in the sense that it places the output in an interval of length M, which is the smallest interval containing a member of each residue class. Indeed, it determines the output P uniquely up to its residue class. Without some knowledge of the residue class of AB and hence of the digits q, this could not be improved.

Notice that the digits of A are processed from least to most significant. So, if it is initially in redundant form, A can be converted on-line to the non-redundant form required. Also, we could represent B and M with radix 2^l rather than 2^k and use a $k{\times}l$-bit multiplier for computing the new value for P. Then the bits of B and M are consumed l bits at a time, propagating carries, and yielding a non-redundant output. Usually dedicated hardware does either that or uses $2k$ full length digit-parallel additions to compute P in a single clock cycle giving a redundant output.

3   Stability

In the context of exponentiation, the output needs to satisfy the same upper bounds as are required for the inputs so that it can be fed straight back in as another input. Under such conditions, the output will remain bounded. Let this bound be $(1+\lambda)M$. Then $P < M+ABr^{-n} \le (1+\lambda)M$ delivers the requirement for stability, namely $M + (1+\lambda)^2M^2r^{-n} \le (1+\lambda)M$, i.e.

$$(1+\lambda)^2 Mr^{-n} \;\le\; \lambda \tag{1}$$

Solving for λ we obtain

$$2^{-1}r^n(1-\sqrt{1-4Mr^{-n}}) \;\le\; (1+\lambda)M \;\le\; 2^{-1}r^n(1+\sqrt{1-4Mr^{-n}}) \tag{2}$$

In particular, this needs to have real roots, i.e. $4M \le r^n$. But this suffices for the existence of a suitable λ and so for the stability of its use in exponentiation. Hence

Theorem 2. MMM *preserves an input bound* $(1+\lambda)M$ *at output whenever* $4M < r^n$ *and* λ *satisfies* (1).

This theorem therefore determines the minimal number of iterations n which MMM must perform in order to obtain a stable exponentiation process, namely the least value of n which satisfies the given inequality. Thus, assuming an initial input less than M,

– Exponentiation using MMM will converge if MMM consumes at least two more bits of A than there are bits in M.

---

[1]   Incidentally, if M_i is chosen such that $M_iM \equiv 1 \bmod r^i$ then $\gamma_i r^i = ((-\alpha_i r^i){\times}\; B{\times}M_i) \bmod r^i$.

Now it is clear that if the top of M is not close to a word boundary (i.e. at least 2 bits away from the next one) then the extra iteration with a zero digit of A, which was applied in [12], is unnecessary. Specifically, the condition used there was $2M < r^{n-1}$. That condition was improved in [6] to $M < r^{n-1}$ when $r \geq 4$. The new bound represents an improvement on both: on [12] when $r > 2$, and on [6] when $r > 4$.

Typically $r = 2^8$, 2^{16} or 2^{32}. In the last case, with a standard 512-bit RSA modulus, 17 iterations are performed on 17 digit numbers, whereas with a 510-bit modulus only 16 iterations would be performed on 16-digit numbers. In both cases, the intermediate calculations actually involve 17 digit numbers, but one might still question whether the extra 2 bits in the modulus to achieve a standard key length are worth the probable extra $6\frac{1}{4}\%$ ($= 17/16 - 1$) processing time. The most efficient cases of MMM occur when the key length has two fewer bits than a multiple of the word or digit size because an extra iteration is not required.

A standard RSA configuration has a modulus which satisfies $\frac{1}{2}r^{n-1} < M < r^{n-1}$. Prior to concerns about DPA, MMM for this case would have been terminated after $n-1$ iterations and, if the output had $P[n-1] \neq 0$, a subtraction of M would be performed. Consequently, inputs bounded above by r^{n-1} would generate outputs also bounded above by r^{n-1} because of the loop invariant $P < M+B$. We will show that, although the extra iteration is still necessary in order to remove the conditional subtraction and the extra digit is needed for intermediate calculations, at least the extra digit can be avoided in the output.

4 Partial Product Bounds

The extreme values for the I/O bound $(1+\lambda)M$ may be of interest i) for designing hardware and ii) for determining allowable input for the multiplier. Assume M satisfies the inequality of Theorem 2. It is readily verified from (1) that $\lambda = 1$ is always an acceptable value, so that if A and B are bounded by $2M$ then so will be the output P. This viewpoint, using λ, provides a very tight bound on the number of bits required for P in terms of those used in M. In a similar vein, it is easy to check that if A and B are bounded by $\frac{1}{2}r^n$ then the associated value of λ satisfies (1) and so the output P is also bounded by $\frac{1}{2}r^n$. This alternative viewpoint provides bounds on the number of digits required for P. Since the extreme values of λ in (2) coincide for $4M = r^n$ at $\lambda = 1$ where $(1+\lambda)M = \frac{1}{2}r^n$, these values are in some sense the only ones which will work for every acceptable modulus satisfying the condition of Theorem 2.

Intermediate values of P are bounded by $M+\alpha_i B$, and hence by $M+B$ and therefore by $(2+\lambda)M$. So,

Theorem 3. *Suppose the modulus M in MMM satisfies $4M < r^n$. If inputs A and B are bounded above by $2M$ (resp. $\frac{1}{2}r^n$), then the output P is similarly bounded above by $2M$ (resp. $\frac{1}{2}r^n$). Moreover, intermediate values of P are bounded above by $\frac{3}{4}r^n$.*

For n satisfying the condition given in the theorem, this determines n as the maximum number of digits needed to represent the various numbers. If n

is also the number of digits in M then n is the exact number of digits required for intermediate and final values of P in MMM. Otherwise, the computations require one more digit than M has. So M lies between $\frac{1}{4}r^{n-1}$ and r^{n-1}. This is the case which usually arises for standard RSA key lengths. In the next section we consider *inter alia* whether the extra nth digit is really necessary in such cases. The answer is surprising.

5   Differential Power Analysis Attacks

Cryptosystems may be vulnerable to timing or differential power attacks [8]. Having eliminated any conditional statements from MMM, it is now necessary to consider transition cases where digits might acquire particular values with frequencies which are measurable using power variations. When the value of n has had to be increased because $4M$ has just exceeded a power of r, a large proportion of very small top digit values can be expected to occur in intermediate results and in final products. These might usefully be partitioned into zero and non-zero values as a basis for an attack. Of particular interest are the cases of standard key sizes, such as 512, 768 or 1024 bits when, as usual, they are multiples of the word length k. Since $2M$ is a bound on the output, we find that the top digit is either 0 or 1. This might well provide exactly the handle that an attacker needs to break the cryptosystem.

Using the same method as described in [14], power or EMR traces from a number of digit-by-digit products can be combined from the same long integer multiplication in order to determine whether or not the leading digit of either input is zero. Alternatively, the Hamming weight of the digit might be measured as it travels along the bus to or from memory. The frequencies of zero and non-zero top digits might then be obtained for each long integer multiplication in an exponentiation. The theory presented by Walter & Thompson in [13] shows that the frequency of a zero top digit will be different for squares and multiplications. (Integrating over the probability density functions for the values of the inputs provides a coefficient $\frac{1}{2} \times \frac{1}{2} = \frac{1}{4}$ in the case of a multiplication, and a coefficient $\frac{1}{3}$ for a square.) This enables squares and multiplications to be distinguished and hence secret key bits to be read from the frequency chart if a sufficiently naïve implementation of exponentiation is used, even if Rivest's method is applied to blinding the input text [1,10]. However, this attack depends on non-zero top digits appearing with sufficient frequency for this difference to be easily measured and for the result to be very reliable for each decision.

Software was constructed for MMM in order to check the theory and gauge the quantity of data that might leak if very small top digit values could be detected. In particular, the frequency of the top digit being non-zero was of interest when M satisfied the most common situation, namely $M < r^{n-1} < 2M$. This revealed that the top digit does not behave as if the distribution of outputs were fairly uniform in the range $[0..2M]$, as one might have expected from the usual $2M$ bound on I/O. Indeed, we will now show that

- in such standard circumstances the top output digit from MMM is non-zero only very exceptionally.

In fact, the final iteration, which was only just necessary, performs in effect an extra shift down of the AB term in the upper bound $M+ABr^{-n}$, making it much smaller than the M component. Thus, from the property $P < M+ABr^{-n} < M+4M^2r^{-n}$ of inputs less than $2M$, one can deduce that the output satisfies $P < r^{n-1}$ if $M+4M^2r^{-n} < r^{n-1}$. Using the Taylor series expansion for $(1+16r^{-1})^{1/2}$ shows that this condition is satisfied when $M < r^{n-1}(1-4r^{-1})$. So,

Theorem 4. *For inputs less than $2M$ and n satisfying the condition of Theorem 2, the property $M < r^{n-1}(1-4r^{-1})$ guarantees that the output of* MMM *satisfies $P < r^{n-1}$.*

Of course, for M close to this upper bound, outputs re-used as inputs satisfy a better bound than $2M$. So it makes sense to see under what conditions $(n-1)$-digit inputs will provide $(n-1)$-digit outputs. In this case, with similar working to that just given, the condition that needs to be satisfied is $M+r^{n-2} < r^{n-1}$. Hence,

Theorem 5. *For inputs of $n-1$ digits and M satisfying the property $M < r^{n-1}(1-r^{-1})$ the output of* MMM *also has at most $n-1$ digits.*

For standard configurations using recommended RSA key lengths and typical off-the-shelf multipliers, n satisfies $\frac{1}{2}r^{n-1} < M < r^{n-1}$, giving the most significant non-zero digit of M in the interval $[\frac{1}{2}r, r-1]$. So the condition in Theorem 5 will hold almost always because the exception requires the top digit of M to be $r-1$ and the probability of this is only 1 in 2^{k-1}. This is certainly unlikely for 16- or 32- bit digits, though occasionally the case if only 8-bit words are employed.

Hence, under standard conditions, a DPA attack which attacks "overflow" output digits in the way described above is unlikely to succeed because the top non-zero digit of M will usually have an unsuitable value. Of course, the same cannot be claimed if M occupies n digits and its topmost digit is just 1. As in the standard case, $2M$ is a poor upper bound for the inputs and outputs during an RSA exponentiation. The width of the interval in inequality (2) is still reasonably large, enabling λ to be chosen fairly small. This will again provide an upper bound only marginally larger than M and average output values close to $M/2$ so that in the region of half the outputs will be expected to have a non-zero top digit. Hence the DPA attack might become feasible.

Between these two situations, consider the exceptional case with $r^{n-1}(1-r^{-1}) < M < r^{n-1}$. This is, in fact, the only case for which the inputs and outputs will have a different number of digits from the modulus M. By taking the maximum lower bound for $(1+\lambda)M$ given in inequality (1), namely when $M = r^{n-1}$, we know that $2^{-1}r^n(1-\sqrt{1-4r^{-1}})$ is an upper bound which will hold for the output if it holds for the inputs. Using the Taylor series expansion for $\sqrt{1-x}$, we obtain $r^{n-1}+r^{n-2}+2r^{n-3}+5r^{n-4}+14r^{n-5}+...$ as a suitable bound on inputs which is preserved. Since the residues $\bmod M$ are expected to be uniformly distributed, we can expect at most $r^{-1}+2r^{-2}+5r^{-3}+14r^{-4}+...$ of the outputs to be greater than r^{n-1}, i.e. to have a non-zero nth digit. Probably less than half this number will occur because such outputs duplicate residues $\bmod M$ which are less than r^{n-1}.

So, for an 8-bit multiplier, $r = 256$ would enable such non-zero digits to be detected with sufficient frequency for squares and multiplications to be distinguished only if at least several thousand exponentiations were performed. Even then, for a given multiplicative operation in the exponentiation scheme, only several tens of top digits will be non-zero over the whole sample. The natural variance will mean that the frequency of these non-zero digits will not be sufficiently stable to distinguish reliably between squares and multiplies in the exponentiation. Hence it will be very hard to determine the bits of the secret exponent unless other data is available. For a 12-bit or larger multiplier, a standard lifetime bound of 10k exponentiations would certainly make such distinctions impossible. In such cases, one could afford to re-introduce a conditional subtraction to keep the result of modular multiplication less than r^{n-1}:

    If M[n-1] = 0 and P[n-1] $\neq$ 0 then P <- P-M

Although this leads to rare timing variations, little is lost if power or EMR variations reveal these cases anyway. Of course, if the modulus is known to the attacker and is seen to be one of these exceptional cases, and the input text is not blinded and individual conditional subtractions can be observed, then the attacker simply has to simulate the exponentiation, determine whether a square or a multiplication would generate an observed subtraction and progressively reconstruct the exponent from this information. Hence, as a matter of course, the input should be blinded by, for example, multiplying it by a random number before the exponentiation and performing an appropriate division afterwards [10].

One can conclude that, in general, key lengths which are multiples of the multiplier word length are close to the least efficient as far as Montgomery multiplication is concerned, but that they are usually no less safe. Potentially unsafe moduli have been identified, but any lack of security should only affect poorly designed implementations with very small multipliers, such as 8-bit multipliers. If these exceptional moduli are excluded or an appropriate conditional subtraction included for them, then hardware savings can be made by not implementing the digit of index n for the output register.

6   Exponentiation

A consequence of the extra, unwanted factor r^{-n} in the MMM output is that every exponentiation has to perform pre- and post- processing. A pre-computed value for r^{2n} mod M is Montgomery-multiplied with the initial text T requiring exponentiation. This introduces an extra factor of r^n mod M to the result of every subsequent multiplication compared to the result obtained if classical modular multiplication were performed instead. To remove this extra factor, a final Montgomery multiplication by 1 is performed. In [12] it was observed that taking $A = 1$ in the version of MMM there produced a result less than M, which therefore needed no further conditional subtraction.

Here, taking $A = 1$ gives a bound $P < M + Br^{-n}$ which, when combined with $B < 2M < \frac{1}{2}r^n$, yields $P \leq M$, as before, because P is an integer. Of course, $P = M$ means the initial text T must have satisfied $T \equiv 0$ mod M. In the context of cryptography, T must be less than M. Undoubtedly it should not equal 0, but if it were, then every intermediate result would also be 0, giving 0 as the final output. So here also, with the better value for n than [12] or [6],

Theorem 6. *For inputs less than $2M$ and n satisfying the condition of Theorem 2, a conditional final subtraction to obtain a result in the interval $[0, M-1]$ is unnecessary in exponentiation using the above version of the MMM algorithm with the usual pre- and post- processing.*

7   Conclusion

By obtaining the best possible bounds on the output, we have shown exactly how many iterations are necessary in an implementation of Montgomery's modular multiplication algorithm in order to avoid the conditional statements which may be subject to timing attacks in cryptographic hardware. This includes eliminating the final conditional subtraction from an exponentiation in which the usual pre- and post- processing of the Montgomery constant $r^n \bmod M$ occurs.

The investigation suggested that it might be both cryptanalytically unsafe and inefficient to use certain moduli which have standard key lengths, as these are normally equal to a multiple of the multiplier word length. However, the potentially weak moduli have been identified as a very small set and the weakness shown to be very slight: the security of these weaker moduli is only in question from DPA attacks where a small multiplier is used in an implementation without input blinding.

For standard RSA configurations, the conditional statements are avoided by a single extra iteration of the multiplication algorithm. If the weaker moduli can be avoided completely then the output has the same number of digits as the modulus and some hardware savings can be made. But, if the weaker moduli cannot be avoided then, in combination with input blinding, a conditional statement can be safely re-introduced to keep the output to the same number of digits as the modulus so that the hardware savings can still be made.

References

1. D. Chaum, *Blind Signatures for Untraceable Payments*, Advances in Cryptology – CRYPTO '82, R. L. Rivest, A. T. Sherman & D. Chaum (editors), Plenum Press, New York, 1982, 199–203
2. W. Diffie & M. E. Hellman, *New Directions in Cryptography*, IEEE Trans. Info. Theory, **IT-22**, no. 6 (1976), 644–654
3. S. E. Eldridge, *A Faster Modular Multiplication Algorithm*, Intern. J. Computer Math., **40** (1991), 63–68
4. S. E. Eldridge & C. D. Walter, *Hardware Implementation of Montgomery's Modular Multiplication Algorithm*, IEEE Trans. Comp. **42** (1993), 693-699
5. T. El-Gamal, *A Public-Key Cryptosystem and a Signature Scheme Based on Discrete Logarithms*, IEEE Trans. Info. Theory, **IT-31**, no. 4 (1985), 469–472
6. G. Hachez & J.-J. Quisquater, *Montgomery exponentiation with no final subtractions: improved results*, Cryptographic Hardware and Embedded Systems (Proc CHES 2000), C. Paar & Ç. Koç (editors), Lecture Notes in Computer Science, **1965**, Springer-Verlag, 2000, 293–301

7. P. Kocher, *Timing attack on implementations of Diffie-Hellman, RSA, DSS, and other systems*, Advances in Cryptology – CRYPTO '96, N. Koblitz (editor), Lecture Notes in Computer Science, **1109**, Springer-Verlag, 1996, 104–113

8. P. Kocher, J. Jaffe & B. Jun, *Differential Power Analysis*, Advances in Cryptology – CRYPTO '99, M. Wiener (editor), Lecture Notes in Computer Science, **1666**, Springer-Verlag, 1999, 388–397

9. P. L. Montgomery, *Modular multiplication without trial division*, Mathematics of Computation, **44** (1985), no. 170, 519–521

10. R. L. Rivest, *Timing cryptanalysis of RSA, DH, DSS*, Communication to sci.crypt Newsgroup, 11 Dec 1995

11. R. L. Rivest, A. Shamir and L. Adleman, *A method for obtaining digital signatures and public-key cryptosystems*, Comm. ACM, **21** (1978), 120–126

12. C. D. Walter, *Montgomery Exponentiation Needs No Final Subtractions*, Electronics Letters, **35**, no. 21, October 1999, 1831–1832

13. C. D. Walter & S. Thompson, *Distinguishing Exponent Digits by Observing Modular Subtractions*, Topics in Cryptology – CT-RSA 2001, D. Naccache (editor), Lecture Notes in Computer Science, **2020**, Springer-Verlag, 2001, 192–207

14. C. D. Walter, *Sliding Windows succumbs to Big Mac Attack*, Cryptographic Hardware and Embedded Systems – CHES 2001, Ç. Koç, D. Naccache & C. Paar (editors), Lecture Notes in Computer Science, **2162**, Springer-Verlag, 2001, 286–299

Montgomery in Practice:
How to Do It More Efficiently in Hardware

Lejla Batina and Geeke Muurling

Pijnenburg SECUREALINK B.V.,
Boxtelseweg 26, 5261 NE,
Vught, The Netherlands
{l.batina,g.muurling}@securealink.com

Abstract. This work describes a fully scalable hardware architecture for modular multiplication which is efficient for an arbitrary bit length. This solution uses a systolic array implementation and can be used for arbitary precision without any modification. This notion of scalability includes both, freedom in choice of operand precision as well as adaptability to any desired gate complexity. We present modular exponentiation based on Montgomery's method without any modular reduction achieving the best possible bound according to C. Walter. Even more, this tight bound appeared to be practical in our architecture. The described systolic array architecture is unique, being scalable in several parameters and resulting in a class of exponentiation engines. The data provided in the figures and tables are believed to be new, providing a practical dimension of this work.

Keywords: Montgomery multiplication, modular exponentiation, systolic array, performance model, scalability

1 Introduction

In 1985 Peter Montgomery introduced a new method for modular multiplication ([17]). This operation is widely used in most cryptographic protocols (public and private key cryptography). Modular multiplication is also the basis of modular exponentiation, which is used in RSA protocols. The approach of Montgomery avoids the time consuming trial division that is a common bottleneck of other algorithms. His method proved to be very efficient and is the basis of many implementations of modular multiplication, both in software and hardware. In this paper we look at a hardware implementation.

RSA is the most popular public key cryptosystem nowadays. However, one cannot deny the attractiveness of Elliptic Curve Cryptography (ECC) when compared to other public-key cryptosystems. ECC allows shorter certificates and higher speed and appears to be less vulnerable to side-channel attacks than RSA cryptosystem. For these and many other reasons a vast amount of research is focused on more efficient and more secure implementation of modular multiplication in hardware.

B. Preneel (Ed.): CT-RSA 2002, LNCS 2271, pp. 40–52, 2002.
© Springer-Verlag Berlin Heidelberg 2002

Efficient implementation in hardware was considered by many authors such as [10], [2], [5] and [18]. A systolic array architecture is one possibility of doing public key cryptography in hardware. Various solutions for systolic arrays were proposed in the past ten years, for example [22], [4], [10], [24] and [20], but no implementations have been reported to our knowledge (see also [3]).

Our contribution is in combining a systolic array architecture, which is assumed to be the best choice for hardware on current ICs, with a Montgomery based RSA implementation, achieving the same notion of scalability as introduced in [2]. Namely, the architecture presented here, when compared with [2] is scalable in both design and implementation. In practice this means that even a once fixed architecture always remains with completely flexible precision. Furthermore, the design explained in [2] is not purely a systolic array, which makes this architecture unique.

Nevertheless, unlike most of the previously mentioned work we are also presenting a missing chain between theory and practice. The data provided in Section 6 present actual state of art in the scalable systolic array architecture doing Montgomery Method. This design has proven to be a widely applicable and secure cryptographic device.

The remainder of this paper is organized as follows. Section 2 gives a survey of previous work on systolic arrays and Montgomery based operations in hardware. In Section 3, we introduce the architecture of the targeted platform. Section 4 gives the underlying methods invented by Peter Montgomery in detail and some comments on the bound condition for avoiding subtraction at the end of every exponentiation step introduced by C. Walter ([23]). Section 5 describes a performance model and explains some trade-offs between area and performance. In Section 6 we illustrate the theory with practical results by providing performance figures. These data stand in favour of this new notion of scalability, introduced in this work. Remarks on security are given in Section 7. Conclusions and benchmarks for future work conclude the paper.

2   Previous Work

This section reviews some of the most relevant previous contributions in this area. Hardware for implementing the Modular Multiplication Method (MMM) is presented in [4] and [14]. The authors have shown how to use hardware much more efficient in order to perform this operation relevant for most cryptographic algorithms. However, the work of Iwamura et al. ([8], [9]) are the first ones to our knowledge presenting the systolic array which can execute a modular exponentiation operation using Montgomery modular multiplication. The same authors were also first to introduce the idea of a scalable architecture. They did not name this idea scalability, but they suggest the design's ability to cope with various precision in bits.

The first definition of scalability is given in [2]. The authors introduced a pipelined Montgomery multiplier, which has the ability to work on any given operand precision and is adjustable to any chip area. The first feature they call scalability and treat as unique in comparison to other designs. Apparently, the architecture described in that work is not purely systolic and has a flavor of serial-parallel implementation ([18]). On the other hand, the architecture presented here is the first purely systolic and scalable design for Montgomery multiplication. In [5] the same design methodology is used to obtain a dual-field mulitplier without compromising scalability. That multiplier would have obvious benefits for vast applications of public key cryptography. In our architecture various lengths in operands are all efficient in practice. This means that ECC as well as RSA can be performed very fast on the same piece of hardware. We provide data to support this claim in tables and figures of Section 6.

Iwamura et al. [10] considered the usual bottleneck for hardware implementations of Montgomery's algorithm, i.e. the fact that the number of output bits may exceed the number of input bits. They derived the bound $R > 2^{n+2}$ for $R = 2^r$ and concluded that $r = n + 2$, is the minimum possible value for which the examination of the size of the output each time the Montgomery method is executed, may be omitted. Here, n is the maximal number of bits for N, so $N < 2^n$. This bound can be further improved to the condition $R > 4N$, which is according to work presented in [23], assumed to be the best possible bound in practice. In Section 4 this will be explained in detail. The idea of splitting both input numbers X and Y into digits is also from [10]. The work of C. Walter offers many useful results for Montgomery's techniques. In [21], which is further improved in [23], he showed that the Montgomery exponentiation method requires no final subtraction, which is very important for fast implementation. Another benefit is that conditional statements, which may be subject to side-channel attacks such as timing attack, power analysis attack etc. may be omitted. Some other results considering constant time implementations which is presumed a first step towards secure hardware solutions are proposed in [7]. However, the result presented in [23] is the best achieved for the bound issue. In this work we prove that it is also practical.

3   Systolic Array and Scalability

As an example of this architecture we introduce the PCC-ISES ([15], [1]) as an integrated circuit with a design which is very suitable for modular multiplication. This design contains two identical Large Number Arithmetic Units (LNAU), each designed as a systolic array. This array is one-dimensional and consists of a fixed number of Processing Elements (PEs). In the remainder of this paper P denotes the number of PEs. For the sake of completeness a short description of this platform is given.

3.1   32-Bit Platform: PCC-ISES

PCC-ISES (Figure 1) has the following characteristics: Embedded Cryptographic Accelerators with 2 LNAU's (Figure 2) capable of performing up to 2048-bit modular arithmetic, Embedded microprocessor ARM7, 128 KB embedded RAM, and other features required for various cryptographic applications.

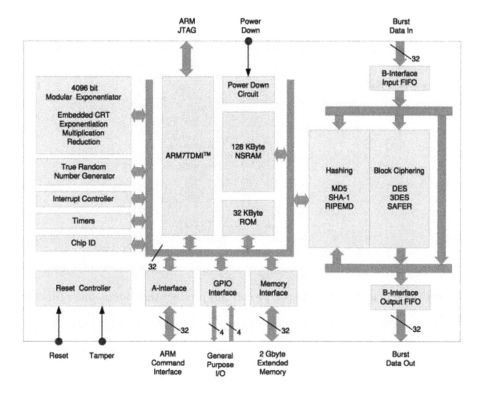

Fig. 1. Architecture of PCC-ISES.

It is well known that modular multiplication in a large prime field is the most time consuming operation in RSA and elliptic curve arithmetic. The two LNAUs of the PCC-ISES provide extremely fast modular exponentiation. Specific commands are defined for modular multiplication and exponentiation, which can be used by the ARM7 processor to access the modular exponentiator. This hardware accelerator can perform two modular multiplication operations at the same time, which provides the possibility of implementing RSA algorithms in parallel. The same holds for ECC.

In addition, PCC-ISES has a true random number generator (TRNG) and SHA-1, which are necessary components for implementing various cryptographic algorithms such as ECDSA, DSA and RSA signature. TRNG, SHA-1, and the modular multiplication are implemented in separate hardware modules, so the ARM7 processor is also fully available for embedded application software.

3.2 Systolic Array

For modular multiplication Montgomery's method is chosen and the notation is as follows:

$$Mont(X, Y) = XYR^{-1} \mod N$$

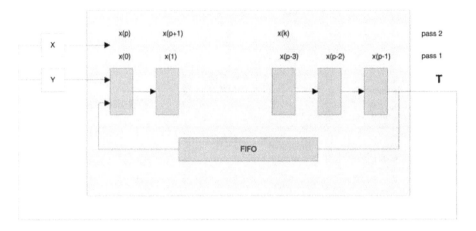

Fig. 2. Large Number Arithmetic Unit of the PCC-ISES.

For the remainder of the paper we call this product the Montgomery function with arguments X and Y. The systolic array consists of a fixed number of PEs (Figure 2). A FIFO memory is added to the design for scalability. A PE contains some adders and multipliers that can process α bits of X and β bits of Y (α and β are not necessarily of the same length) in one clock cycle. So, in one clock cycle a number of additions and multiplications can be performed which differs from relevant work of other authors. More precisely, in each PE, within this architecture, one loop of the multiplication algorithm is performed in one cycle. Each PE calculates $\frac{T_j + x_i y_j + m_i N_j}{2^\alpha}$ in each clock cycle (Algorithm 1). Other authors usually consider one addition or multiplication in one cycle.

3.3 Scalability

An arithmetic unit is scalable if: "the unit can be reused or replicated in order to generate long-precision results independly of the data path precision for which the unit was originally designed" ([2]). In further work the authors maintain this definition of scalability.

Our architecture is also scalable according to the previous definition. If the operands are "too large" to fit in the available number of PEs the intermediate result of the last PE is fed into the first PE. These intermediate results are temporarily stored in a FIFO memory, if necessary.

4   Montgomery Based Operations

Modular multiplication forms the basis of the RSA cryptosystems and also for many other cryptographic protocols. For a word base $b = 2^\alpha$, R has to be chosen such that $R = 2^r = (2^\alpha)^l > N$. Then, there is a one-to-one correspondence between each element $x \in \mathbb{Z}_N$ and its representation $xR \bmod N$. This

Montgomery representation allows very efficient modular arithmetic especially for multiplication. Montgomery's method for multiplying two integers x and y (called N-residues) modulo N, avoids division by N which is the most expensive operation in hardware. The method requires conversion of x and y to an N-residue domain and conversion of the calculation result back to the integer domain. The procedure is as follows. To compute $Z = xyR \bmod N$, one first has to compute the Montgomery function of x and $R^2 \bmod N$ to get $Z' = xR \bmod N$. $Mont(Z', y)$ gives the desired result. When computing the Montgomery product $T = XYR^{-1} \bmod N$, the following procedure is performed ([16]):

Algorithm 1. Montgomery modular multiplication
INPUT: Integers $N(odd), x \in [0, N-1], y \in [0, N-1], R = 2^r$, and $N' = -N^{-1} \bmod 2^\alpha$
OUTPUT: $xyR^{-1} \bmod N$
1. T $\leftarrow$ 0. (Notation $T = (t_l t_{l-1}...t_0)$
2. For i from 0 to (l-1) do:
2.1 $m_i \leftarrow (t_0 + x_i y_0) N' \bmod 2^\alpha$
2.2 $T \leftarrow (T + x_i y + m_i N)/2^\alpha$
3. If $T \geq N$, then $T \leftarrow T - N$
4. Return (T)

In the original notation of Montgomery after each multiplication a reduction was needed (step 3 in the algorithm above). The input had the restriction $X, Y < N$ and the output T was bounded by $T < 2N$. The result of this is that in the case $T > N$, N must be subtracted so that the output can be used as input of the next multiplication. To avoid this subtraction a bound for R is known ([23]) such that for inputs $X, Y < 2N$ also the output is bounded by $T < 2N$.

In [23] the need of avoiding reduction after each multiplication is addressed. In practice this means that the output of the multiplication can be directly used as an input of the next Montgomery multiplication. We want to find a bound on R such that with $X, Y < 2N$ the output of the Montgomery multiplication $T < 2N$. Write $R \geq kN$, then:

$$T = \frac{XY + mN}{R} = \frac{XY}{R} + \frac{m}{R}N < \frac{4}{k}N + N \tag{1}$$

where, $m = (XY \bmod R)N' \bmod R$.

Hence, $T < 2N$ for $k \geq 4$, implying: $4N \leq R$. To guarantee the existence of the modular inverse of R, R and N should be relatively prime. This excludes $4N = R$.

Theorem 1. *The result of a Montgomery multiplication* $XYR^{-1} \bmod N < 2N$ *when* $X, Y < 2N$ *and* $R > 4N$.

This is the same result as in [23]. The final round in the modular exponentiation is the conversion to the integer domain, i.e. calculating the Montgomery function of the last result and 1. The same arguments as above prove that this final step remains within the following bound: $Mont(T, 1) \leq N$. In practice, A^B mod $N = N$ will never occur since $A \neq 0$.

5   Performance Model

In the previous sections the building blocks of the PCC-ISES are given: the architecture and the algorithms. The parameters of the architecture can be chosen according to the targeted application. There is always a trade-off between the size of the IC and the performance. In this Section we define a performance model and an area estimation, showing the influence of the different parameters on these criteria. The modular exponentiation is implemented as a repeated Montgomery multiplication. The number of multiplications for the exponentiation and thus the performance depends on the exponent. Re-coding of the exponent reduces the number of multiplications but this number stays of order of the exponent length [16]. Furthermore, the performance is determined by the performance of a multiplication. The operands of the multiplication are divided into words, not necessarily of the same length as described in Section 3. The words of X are divided over the PEs. When there are not enough PEs more rounds are needed. So the number of rounds is $\lceil \frac{n'}{P} \rceil$, where n' is the number of words X of length α bits. In one round each word of Y has to pass all PEs. Each PE calculation takes one clock-cycle. So, in P clock-cycles a word of Y passes all PEs in the array. When the number of words of Y is larger than the number of PEs the FIFO memory is used to store the intermediate results of the last PE. Now each round costs $max(P, n'')$ cycles, where n'' is the number of words of Y of length β bits. The performance (number of clock-cycles for one exponentiation) is now modeled as:

$$\left\lceil \frac{n'}{P} \right\rceil \cdot max(P, n'') \cdot c_1 \cdot n$$

$$n' = \left\lceil \frac{n}{\alpha} \right\rceil, \; n'' = \left\lceil \frac{n}{\beta} \right\rceil \tag{2}$$

where n is the length of the modulus in bits and $c_1 \cdot n$ is the number of multiplications needed for the exponentiation. The value for c_1 is typically 1.5 for the well known left to right square and multiply exponentiation method. When using exponent re-coding the value of c_1 is usually estimated with 1.2. In both estimations an exponent with a weight of approximately half of the length of the exponent is considered. From (2) is easy to conclude that for modulus lengths with $n'' < P$ the performance is quadratic in n. For larger modulus sizes the performance is cubic in the modulus length. The fixed word size of α and β bits (the ceiling function in the model) causes jumps in the performance curve (see Figure 3). The model shows that more PEs does not always lead to a better performance. It is most efficient to have a minimum number of PEs for which only one round is needed for a certain modulus length. As the number of PEs is fixed and the input of the LNAU is not of a fixed size, the chosen number of PEs cannot be optimal for all modulus lengths. Another restriction is that

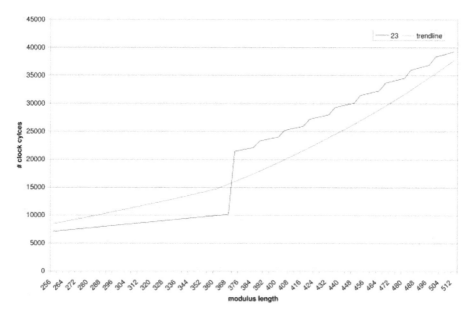

Fig. 3. Modular exponentiation performance curve for 23 PEs and its trendline.

usually the number of PEs is restricted because of a maximal implementation complexity. The performance can be further improved by implementing more than one LNAU. This is easy to achieve because each LNAU has a relatively low power consumption. The LNAUs perform exponentiations separately, so the performance is linear in the number of LNAUs. Hence the following model for the area occupied by the exponentiation engine is obtained:

$$L \cdot (c_2 \cdot \alpha \cdot \beta \cdot P + c_3) \tag{3}$$

where L is the number of LNAUs, c_2 is the area of one PE, capable of processing one bit of both X and Y, and c_3 denotes the area of the remaining components, like the control module and the FIFO.

In both of the models the following parameters: α, β, P, L are used. It can be concluded that the values of α and β must be larger for a better performance but also yield a larger area. Smaller α and β resume in a better overall performance. A larger number of LNAUs (L) improves the performance but does not change the latency, i.e. the time needed for one modular exponentiation.

6   Performance Figures

Figure 3 shows the performance curve for a modular exponentiation on a LNAU with 23 PEs and $\alpha = \beta = 16$. The effect of the ceiling functions can be observed as jumps in the graph.

Figure 4 shows the performance as function of the number of PEs. In this graph the peaks represent the extreme values of the ceiling function for the

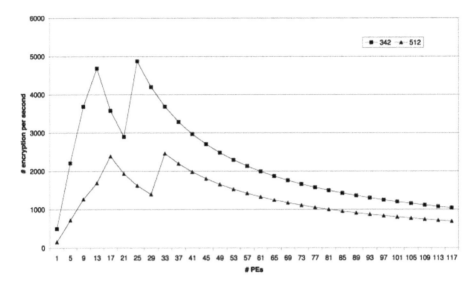

Fig. 4. Influence of the number of PEs

number of rounds. For example the maximal value between 5000 and 6000 encryptions per second stands for the beginning of the interval in which, for the given modulus size, only one round is needed. The second peak in the range of 5000-6000 marks the beginning of the interval for which 2 rounds are required. The modulus lengths are smaller than 1024 bits (512 and 342 bits) because of a computation which is using the Chinese Remainder Theorem (CRT). For example Compaq's $MultiPrime^{TM}$ with three primes is using this type of calculation ([19]). This is an example of the RSA crypto-system for which this platform is optimized. In this scheme instead of factors p, q of the composite N, three or more factors can be used.

Similar behavior of the performance figures can be found in [2]. Jumps in the graphs can be also observed. However, it is difficult to compare figures of two completely different architectures. Figure 5 presents the performance figures for various number of PEs. Here is shown that for different modulus lengths the number of PEs can have a different optimal value.

In ([23]) and ([7]) a remark is made on the fact that a 510 bit modulus has a more beneficial number of iterations than a 512 bit modulus. The assumption was that the difference in performance could go up to 13%. In our implementation the difference in performance between a 510 and 512 bit modulus is only 0.4%. In Figure 3 a jump of almost a factor of 2 can be observed between a modulus length of 368 and 372. However, for other number of PEs the difference between the 368 and 372 bit modulus is only 1%. So one should always try to choose a number of PEs for which the corresponding performance for the targeted modulus lengths lies under the trendline.

Table 1 presents the timings of the ISES-core (2 LNAUs) for modular exponentiation for various key-lengths and with/without CRT. This includes all over-

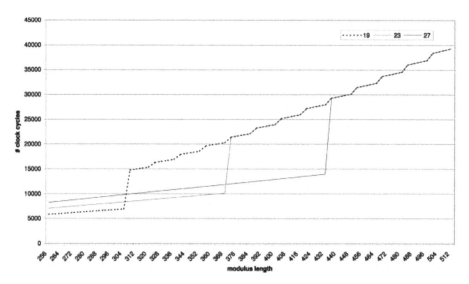

Fig. 5. Modular exponentiation for three different number of PEs

Table 1. Timings on the PCC-ISES with 2 LNAUs and clock-frequency of 50 Mhz.

	CRT (exp/s)	no CRT (exp/s)
512 bits	1879	1418
1024 bits	688	337
2048 bits	166	43

Table 2. Timing differences (in μs) between [5] and PCC-ISES for modular multiplication (80 MHz)

precision	Tenca and Koç ([5]) (μs)	This architecture (μs)	ratio
160 bits	4.1	4.51	0.9
192 bits	5.0	4.52	1.1
224 bits	5.9	4.54	1.3
256 bits	6.6	4.55	1.5
512 bits	-	5.95	-
1024 bits	61	9.41	6.5
2048 bits	-	26.55	-

head of the controller and pre-calculations for Montgomery, hence these timings represent the real-time performance of ISES.

In [5] a table with timings for multiplications in $GF(p)$ can be found. In Table 2 we compare these timings with the timings on 1 LNAU, even with controller overhead included, for modular multiplication. Improvement in timings when comparing this design with [5] is evident, except for 160 bits, where the performance is similar for both designs.

7   Remarks on Security

When considering efficient implementation one has to focus on the side channel attacks, first of all timing ([6]) and power analysis attacks ([13]). Ever since P. Kocher introduced a new type of attack ([12]), the so-called Differential Power Analysis attack (DPA for short), reasonable amount of research has been done on this subject. The same author published in 1996 ([11]) one of the relevant papers on time-difference based attacks. Namely, computations performed in non-constant time (usually because of performance optimisations) may leak secret key information. This observation is the basis for timing attacks. On the other hand, power analysis based attacks use the fact that the power consumed at any particular time during a cryptographic operation is related to the function being performed and (possibly sensitive) data being processed.

RSA appeared to be more vulnerable to side-channel attacks than ECC because it looks easier to protect ECC from these threats. However, as results of several contributions, some recommendations have been given for both types of public key cryptography, such as the benefit of using the method of Montgomery for multiplication. One of the reasons for that is the following: a modular reduction step may also be vulnerable and in the case of MMM at most one modular reduction is introduced for every multiplication or per step of exponentiation. In our implementation even these reductions are excluded. The weaknesses in the conditional statements of the algorithm (used for realization of the reduction step) are time variations and therefore these should be omitted. By use of an optimal upper bound the number of iterations required in the MMM can be reduced ([25]). More precisely, some savings in hardware that have been included in our architecture avoid the conditional statements while performing the exponentiation. In that way, our implementation of modular exponentiation operates in constant time, which is presumed to significantly reduce the potential risk of the timing attack.

When considering power analysis attacks, some precautions have also been introduced. Exponent re-coding is more difficult to attack than standard square-and-multiply and the fact that multiple LNAUs operate in parallel make these types of attacks far less likely to succeed.

8   Conclusions

We have described an efficient and scalable implementation of modular exponentiation in a systolic array architecture. After a number of proposals for this type of architecture in literature this is the first reported implementation, at our knowledge. We use the methods of Montgomery, which where proven to be very secure in hardware. Namely, the optimal bound is achieved which, with some savings in hardware, omits completely all reduction steps that are presumed to be vulnerable to side-channel attacks. Nevertheless the presented implementation has a world class performance in the sense of speed. Because of scalability, this architecture can be further improved on the basis of any desired requirement.

Acknowledgements

We are grateful to Cees Jansen for his contribution to all aspects of his work. We would like to thank Colin D. Walter for his papers and preprints as well. Comments and recommendations of anonymous referees influenced the final version of this paper substantially.

References

1. PCC-ISES datasheet, www.secure-a-link.com /pdfs/isespdf/ isesdata.pdf.
2. A.F.Tenca and Ç.K. Koç. A scalable architecture for Montgomery multiplication. *Lecture Notes in Computer Science, Springer-Verlag*, (1717):94–108, 1999. Cryptographic Hardware and Embedded Systems-CHES 1999.
3. T. Blum. Modular exponentiation on reconfigurable hardware. Master's thesis, Worcester Polytechnic Institute, April 1999.
4. S.E. Eldridge and C.D. Walter. Hardware implementation of Montgomery's modular multiplication algorithm. *IEEE Transactions on Computers*, (42):693–9, 93.
5. E.Savaç, A.F.Tenca, and Ç.K. Koç. A scalable and unified multiplier architecture for finite fields GF(p) and GF(2^m). *Lecture Notes in Computer Science, Springer-Verlag, Cryptographic Hardware and Embedded Systems-CHES 2000*, (1965):281–296, 2000.
6. G. Hachez, F. Koeune, and J.-J. Quisquater. Timing attack: what can be achieved by a powerful adversary? *Proceedings of the 20th symposium on Information Theory in the Benelux*, pages 63–70, May 1999.
7. G. Hachez and J.-J. Quisquater. Montgomery exponentiation with no final subtractions: Improved results. *Lecture Notes in Computer Science, Springer-Verlag*, (1965):293–301, 2000. Cryptographic Hardware and Embedded Systems-CHES 2000.
8. K. Iwamura, T. Matsumoto, and H. Imai. Systolic-arrays for modular exponentiation using Montgomery method. *Lecture Notes in Computer Science, Springer-Verlag*, 1440:477–481, 1981-1996. presented in Rumpsession of Eurocrypt 1992, May 24-28.
9. K. Iwamura, T. Matsumoto, and H. Imai. High-speed implementation methods for RSA scheme. *Lecture Notes in Computer Science, Springer-Verlag*, 658:221–238, 1992. Advances in Cryptology-EUROCRYPT 92.
10. K. Iwamura, T. Matsumoto, and H. Imai. Montgomery modular multiplication method and systolic arrays suitable for modular exponentiation. *Electronics and Communications in Japan*, 77(3):40–50, 1994.
11. P. Kocher. Timing attacks on implementations of Diffie-Hellman, RSA, DSS and other systems. *Lecture Notes in Computer Science, Springer-Verlag*, pages 104–113, 1996. Advances in Cryptology-CRYPTO 96.
12. P. Kocher, J. Jaffe, and B. Jun. Introduction to differential power analysis and related attacks. www.cryptography.com/dpa/technical, 1998.
13. P. Kocher, J. Jaffe, and B. Jun. Differential power analysis. *Lecture Notes in Computer Science, Springer-Verlag*, pages 388–397, 1999. Advances in Cryptology-CRYPTO 99.
14. Peter Kornerup. A systolic, linear-array multiplier for a class of right-shift algorithms. *IEEE Transactions on Computers*, 43(8):892–898, August 1994.

15. Erwin Kuipers. Design of an RSA crypto-processor using a systolic array. Master's thesis, Technical University of Eindhoven, The Netherlands, June 1996.
16. A. Menezes, P. van Oorschot, and S. Vanstone. *Handbook of Applied Cryptography.* CRC Press, 1997.
17. P. Montgomery. Modular multiplication without trial division. *Mathematics of Computation*, Vol. 44:519–521, 1985.
18. A. F. Tenca, Georgi Todorov, and Ç. K. Koç. High-radix design of a scalable modular multiplier. *Lecture Notes in Computer Science, Springer-Verlag*, (2162):189–205, 2001. Cryptographic Hardware and Embedded Systems-CHES 2001.
19. COMPAQ. Cryptography using Compaq MultiPrime technology in a parallel processing environment. www.compaq.com, Electronic Commerce Technical Brief, January 2000.
20. E. Trichina and A. Tiountchik. Scalable algorithm for Montgomery multiplication and its implementation on the coarse-grain reconfigurable chip. *Lecture Notes in Computer Science, Springer-Verlag*, (2020):235–249, 2001. Topics in Cryptology - CT-RSA 2001.
21. C.D. Walter. Montgomery exponentiation needs no final subtraction. *Electronic letters*, 35(21):1831–1832, October 1999.
22. C.D. Walter. Improved linear systolic array for fast modular exponentiation. *IEEE Computers and Digital Techniques*, 147(5):323–328, September 2000.
23. C.D. Walter. Precise bounds for Montgomery modular multiplication and some potentially insecure RSA moduli. *Lecture Notes in Computer Science, Springer-Verlag*, RSA 2002 Cryptographers' track (2271):30–39, 2002, (This Volume).
24. C.D. Walter. Systolic modular multiplication. *IEEE Transactions on Computers*, (42):376–378, 93.
25. C.D. Walter and S. Thompson. Distinguishing exponent digits by observing modular subtractions. *Lecture Notes in Computer Science, Springer-Verlag*, (2020):192–207, 2001. Topics in Cryptology – CT-RSA 2001.

Mist: An Efficient, Randomized Exponentiation Algorithm for Resisting Power Analysis

Colin D. Walter

Comodo Research Laboratory
10 Hey Street, Bradford, BD7 1DQ, UK
www.comodo.net

Computation Department, UMIST
PO Box 88, Sackville Street, Manchester, M60 1QD, UK
www.co.umist.ac.uk

Abstract. The MIST algorithm generates randomly different addition chains for performing a particular exponentiation. This means that power attacks which require averaging over a number of exponentiation power traces becomes impossible. Moreover, attacks which are based on recognising repeated use of the same pre-computed multipliers during an individual exponentiation are also infeasible. The algorithm is particularly well suited to cryptographic functions which depend on exponentiation and which are implemented in embedded systems such as smart cards. It is more efficient than the normal square-and-multiply algorithm and uses less memory than 4-ary exponentiation.

Key words: Mist exponentiation algorithm, division chains, addition chains, power analysis, DPA, blinding, smart card.

1 Introduction

Recent progress in side channel attacks [4], [5] on embedded cryptographic systems has exposed the need for new algorithms which can be implemented in more secure ways than those currently in use. This is particularly true for exponentiation, which is a major process in many crypto-systems such as RSA and Diffie-Hellman. Timing attacks on modular multiplication can usually be avoided easily by removing data-dependent conditional statements [12]. But, with timing variations removed, power attacks on exponentiation become easier. Initial power attacks required averaging over a number of exponentiations [5]. Although the necessary alignment of power traces can be made more difficult by the insertion of obfuscating, random, non-data-dependent operations, the data transfers between operations usually reveal the commencement of every long integer operation very clearly. Fortunately, such attacks can be defeated by modifying the exponent e to $e+rg$ where r is a random number and g is the order of the (multiplicative) group in which the exponentiation is performed [4]. This results in a different exponentiation being performed every time.

However, the author showed recently [11] that there were strong theoretical grounds for believing that, given the right monitoring equipment [2], it would

B. Preneel (Ed.): CT-RSA 2002, LNCS 2271, pp. 53–66, 2002.
© Springer-Verlag Berlin Heidelberg 2002

be possible to break the normal m-ary exponentiation method [3] and related sliding windows techniques using a single exponentiation. This method relies on being able to recognise the same multipliers being reused over and over, namely the pre-computed powers of the initial text, and requires no knowledge of the modulus, input text or output text. It renders useless the choice of $e+rg$ as a counter-measure, even for the case of $m = 2$, namely the standard square-and-multiply algorithm.

In an embedded system, the re-use of the same multipliers is useful because it reduces data movement. However, if such re-use is dangerous, other methods must be employed. Performing square-and-multiply in the opposite direction, namely consuming exponent bits from least to most significant, is the obvious starting point. With a random variation in the exponent as a counter-measure, this seems to defeat the attacks mentioned so far. Unfortunately, long integer squares and multiplications are among the easiest operations to distinguish in integer RSA [7]. Even the movement of data is different for the two operations. Thus an exponentiation algorithm is required which does not reveal its exponent through knowledge of the sequence of squares and multiplies.

A novel exponentiation algorithm is presented here which avoids all of the above-mentioned pitfalls. It can also be combined with most other counter-measures, such as using $e+rg$ instead of e. It relies on the generation of random addition chains [3] which determine the operations to be performed, and is based on previous work by the author [9] for finding efficient exponentiation schemes. This earlier work required extensive computation to establish near optimal addition chains, and so it is by no means obvious that the method can be used on-the-fly without an impractical overhead. Clearly, such computation can be performed in the factory prior to the production of an embedded crypto-system, and then each individual item can be issued with a different embedded addition chain which, although efficient, must be re-used on each exponentiation and is incompatible with the $e+rg$ counter-measure [1].

Here we develop that algorithm to the point where it is possible to efficiently generate fresh, efficient addition chains for any exponent and every exponentiation. The main aim here, besides presentation of the algorithm itself, is to establish that it has a time complexity better than square-and-multiply, and to note that, as far as space is concerned, only three long integer registers are required for executing the addition chain operations.

2   The MIST Algorithm

For notation, let us assume that M^E has to be computed. D will always represent a "divisor" in the sense of [9], and R a residue modulo D. A key ingredient of this algorithm is that we can efficiently compute both A^R and A^D from A using a single addition chain. A set of divisors is chosen in advance, and an associated table of these addition chains is stored in memory. Several variables are used in the code below. There are three which contain powers of M, namely $StartM$, $TempM$ and $ResultM$. Rather than destroy the initial value of E, we also use a variable called $RemE$ which represents the power to which $StartM$ still has to be raised before the exponentiation is complete. When the divisor set consists of the single divisor 2, the algorithm simplifies to the following right-to-left binary exponentiation algorithm, that is, to the standard version of square-and-multiply in which the least significant exponent bit is processed first:

RIGHT-TO-LEFT SQUARE-AND-MULTIPLY EXPONENTIATION ALGORITHM

```
{ Pre-condition: E ≥ 0 }
RemE    := E  ;
StartM  := M  ;
ResultM := 1  ;
While RemE > 0 do
Begin
    If (RemE mod 2) = 1 then
        ResultM := StartM×ResultM ;
    StartM  := StartM² ;
    RemE    := RemE div 2 ;
    { Loop invariant: M^E = StartM^{RemE} × ResultM }
End ;
{ Post-condition: ResultM = M^E }
```

Thus $StartM$ contains the initial value M raised to a power of 2, and it is the starting point for each loop iteration. $ResultM$ is the partial product which contains M raised to the power of the processed suffix of E and ends up containing the required output.

Without the *essential* random feature made possible by a larger divisor set, this is an insecure special case of MIST. In fact, we are assuming that the attacker can distinguish between squares and multiplies. Hence he can read off the bits of the exponent in the above from the sequence of long integer instructions which he deduces. However, unlike the left-to-right version of square-and-multiply, both arguments in the conditional product are now changed for every multiplication. This, at least, makes an attack of the type [11] impractical.

Next is the generalisation of this, which is named "MIST" because of its use in obscuring the exponent from side channel attacks. For convenience and comparison, the processing of the exponent E is presented as being performed within the main loop. In practice, it *must* be scheduled differently. As will be pointed out below, the illustrated processing order is insecure from the point of differential power analysis because it can reveal the random choice of the divisor D, which *must* remain secret. In fact, the choices of divisors and computations with the exponent must be done initially in a secure way, or at least interleaved with the long integer operations in a more considered way. To be even more precise, the complete schedule of long integer operations, i.e. the addition chain which determines the exponentiation scheme, should be computed with great care in order to hide the choice of divisors.

THE MIST EXPONENTIATION ALGORITHM
{ before proper re-scheduling of addition chain choices }

```
{ Pre-condition: E ≥ 0 }
RemE    := E  ;
StartM  := M  ;
ResultM := 1  ;
While RemE > 0 do
Begin
    Choose a random "divisor" D ;
    R := RemE mod D ;
    If R ≠ 0 then
        ResultM := StartMᴿ×ResultM ;
    StartM  := StartMᴰ ;
    RemE    := RemE div D ;
    { Loop invariant: Mᴱ = StartMᴿᵉᵐᴱ×ResultM }
End ;
{ Post-condition: ResultM = Mᴱ }
```

Observe that there are no powers of M which are repeatedly re-used during the exponentiation, so that the attack described in [11] on a single exponentiation is inapplicable in its current form. Also, the random choice of divisors achieves different exponentiation schemes on successive runs and so makes impossible the usual averaging process required for differential power analysis [5].

For the proof of correctness, it is immediate that the stated invariant

$$M^E \quad = \quad StartM^{RemE} \times ResultM$$

holds at the start of the first iteration of the loop. Because initial and final values of $RemE$ for one loop iteration satisfy

$$RemE_{Initial} \quad = \quad D \times RemE_{Final} + R$$

it is easy to check that if the invariant holds at the start of an iteration then it holds again at the end of the iteration. Consequently, the invariant holds at the end of every iteration. In particular, at the end of the last iteration, we have $RemE = 0$ and hence, by simplifying the invariant, $ResultM = M^E$. So $ResultM$ yields the required result on termination. Needless to say, termination is guaranteed because only divisors greater than 1 are allowed, and so RemE decreases on every iteration.

The choices of divisor set and associated addition chains for each residue R are made with security and efficiency in mind. In particular, for efficiency the choice of addition chain for raising to the power D always includes R so that the computation of $StartM^D$ provides $StartM^R$ *en route* at little or no extra cost. Thus, these two power computations are not performed independently, as might be implied by the code. They are to be implemented so that $StartM^D$ uses all the work done already to compute $StartM^R$. So, in the case of RSA, the main cost of a loop iteration is only the cost of computing $StartM^D$ plus the conditional extra multiplication involving $ResultM$. In terms of space, an extra long integer variable $TempM$ is required to enable the addition chain operations to be carried out. Its value is always a power of the value of $StartM$ at the beginning of the main loop above, but its value does not need to persist between successive iterations of the loop. Of course, on the final iteration, once $StartM^R$ has been computed there is no need to continue to compute the rest of $StartM^D$. Likewise, the initial multiplication of $ResultM$ by 1 can be easily omitted.

For security reasons, some care is necessary in the initial choice of divisor set and addition chains, and in how and when the divisors are chosen for each exponentiation. The selection of the divisor and associated addition chain instructions can be performed by the CPU on-the-fly while the exponentiation is performed in parallel by the co-processor, or it can be done in advance when there is no co-processor. At any rate, these computations must be scheduled so as not to reveal the end points of each iteration of the main loop. Otherwise, the number and type of long integer operations during the loop iteration may leak the values of D and R, enabling E to be reconstructed.

A typical safe set of divisors is $\{2,3,5\}$. If the exponent is represented using a radix which is a multiple of every divisor (say 240 in this case if an 8-bit processor is being used) then the divisions of the exponent by D become trivial. So, over and above the operations for executing the addition chain, the cost of the algorithm is little more than that for calling the random number generator to select each D.

3   The Addition Sub-chains and Space Requirements

When divisor choices are made during pre-processing, the sequence of operations to perform the exponentiation is stored as an *addition chain* [3]. For a safe implementation, the pattern of squares and multiplies in this chain must not reveal too much about the divisors and residues. However, for the complexity considerations of this and the following section, we only need to look at the subchain associated with a single divisor. Such subchains can be concatenated to yield an addition chain for the whole exponentiation, if desired.

Let us choose the divisor set to be $\{2,3,5\}$. The full list of minimal addition subchains can be represented as follows:

1+1=2	for divisor 2 with any residue R
1+1=2, 1+2=3	for divisor 3 with any residue R
1+1=2, 1+2=3, 2+3=5	for divisor 5 with any residue except 4
1+1=2, 2+2=4, 1+4=5	for divisor 5 with any residue except 3

The third case corresponds to using an initial M^1 to compute M^2 with one multiplication, then M^3 with another multiplication, and finally M^5 with a third multiplication. The first three addition chains provide M^R when R is 0, 1, 2 or 3: for $0 < R < D$ the chain already contains the value of R, while the case $R = 0$ requires no multiplication and so 0 does not need to appear. The last addition chain can be used when $R = 4$. *Minimal* here means that any other addition chains which give a power equal to the divisor are longer. The subchains above are minimal. To achieve the fastest exponentiation, we will not include longer chains. However, there may be extra cryptographic strength in extending the choice in such a way.

There is no instruction which updates the value of *ResultM* in the above addition subchains, but it can be represented explicitly using the following notation. Suppose we number the registers 1 for *StartM*, 2 for *TempM* and 3 for *ResultM*. Then the subchains can be stored as sequences of triples $ijk \in \{1, 2, 3\}^3$, where ijk means read the contents from registers i and j, multiply them together, and write the product into register k. In particular, *ResultM* will always be updated using a triple of the form $i33$ and 3 will not appear in triples otherwise. Now, adding in the instruction for updating *ResultM* yields the following as a possible list of subchains, with one representative for each divisor/residue pair $[D, R]$. Such a table requires only a few bytes of storage.

Table 3.1. A Choice for the Divisor Sub-Chains.

$[2, 0]$	(111)
$[2, 1]$	(112, 133)
$[3, 0]$	(112, 121)
$[3, 1]$	(112, 133, 121)
$[3, 2]$	(112, 233, 121)
$[5, 0]$	(112, 121, 121)
$[5, 1]$	(112, 133, 121, 121)
$[5, 2]$	(112, 233, 121, 121)
$[5, 3]$	(112, 121, 133, 121)
$[5, 4]$	(112, 222, 233, 121)

Many other choices are possible. In particular, we might prefer to preserve the location of *StartM* to be in register 1 from one divisor to the next. This could be achieved by choosing (133, 111) instead of (112, 133) for [2,1]. Then subchains of triples ijk could be concatenated without modification to provide the complete exponentiation scheme. However, for the given subchain, the new value for *StartM* is in register 2 instead of register 1. Rather than waste time copying, the computation should continue with 2 as the address for *StartM*, and the next subchain then has to be updated with the register addresses 1 and 2 interchanged. However, as noted in the penultimate section, this apparently less obvious choice for [2,1] makes the implementation more secure against differential power analysis.

Following this last remark, it is clear that we could provide an additional source of randomness by writing the product into any register whose current contents are no longer required [6]. Thus the purpose of each register can be changed. If we include every possible minimal addition subchain with such variations, then we obtain 2, 6, 2, 6, 4, 4, 16, 8, 4 and 4 possibilities respectively for the 10 divisor/residue cases. The storage is still only a few bytes, but it provides an extra source of randomness which may provide added security.

Thus the space order required for MIST is low. In the next section we turn to the time complexity, which we measure in terms of the number of multiplications in the exponentiation scheme.

4 The Time Complexity

The usual square-and-multiply algorithm uses $1.5 \times \lfloor \log_2 E \rfloor$ multiplicative operations (including squarings) on average and a maximum of at most $2 \times \lfloor \log_2 E \rfloor$. In this section almost the same upper bound will be established for MIST, and an improved average. In practice the variance is very small. Consequently, if a pre-computed scheme involved more than $1.5 \times \lfloor \log_2 E \rfloor$ multiplications, say, it could be abandoned and an alternative scheme computed.

> **Theorem 4.1** Assume minimal subchains are provided as above for the divisor set $\{2,3,5\}$ and $E > 0$, and unnecessary initial and final multiplications have been omitted. Then the maximum number of operations in an addition chain for E is at most $2 \log_2 E$ with equality possible only for $E = 1$.

Proof. We use a proof by descent: assuming E is a smallest counter-example, we will construct a smaller counter-example E' for which the theorem fails. Such a contradiction will prove the theorem.

First assume that at least two divisors are required to reduce the minimal counter-example E to 0. We will consider the three choices for the first divisor of E in turn.

Suppose the first divisor is 3. Then $E' = E$ div 3 is the next value of $RemE$ after E and $3E' \le E$. Let m be the number of multiplications for E, and k the number for this first divisor 3. Then $k = 2$ or 3, so that $k < 2(\log_2 3) \approx 2 \times 1.585$. By assumption, $m \ge 2 \log_2 E$. So $m \ge 2 \log_2(3E') = 2 \log_2(E') + 2(\log_2 3)$ gives $m - k > m - 2(\log_2 3) \ge 2 \log_2(E')$. But $m - k$ is the number of multiplications for E'. Hence we obtain a smaller E for which the theorem does not hold. This is a contradiction unless $E' = 0$. However, that is impossible as there is another divisor in the exponentiation scheme.

Similarly, suppose the first divisor is 5 and E' is the next value for $RemE$. Then the associated subchain requires $k = 3$ or 4 multiplications, so that $k < 2(\log_2 5) \approx 2 \times 2.322$. Thus, a similar argument yields a contradiction again.

Lastly, suppose the first divisor is 2 and E' is defined again as the next value for $RemE$. If E is odd, then $2E' < E$ and the corresponding division requires $k = 2 = 2(\log_2 2)$ multiplications. By assumption, $m \ge 2 \log_2 E > 2 + 2 \log_2 E'$. So the number of multiplications used by E' is $m - 2 > 2 \log_2(E')$ and E' is also a failing instance unless $E' = 0$, which is a contradiction as there is still another divisor to come. On the other hand, if E is even, then the corresponding division requires $k = 1$ multiplication. By assumption, $m \ge 2 \log_2 E = 2 + 2 \log_2 E'$. So the number of multiplications used by E' is $m - 1 > 2 \log_2(E')$ and E' is also a failing instance unless $E' \le 0$, which, as before, is a contradiction.

It remains to consider the cases of a single divisor reducing E to 0. For all of these, the chosen divisor satisfies $D > E$ since E div $D = 0$ and so $R = E$. The only multiplications are those which construct M^E. For $E = 1, 2, 3, 4$ the number of multiplications are 0, 1, 2 and 2 respectively, all of which satisfy the theorem. Hence, by the method of descent, there are no failing instances, and the theorem holds.

It may be worth noting that removal of a divisor 3 or 5 at any point from the addition chain for E (other than the last) yields a chain for which the bound in the theorem is tighter. The same is true when the divisor is 2 and the residue 0. Thus, the tightest bounds (i.e. the most multiplications for given E) will occur when only divisor 2 is chosen and the residue is always 1. Then $E = 2^n - 1$ for some n and E requires $n-1$ subchains of 2 operations to reduce $RemE$ to $2^1 - 1$, leading to $2n - 2$ operations in all. So $2 \times \lfloor \log_2 E \rfloor$ would normally hold as an upper bound. However, it requires more care to establish the exceptions for this slightly better bound.

5   A Weighting for the Choice of Divisor

Suppose k is the number of multiplications required for the divisor/residue pair (D, R). The result of picking divisor D is that $RemE$ is reduced by a factor which is very close to D, especially when $RemE$ is large. So, we would expect the total number of multiplications to be close to $k \log_D E$ if (D, R) occurred for every divisor. Consequently, the cost of using (D, R) is proportional to $k/\log D$. We might use this to bias the choice of divisor in an attempt to decrease the number of multiplications. However, this is not the best measure because the effect of larger divisors is longer lasting than that of smaller divisors. A small divisor with many multiplications, which is expected to be followed by divisors with an average number of multiplications, may be a better choice than picking a larger divisor with a better, but poorer than average, ratio $k/\log D$ (see [9]).

Suppose α is the average number of multiplications required to reduce E by a factor of 2. Then reducing E by a factor F will require $\alpha \log_2 F$ multiplications, on average. For the usual square-and-multiply algorithm $\alpha = 1.5$, for 4-ary exponentiation $\alpha = 1.375$, and for the sliding window version of 4-ary exponentiation $\alpha = 1.333...$ Here we can manage just a little better than $\alpha = 1.4$ with appropriate choices for the divisors[1]. Comparing (3,0) with (5,0) we find 3 multiplications will reduce E by a factor of 5 when (5,0) is chosen, whereas the same factor of 5 takes $2 + \alpha \log_2(5/3)$ multiplications on average if (3,0) is chosen instead. Equating these costs, $3 = 2 + \alpha \log_2(5/3)$, provides the cross-over point $\alpha' = 1/\log_2(5/3) = 1.357$ at which one becomes a better choice than the other. Since $\alpha' < \alpha$, for speed we should choose (5,0) in preference to (3,0) although the ratios $k/\log D$ suggested the opposite preference. In a similar way, for the expected range of α, (3,0) is preferred to (5, R) for $R \neq 0$; (5, R) to (3, S) for R,

---

[1] Choose $D = 2$ if the residue is 0, else $D = 3$ if the residue is 0, else $D = 5$ if the residue is 0, else $D = 2$. However, the probabilities of these choices need reducing to less than 1 in order to yield randomly different exponentiation schemes.

$S \neq 0$; $(3, R)$ to $(2,1)$ for $R \neq 0$; and $(5, R)$ to $(2,1)$ for $R \neq 0$. This yields the following order of desirability for the pairs (D, R):

$$(2,0) < (5,0) < (3,0) < (5,1) = (5,2) = (5,3) = (5,4) < (3,1) = (3,2) < (2,1)$$

So $(2,0)$ will lead to the shortest chains, and $(2,1)$ to the longest.

Once more we caution that this is only a better approximation than the previous one. Successive divisors are not independent, so that the above argument is not quite accurate. This is clear from an example. We consider the extreme case where divisibility by a divisor always leads to the choice of that divisor. If 5 is chosen as the divisor and its residue is 0 then its residue mod 30 was 5 or 25. Residue 5 mod 30 leads to the next residue mod 30 being 1, 7, 13, 19 or 25, and residue 25 mod 30 leads to it being 5, 11, 17, 23 or 29. Hence 5 becomes the most likely choice for the next divisor as divisibility by 2 or 3 cannot to occur. Indeed, the dependence is inherited by the next residue beyond this as well, since these residues favour 5 as the next divisor and so 5 or 25 as the following residue.

Choosing one of the three divisors with equal probability leads to an inefficient process. So we make a choice which is biased towards the better pairs (D, R). For non-trivial reasons outlined in the penultimate section, it is less safe cryptographically to make a deterministic choice of divisor even if the exponent is modified randomly on every occasion. So 2 is not chosen whenever its residue is 0. Instead, we might use code such as the following for choosing the divisor non-deterministically. (*Random* returns a fresh random real in the range $[0,1]$.)

```
D := 0 ;
If Random(x) < 7/8 then
      If 0 = RemE mod 2 then D := 2 else
      If 0 = RemE mod 5 then D := 5 else
      If 0 = RemE mod 3 then D := 3 ;
   If D = 0 then
   Begin
      p := Random(x) ;
      If p < 6/8 then D := 2 else
      If p < 7/8 then D := 3 else
                      D := 5 ;
   End ;
```

The parameters, such as 6/8 and 7/8, are free for the implementor to choose, and might even be adjusted dynamically. This is the code which will be used for calculating the various parameters of interest in the rest of this article, such as α, which turns out to be 1.4205 here. So a good implementation of MIST can be expected to have a time efficiency midway between the square-and-multiply and 4-ary exponentiation methods.

6   A Markov Process

As the successive residues of $RemE$ mod 30 are not independent, we must study them as forming a Markov process. By forming the probability matrix of output

Table 6.1. The limit probabilities of residues mod 30.

0	0.02914032	1	0.03448691	2	0.01660770
3	0.05345146	4	0.01884630	5	0.03655590
6	0.02920902	7	0.04100675	8	0.02436897
9	0.04985984	10	0.02011853	11	0.04919923
12	0.02014301	13	0.04214864	14	0.03407472
15	0.03681055	16	0.01526795	17	0.03368564
18	0.03160194	19	0.03600811	20	0.03187408
21	0.04915191	22	0.02166433	23	0.04323007
24	0.03197409	25	0.02706484	26	0.03224476
27	0.04020936	28	0.02102305	29	0.04897205

Table 6.2. The limit probabilities $p_{D,R}$ of the divisor/residue pairs (D, R)

(D, R)	0	1	2	3	4
2	0.35012341	0.27101192	-	-	-
3	0.18867464	0.02086851	0.02419582	-	-
5	0.09792592	0.01202820	0.01060216	0.01227849	0.01229092

residues against input residues and iterating a number of times, it is possible to obtain the limiting relative frequencies of the residues. These are given in Table 6.1, from which it is apparent that the residues do not occur with equal frequency, and so, as one would expect, the probabilities of divisors and divisor/residue pairs are also not what we might initially expect from the code above.

From this table the probability $p_{D,R}$ of each divisor/residue pair (D, R) can be obtained as well as the probability $p_{\sim 0}$ of selecting a divisor with a non-zero residue and the probability p_D of each divisor D:

7   Average Properties of the Addition Chain

The probabilities in the above tables are fairly accurate after only a small number of divisors have been applied. So the following results hold very closely for any exponents related to integer RSA decryption.

Theorem 7.1 The average subchain length is just under 1.89 operations per divisor as $E \to \infty$.

Proof. With probability $p_{2,0} = 0.35012341$ the subchain has length 1, with probability $p_{3,0}+p_{2,1} = 0.45968656$ the subchain has length 2, with probability $p_{5,0}+p_{3,1}+p_{3,2} = 0.14299025$ the subchain has length 3, and with probability $p_{5,1}+p_{5,2}+p_{5,3}+p_{5,4} = 0.04719977$ the subchain has length 4. The average length of a subchain is therefore $1(p_{2,0}) + 2(p_{3,0}+p_{2,1}) + 3(p_{5,0}+p_{3,1}+p_{3,2}) + 4(p_{5,1}+p_{5,2}+p_{5,3}+p_{5,4}) = 1.88726638$.

Theorem 7.2 The average number of subchains in the addition chain for E is approximately $0.75 \times \log_2 E$ as $E \to \infty$.

Table 6.3. The limit probability p_D for each divisor D.

$$p_2 = 0.62113534$$
$$p_3 = 0.23373897$$
$$p_5 = 0.14512570$$

Proof. Using the divisor probabilities listed in Table 6.3, the average decrease in size of $RemE$ due to a single subchain is by the factor $2^{p_2} 3^{p_3} 5^{p_5} = 2\uparrow\{0.6211353 \times \log_2 2 + 0.233739 \times \log_2 3 + 0.1451257 \times \log_2 5\} \approx 2^{1.328574}$. Hence the average number of subchains is about $\log_{2\uparrow(1.328574)} E = 0.75268656 \times \log_2 E$.

Theorem 7.3 The average number of operations in the addition chain for E is approximately $1.42 \times \log_2 E$ as $E \to \infty$. This is about 3% above the $1.375 \times \log_2 E$ of the 4-ary method, and noticeably less than the $1.5 \times \log_2 E$ of the square-and-multiply method.

Proof. Using the results of the last two theorems, the average number of operations for the whole addition chain is approximately

$$1.88726638 \times 0.75268656 \times \log_2 E \quad \approx \quad 1.42052005 \times \log_2 E$$

For small exponents E the approximations are slightly more inaccurate because modular division by the divisor produces a result which differs more from rational division. However, each subchain reduces the exponent by *at least* the divisor. Hence exact calculations here should yield an *upper* bound on the average.

Lastly, we note that the number of multiplications in the exponentiation schemes for a given exponent E does vary between different executions. The variance is usually small but depends on the method of picking divisors. Typically, the choice of divisors is restricted, as here, in order to improve efficiency, and this reduces the variance. Indeed, in the limit, deterministic choices lead to zero variance for fixed E.

8 Data Leakage

Because of the difficulty of successfully hiding all the differences between squaring and non-squaring multiplications, the MIST algorithm has been created in order to make it impossible to deduce the secret exponent E from leaked knowledge of the sequence of square or multiply instructions which perform an exponentiation. A detailed exposition of how easy it is to reconstruct the exponent from this and other related data is beyond the scope of this article. However, we will outline some of the issues. Fuller details will be published elsewhere.

Standard power analysis attacks [5] average traces over a number of exponentiations in order to determine information such as which operations are multiplications and which are squares. This requires the same exponentiation scheme to be used every time. However, assuming the choice of divisors does vary from one exponentiation to the next (or at least changes with sufficient frequency), such an averaging process cannot be carried out here.

Differences between squaring and multiplying are so great that one should assume that they can be correctly distinguished for a single exponentiation. In the case of the standard square-and-multiply methods, such knowledge can be translated immediately into the bit sequence for E. But here the equivalent process requires first parsing the sequence of squares and multiplies into divisor subchains in order to deduce each choice of divisor/residue pair. Then E is reconstructed by working backwards from the final value of 0. However, the subsequences are identical for various pairs, such as for $[2,1]$ and $[3,0]$, and for $[3,1]$, $[3,2]$ and $[5,0]$. It is easy enough to verify that these ambiguities put the number of possibilities for E far outside the limits of feasible computation, making MIST secure against such an attack. This was one reason for choosing the less obvious subchain for $[2,1]$ given in Table 3.1.

The m-ary and sliding windows methods of exponentiation can be defeated for a single exponentiation if the power attack described in [11] can be applied. That attack depends on identifying the re-use of multiplicands. In particular, whenever a digit i is encountered in the exponent, a multiplication is performed using the pre-computed value M^i. Then careful averaging of the power trace subsections enables multiplications which involve the same power M^i to be identified, and this leads to recovery of the exponent E.

The same attack has an analogue here. Every multiplication here involves a new power of M. More precisely, $StartM$ and $ResultM$ represent higher powers of M every time they are updated. So the possibilities of two multiplications sharing the same multiplicand are mostly limited to local considerations. If operations which share a common argument can be identified, it usually becomes possible to determine uniquely the subchains which correspond to each divisor. Of course, a careless implementation of MIST might do this anyway: for example, if the exponentiation is halted in every iteration of the main loop while determination of the next divisor is made. However, although knowledge of argument sharing now distinguishes $[3,1]$, $[3,2]$ and $[5,0]$, it does not distinguish $[2,1]$ from $[3,0]$. Thus it does not necessarily determine the divisor which was chosen. A more careful estimate of the number of exponents E which satisfy all the operand sharing requirements shows that such an attack is still infeasible.

Finally, it would be nice to make deterministic choices of the divisor. In particular, one would like to pick $[2,0]$ when possible because of its higher efficiency. However, this is generally a bad idea, and was deliberately avoided in the code suggested in Section 5. This is because when such choices are used to prune the possible values of E which are deduced from knowledge of operand sharing, it may become feasible to recover E.

The choices of the divisor set $\{2,3,5\}$ and the addition subchains in Table 3.1 have been made with some care to ensure such attacks as the above leave an infeasible number of values which E might be. Small divisors lead to short addition subchains and hence more ambiguity over which divisor occurred, but large divisors may lead to characteristic patterns which identify conditions under which the search tree for E can be pruned to a computationally feasible size.

9   Conclusion

An exponentiation algorithm "MIST" has been presented which has a variety of features which make it much more resilient to attack by differential power analysis than the normal m-ary or sliding window methods. MIST uses randomly different multiplication schemes on every run in order to avoid the averaging which is normally required for power analysis attacks to succeed. It also avoids re-using multiplicands within a single exponentiation, thereby defeating some of the more recent power analysis attacks.

There are many different ways in which to program the random selection of the so-called divisors of the algorithm. For the code presented here, about $1.42 \log_2 E$ multiplications are required for the exponent E, thereby making it more efficient than square-and-multiply. Three read/write registers are required for storing intermediate powers, together with somewhat less memory for storing a pre-computed addition chain which determines the exponentiation scheme. Otherwise there is little extra overhead in terms of either space or time. In a processor/co-processor set-up, the additional minor computing associated with the exponent can be carried out on the processor in parallel with, and without holding up, the exponentiation on the co-processor.

As with all algorithms, poor implementation can lead to data leakage. The main sources of such weakness have been identified. In particular, divisor sets and addition subchains must be chosen carefully and a predictable choice of any divisor must be avoided.

MIST is compatible with most other blinding techniques, and independent of the methods used for other arithmetic operators. So, in conjunction with existing methods, it offers a sound basis for high specification, tamper resistant crypto-systems.

References

1. C. Clavier & M. Joye, *Universal Exponentiation Algorithm*, Cryptographic Hardware and Embedded Systems – CHES 2001, Ç. Koç, D. Naccache & C. Paar (editors), Lecture Notes in Computer Science, **2162**, Springer-Verlag, 2001, 300–308.
2. K. Gandolfi, C. Mourtel & F. Olivier, *Electromagnetic Analysis: Concrete Results*, Cryptographic Hardware and Embedded Systems – CHES 2001, Ç. Koç, D. Naccache & C. Paar (editors), Lecture Notes in Computer Science, **2162**, Springer-Verlag, 2001, 251–261.
3. D. E. Knuth, *The Art of Computer Programming*, vol. **2**, "Seminumerical Algorithms", 2nd Edition, Addison-Wesley, 1981, 441–466.
4. P. Kocher, *Timing Attack on Implementations of Diffie-Hellman, RSA, DSS, and other systems*, Advances in Cryptology – CRYPTO '96, N. Koblitz (editor), Lecture Notes in Computer Science, **1109**, Springer-Verlag, 1996, 104–113.

5. P. Kocher, J. Jaffe & B. Jun, *Differential Power Analysis*, Advances in Cryptology – CRYPTO '99, M. Wiener (editor), Lecture Notes in Computer Science, **1666**, Springer-Verlag, 1999, 388–397.

6. D. May, H.L. Muller & N.P. Smart, *Random Register Renaming to Foil DPA*, Cryptographic Hardware and Embedded Systems – CHES 2001, Ç. Koç, D. Naccache & C. Paar (editors), Lecture Notes in Computer Science, **2162**, Springer-Verlag, 2001, 28–38.

7. T. S. Messerges, E. A. Dabbish & R. H. Sloan, *Power Analysis Attacks of Modular Exponentiation in Smartcards*, Cryptographic Hardware and Embedded Systems (Proc CHES 99), C. Paar & Ç. Koç (editors), Lecture Notes in Computer Science, **1717**, Springer-Verlag, 1999, 144–157.

8. E. Oswald & M. Aigner, *Randomized Addition-Subtraction Chains as a Countermeasure against Power Attacks*, Cryptographic Hardware and Embedded Systems – CHES 2001, Ç. Koç, D. Naccache & C. Paar (editors), Lecture Notes in Computer Science, **2162**, Springer-Verlag, 2001, 39–50.

9. C. D. Walter, *Exponentiation using Division Chains*, IEEE Transactions on Computers, **47**, No. 7, July 1998, 757–765.

10. C. D. Walter & S. Thompson, *Distinguishing Exponent Digits by Observing Modular Subtractions*, Topics in Cryptology – CT-RSA 2001, D. Naccache (editor), Lecture Notes in Computer Science, **2020**, Springer-Verlag, 2001, 192–207.

11. C. D. Walter, *Sliding Windows succumbs to Big Mac Attack*, Cryptographic Hardware and Embedded Systems – CHES 2001, Ç. Koç, D. Naccache & C. Paar (editors), Lecture Notes in Computer Science, **2162**, Springer-Verlag, 2001, 286–299.

12. C. D. Walter, *Precise Bounds for Montgomery Modular Multiplication and Some Potentially Insecure RSA Moduli*, Topics in Cryptology – CT-RSA 2002, B. Preneel (editor), Lecture Notes in Computer Science, **2271**, Springer-Verlag, 2002, 30–39, This Volume.

An ASIC Implementation of the AES SBoxes*

Johannes Wolkerstorfer[1], Elisabeth Oswald[1], and Mario Lamberger[2]

[1] Institute for Applied Information Processing and Communications,
Graz University of Technology, Inffeldgasse 16a, A-8010 Graz, Austria
Johannes.Wolkerstorfer@iaik.at, http://www.iaik.at

[2] Department of Mathematics
Graz University of Technology, Steyrergasse 30, A-8010 Graz, Austria

Abstract. This article presents a hardware implementation of the S-Boxes from the Advanced Encryption Standard (AES). The SBoxes substitute an 8-bit input for an 8-bit output and are based on arithmetic operations in the finite field $GF(2^8)$. We show that a calculation of this function and its inverse can be done efficiently with combinational logic. This approach has advantages over a straight-forward implementation using read-only memories for table lookups. Most of the functionality is used for both encryption and decryption. The resulting circuit offers low transistor count, has low die-size, is convenient for pipelining, and can be realized easily within a semi-custom design methodology like a standard-cell design. Our standard cell implementation on a 0.6 μm CMOS process requires an area of only 0.108 mm^2 and has delay below 15 ns which equals a maximum clock frequency of 70 MHz. These results were achieved without applying any speed optimization techniques like pipelining.

Keywords: Advanced Encryption Standard (AES), finite field arithmetic, inversion, Application Specific Integrated Circuit (ASIC), standard-cell design, Very Large Scale Integration (VLSI), scalability, pipelining.

1 Introduction

The Advanced Encryption Standard (AES) is a symmetric encryption algorithm. It will become a FIPS standard in Fall 2001 [1]. AES will replace the DES-algorithm in the coming years since it offers higher levels of security. AES supports key lengths of 128, 192, and 256 bits. It operates on 128-bit data blocks. The major building blocks of the AES algorithm are the non-linear SBoxes (SubByte-operation) and the MixColumn-operation. Both are based on finite field arithmetic and have an inverse function which is used for decryption.

---

* The work described originates from the European Commission funded Project *Secure Terminal IC (SETIC)* established under contract IST-2000-25167 resp. *Crypto Module with USB Interface (USB_CRYPT)* established under contract IST-2000-25169 in the Information Society Technologies (IST) Program.

B. Preneel (Ed.): CT-RSA 2002, LNCS 2271, pp. 67–78, 2002.
© Springer-Verlag Berlin Heidelberg 2002

```
AESROUND () {
    SubByte(State);
    ShiftRow(State);
    MixColumn(State);
    AddRoundKey(State, RoundKey);
}
```

Fig. 1. Round function of the AES-algorithm

The AES-algorithm's operations are performed on a two-dimensional array of bytes called the *State*. The State consists of four columns and four rows of bytes. For both encryption and decryption, the AES-algorithm uses a round function that is composed of four different transformations which modify the State (see Fig. 1). First, the SubByte-function substitutes all bytes of the State using a lookup-table called SBox. SBox-table entries are calculated by inversion in the finite field $GF(2^8)$ followed by a short final transformation. Second, the rows of the State are shifted by different offsets (ShiftRow-function). The MixColumn-function scrambles the columns of the State by multiplying a finite field constant. An addition of the State with the Roundkey – which is derived from the input key – concludes the round function which is executed ten times when 128-bit keys are used. The RoundKey is calculated by operations which are similar to those of the round function and require the SBox functionality too.

The efficiency of an AES hardware implementation in terms of die-size, throughput, and power consumption is mainly determined by the implementation of the MixColumn-operation and the SBoxes. The remaining operations are trivial: ShiftRow is a simple cyclic shift, and AddRoundKey is a XOR-operation of the State and the RoundKey. Up to 20 instances of AES-SBoxes are used to realize hardware for the AES round function. The exact number of SBoxes depends on the architecture's degree of parallelism and is determined by throughput requirements and the desired clock frequency. In case that the AES-module should also decrypt data, it has to be taken into account that the SBoxes used for decryption have a different functionality. The number of SBoxes and their style of implementation has important influence on the size and the speed of an AES hardware. For this reason, V. Rijmen (one of the AES inventors) suggests in [4] an alternative method for the computation of the AES-SBox. It consists essentially of a replacement of the SBox lookup-table by an efficient combinational logic for the computation of the inverse elements in $GF(2^8)$. Therefore, another representation of the finite field $GF(2^8)$ is used. This representation leads to an efficient implementation of the finite field arithmetic and was investigated in connection with the implementation of error correcting codes in C. Paar [7], Soljanin et al. [5], and Mastrovito [6]. In contrast to V. Rijmen's original proposal which additionally suggests the optimal normal basis representation of finite field elements (for a definition see [3]) we use the polynomial representation of finite field elements. The benefit of our method is that we have a far more flexible hardware architecture (in comparison to the possible architectures with a straightforward SBox implementation) without the necessity to do complex conversions from one representation (of finite field elements) to another. The main advantages of our architecture are:

- lower transistor count and die-size than a ROM-based approach,
- a short critical path to achieve a high operational frequency,
- easier implementation within a semi-custom design methodology since all computations can be done with standard-cells,
- flexibility for speed optimization: pipelining techniques can trade throughput for latency.
- suitability for a full-custom implementation: a few leaf-cells using an appropriate logic-style could increase speed or decrease power consumption.

The remainder of this article provides the mathematical background of the finite field arithmetic and the computation of the AES SBoxes in Sect. 2. The building blocks of an SBox and the according formulas are given in Sect. 3. Section 4 presents the implementation.

2   Mathematical Background

This article uses the same notation and conventions as the AES specification [1]. All notations and mathematical operations required for the SBox-operation are presented in a condensed form.

Bytes. The basic data unit of AES are Bytes $a = \{a_7, a_6, a_5, a_4, a_3, a_2, a_1, a_0\}$ each holding eight bits. A Byte can be interpreted as an element of the Galois-Field $GF(2^8)$ in polynomial representation:

$$a(x) = \sum_{i=0}^{7} a_i x^i = a_7 x^7 + a_6 x^6 + a_5 x^5 + a_4 x^4 + a_3 x^3 + a_2 x^2 + a_1 x + a_0.$$

The coefficients a_i of a polynomial $a(x)$ are bits. Bytes can be written in different notations. For example, the binary value $\{01100011\}$ is $\{63\}$ in hexadecimal notation and represents the polynomial $x^6 + x^5 + x + 1$.

Addition. The addition of two Bytes representing polynomials $a(x), b(x) \in GF(2^8)$ is achieved by adding their corresponding coefficients modulo 2 which is a XOR-operation usually denoted with $\oplus$.

$$a(x) \oplus b(x) = \sum_{i=0}^{7} a_i x^i \oplus \sum_{i=0}^{7} b_i x^i = \sum_{i=0}^{7} (a_i \oplus b_i) x^i \qquad (1)$$

The additive inverse of a Byte is the Byte itself: $-b(x) = b(x)$ and therefore subtraction is identical with addition: $a(x) - b(x) = a(x) + b(x)$.

Multiplication. The multiplication of $a(x), b(x) \in GF(2^8)$ – denoted with $a(x) \otimes b(x)$ – requires an irreducible polynomial of degree 8. For the AES-algorithm it is defined as

$$m(x) = x^8 + x^4 + x^3 + x + 1 = 1\{00011011\} \text{ (bin)} = 1\{1b\} \text{ (hex)}.$$

The multiplication $q(x) = a(x) \cdot b(x)$ in $GF(2^8)$ is done by multiplying the polynomials $a(x)b(x)$ which yields a polynomial $p(x)$ with degree less than 15. This step is followed by a modular reduction step $q(x) = p(x) \bmod m(x)$ to ensure that the result is an element of $GF(2^8)$.

A convenient method to multiply in the finite field $GF(2^8)$ is to generate eight partial products: $P_i(x) = a(x) \cdot x^i$ and to add those partial products where the according bit b_i of the multiplier $b(x)$ is 1: $q(x) = \sum_{i=0}^{7} P_i b_i$. The partial products can be calculated efficiently by iterating a multiplication by x: $P_i(x) = P_{i-1}(x) \cdot x \bmod m(x), P_0(x) = a(x)$. A multiplication by x is termed *xtimes* and is given by

$$q(x) = xtimes(a) = a(x)x \bmod m(x) \tag{2}$$

$$q_0 = a_7, \quad q_1 = a_0 \oplus a_7, \quad q_2 = a_1, \quad q_3 = a_2 \oplus a_7$$

$$q_4 = a_3 \oplus a_7, \quad q_5 = a_4, \quad q_6 = a_5, \quad q_7 = a_6.$$

Xtimes can be implemented by a shift left operation of the input Byte a and a conditional addition of the irreducible polynomial $m(x)$ if the most significant bit (a_7) of a is set. This ensures a Byte as result.

Inversion. The multiplicative inverse a^{-1} of an element $a \in GF(2^8)$ has the property that $\forall a \in GF(2^8) \setminus \{0\} : a \otimes a^{-1} = \{1\}$. Calculating the inverse of a Byte is even more costly than multiplying Bytes. A widely used algorithm for inversion is the *extended Euclidean algorithm* described in [2]. Unfortunately, this algorithm is not suitable for a hardware implementation.

2.1 GF(2^8) as an Extension of GF(2^4)

Usually, the field $GF(2^8)$ is seen as a field extension of $GF(2)$ and therefore its elements can be represented as Bytes. An isomorphic – but for hardware implementations far better suited – representation is to see the field $GF(2^8)$ as a quadratic extension of the field $GF(2^4)$. In this case, an element $a \in GF(2^8)$ is represented as a linear polynomial with coefficients in $GF(2^4)$,

$$a \cong a_h x + a_l, \quad a \in GF(2^8), \quad a_h, a_l \in GF(2^4) \tag{3}$$

and will be denoted by the pair $[a_h, a_l]$. Both coefficients of such a polynomial have four bits. All mathematical operations applied to elements of $GF(2^8)$ can also be computed in this representation which we call *two-term polynomials*. Two-term polynomials are added by addition of their corresponding coefficients

$$(a_h x + a_l) \oplus (b_h x + b_l) = (a_h \oplus b_h)x + (a_l \oplus b_l). \tag{4}$$

Multiplication and inversion of two-term polynomials require a modular reduction step to ensure that the result is a two-term polynomial too. The irreducible polynomial needed for the modular reduction is given by

$$n(x) = x^2 + \{1\}x + \{e\}. \tag{5}$$

The coefficients of $n(x)$ are elements in $GF(2^4)$ and are written in hexadecimal notation. Their particular values are chosen to optimize the finite field arithmetic.

The multiplication of two-term polynomials involves multiplication of elements in $GF(2^4)$ which requires an irreducible polynomial of degree 4 which is given by

$$m_4(x) = x^4 + x + 1. \tag{6}$$

Deriving formulas for multiplication in $GF(2^4)$ is similar to Byte-multiplication. Multiplication in $GF(2^4)$ is given by

$$q(x) = a(x) \otimes b(x) = a(x) \cdot b(x) \bmod m_4(x), \quad a(x), b(x), q(x) \in GF(2^4) \tag{7}$$

$$a_A = a_0 \oplus a_3, \quad a_B = a_2 \oplus a_3$$

$$q_0 = a_0 b_0 \oplus a_3 b_1 \oplus a_2 b_2 \oplus a_1 b_3 \quad q_1 = a_1 b_0 \oplus a_A b_1 \oplus a_B b_2 \oplus (a_1 \oplus a_2) b_3$$

$$q_2 = a_2 b_0 \oplus a_1 b_1 \oplus a_A b_2 \oplus a_B b_3 \quad q_3 = a_3 b_0 \oplus a_2 b_1 \oplus a_1 b_2 \oplus a_A b_3.$$

Squaring in $GF(2^4)$ is a special case of multiplication and is given by

$$q(x) = a(x)^2 \bmod m_4(x), \quad q(x), a(x) \in GF(2^4) \tag{8}$$

$$q_0 = a_0 \oplus a_2, \quad q_1 = a_2, \quad q_2 = a_1 \oplus a_3, \quad q_3 = a_3.$$

The inverse a^{-1} of an element $a \in GF(2^4)$ can be derived by solving the equation $a(x) \cdot a^{-1} \bmod m_4(x) = 1$ as follows

$$q(x) = a(x)^{-1} \bmod m_4(x), \quad q(x), a(x) \in GF(2^4) \tag{9}$$

$$a_A = a_1 \oplus a_2 \oplus a_3 \oplus a_1 a_2 a_3$$

$$q_0 = a_A \oplus a_0 \oplus a_0 a_2 \oplus a_1 a_2 \oplus a_0 a_1 a_2$$

$$q_1 = a_0 a_1 \oplus a_0 a_2 \oplus a_1 a_2 \oplus a_3 \oplus a_1 a_3 \oplus a_0 a_1 a_3$$

$$q_2 = a_0 a_1 \oplus a_2 \oplus a_0 a_2 \oplus a_3 \oplus a_0 a_3 \oplus a_0 a_2 a_3$$

$$q_3 = a_A \oplus a_0 a_3 \oplus a_1 a_3 \oplus a_2 a_3.$$

The concatenation of two bits $a_i a_j$ in Equations 7 and 9 represents a binary multiplication which is an AND-operation. In contrast to inversion in $GF(2^8)$, inversion in $GF(2^4)$ is suitable for a hardware implementation using combinational logic since all Boolean equations depend only on four input bits.

Inversion of Two-Term Polynomials. Inversion of two-term polynomials is the equivalent operation to inversion in $GF(2^8)$. A multiplication of a two-term polynomial with its inverse yields the 1-element of the field: $(a_h x + a_l) \otimes (a'_h x +$

$a'_l) = \{0\}x + \{1\}$, $a_h, a_l, a'_h, a'_l \in GF(2^4)$. From this definition the formula for inversion can be derived:

$$(a_h x + a_l)^{-1} = a'_h x + a'_l = (a_h \otimes d)x + (a_h \oplus a_l) \otimes d \qquad (10)$$
$$d = ((a_h^2 \otimes \{e\}) \oplus (a_h \otimes a_l) \oplus a_l^2)^{-1}.$$

Inversion of two-term polynomials involves only operations in GF(2^4) which are suitable for a hardware implementation using combinational logic. Most of the functionality is used to calculate the term d which is used to calculate both coefficients of the inverted two-term polynomial.

Transition between Representations of GF(2^8). The finite field GF(2^8) is isomorphic to the finite field GF($(2^4)^2$) which means that for each element in GF(2^8) there exists exactly one element in GF($(2^4)^2$). The bijection from an element $a \in GF(2^8)$ to a two-term polynomial $a_h x + a_l$ where $a_h, a_l \in GF(2^4)$ is given by the function map:

$$a_h x + a_l = map(a), \qquad a_h, a_l \in GF(2^4), \qquad a \in GF(2^8) \qquad (11)$$

$$a_A = a_1 \oplus a_7, \qquad a_B = a_5 \oplus a_7, \qquad a_C = a_4 \oplus a_6$$
$$a_{l0} = a_C \oplus a_0 \oplus a_5, \qquad a_{l1} = a_1 \oplus a_2, \qquad a_{l2} = a_A, \qquad a_{l3} = a_2 \oplus a_4$$
$$a_{h0} = a_C \oplus a_5, \qquad a_{h1} = a_A \oplus a_C, \qquad a_{h2} = a_B \oplus a_2 \oplus a_3, \qquad a_{h3} = a_B.$$

The inverse transformation (map^{-1}) converts two-term polynomials $a_h x + a_l$, $a_h, a_l \in GF(2^4)$ back into elements $a \in GF(2^8)$. It is given by

$$a = map^{-1}(a_h x + a_l), \qquad a \in GF(2^8), \qquad a_h, a_l \in GF(2^4) \qquad (12)$$

$$a_A = a_{l1} \oplus a_{h3}, \qquad a_B = a_{h0} \oplus a_{h1},$$
$$a_0 = a_{l0} \oplus a_{h0}, \qquad a_1 = a_B \oplus a_{h3}$$
$$a_2 = a_A \oplus a_B, \qquad a_3 = a_B \oplus a_{l1} \oplus a_{h2}$$
$$a_4 = a_A \oplus a_B \oplus a_{l3}, \qquad a_5 = a_B \oplus a_{l2}$$
$$a_6 = a_A \oplus a_{l2} \oplus a_{l3} \oplus a_{h0}, \qquad a_7 = a_B \oplus a_{l2} \oplus a_{h3}$$

Both transformations can be derived by following the procedure given in C. Paar's PhD thesis [7]. They differ from the transformations given in [5] because the irreducible polynomial $m(x)$ for GF(2^8) is different.

3   SBox Building Blocks

The SubByte transformation operates independently on each Byte of the State using a substitution table (SBox). An AES-SBox is composed of two transformations:

1. Calculate the multiplicative inverse in the finite field GF(2^8). The element $\{00\}$ is mapped to itself. Table 1 presents the inversion.

Table 1. Inversion of a Byte $\{xy\} \in GF(2^8)$ in hexadecimal notation.

$x \backslash y$	0	1	2	3	4	5	6	7	8	9	a	b	c	d	e	f
0	00	01	8d	f6	cb	52	7b	d1	e8	4f	29	c0	b0	e1	e5	c7
1	74	b4	aa	4b	99	2b	60	5f	58	3f	fd	cc	ff	40	ee	b2
2	3a	6e	5a	f1	55	4d	a8	c9	c1	0a	98	15	30	44	a2	c2
3	2c	45	92	6c	f3	39	66	42	f2	35	20	6f	77	bb	59	19
4	1d	fe	37	67	2d	31	f5	69	a7	64	ab	13	54	25	e9	09
5	ed	5c	05	ca	4c	24	87	bf	18	3e	22	f0	51	ec	61	17
6	16	5e	af	d3	49	a6	36	43	f4	47	91	df	33	93	21	3b
7	79	b7	97	85	10	b5	ba	3c	b6	70	d0	06	a1	fa	81	82
8	83	7e	7f	80	96	73	be	56	9b	9e	95	d9	f7	02	b9	a4
9	de	6a	32	6d	d8	8a	84	72	2a	14	9f	88	f9	dc	89	9a
a	fb	7c	2e	c3	8f	b8	65	48	26	c8	12	4a	ce	e7	d2	62
b	0c	e0	1f	ef	11	75	78	71	a5	8e	76	3d	bd	bc	86	57
c	0b	28	2f	a3	da	d4	e4	0f	a9	27	53	04	1b	fc	ac	e6
d	7a	07	ae	63	c5	db	e2	ea	94	8b	c4	d5	9d	f8	90	6b
e	b1	0d	d6	eb	c6	0e	cf	ad	08	4e	d7	e3	5d	50	1e	b3
f	5b	23	38	34	68	46	03	8c	dd	9c	7d	a0	cd	1a	41	1c

2. Apply the affine transformation which is given by Equation 13.

$$q = \text{aff\_trans}(a) \qquad (13)$$

$$a_A = a_0 \oplus a_1, a_B = a_2 \oplus a_3,$$
$$a_C = a_4 \oplus a_5, a_D = a_6 \oplus a_7$$

$$q_0 = \overline{a_0} \oplus a_C \oplus a_D$$
$$q_1 = \overline{a_5} \oplus a_A \oplus a_D$$
$$q_2 = a_2 \oplus a_A \oplus a_D$$
$$q_3 = a_7 \oplus a_A \oplus a_B$$
$$q_4 = a_4 \oplus a_A \oplus a_B$$
$$q_5 = \overline{a_1} \oplus a_B \oplus a_C$$
$$q_6 = \overline{a_6} \oplus a_B \oplus a_C$$
$$q_7 = a_3 \oplus a_C \oplus a_D.$$

$$q = \text{aff\_trans}^{-1}(a) \qquad (14)$$

$$a_A = a_0 \oplus a_5, a_B = a_1 \oplus a_4,$$
$$a_C = a_2 \oplus a_7, a_D = a_3 \oplus a_6$$

$$q_0 = \overline{a_5} \oplus a_C$$
$$q_1 = a_0 \oplus a_D$$
$$q_2 = \overline{a_7} \oplus a_B$$
$$q_3 = a_2 \oplus a_A$$
$$q_4 = a_1 \oplus a_D$$
$$q_5 = a_4 \oplus a_C$$
$$q_6 = a_3 \oplus a_A$$
$$q_7 = a_6 \oplus a_B.$$

Overlined bits in Equation 13 and 14 denote inverted bits. Decryption requires the inverse function of SubByte (*InvSubByte*) which reverses the SBox-operation by applying the inverse affine transformation first (Equation 14). Then, the multiplicative inverse in the finite field GF(2^8) is calculated.

Inversion in the finite field GF(2^8) is needed to calculate the SubByte-function as well as InvSubByte. It makes sense, to merge the encryption SBox with the decryption SBox in order to reuse the finite field inversion circuit for decryption. Figure 2 depicts this approach. The control signal *enc* switches between encryption and decryption. If encryption is chosen (*enc*=1), the inverse affine transformation (*aff_trans*$^{-1}$) is bypassed and the input a is directly fed

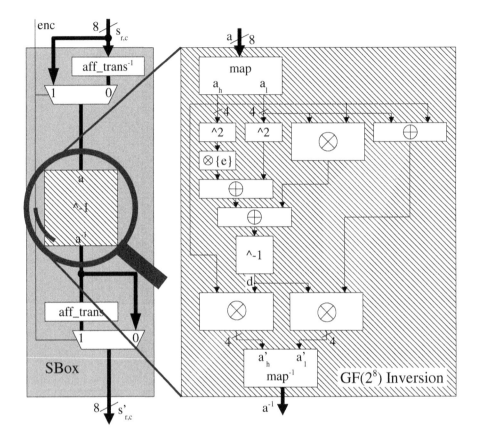

Fig. 2. Architecture of the AES-SBox

into the inversion circuit. The output of the inversion circuit is modified by the affine transformation block which calculates the result of the SubByte-function. During decryption ($enc=0$), the inverse affine transformation is active and the affine transformation is bypassed to calculate InvSubByte. The delay for encryption and decryption is essentially the same because the circuit's complexity for the affine transformation and its inverse are equal.

The circuit for inversion of elements in $GF(2^8)$ covers most of the SBox functionality. In our approach the inversion is calculated with combinational logic and is based on Equation 10 which operates in $GF(2^4)$. The operations occurring in this equation correspond to the function blocks shown in Fig. 2. Furthermore, this function block has to convert data from $GF(2^8)$ to two-term polynomials and vice versa. The blocks *map* resp. map^{-1} provide this functionality based on Equation 11 and 12. Addition of elements in $GF(2^4)$ is accomplished by a bitwise XOR-operation. Squaring relies on Equation 9. Multiplication of an element in $GF(2^4)$ with the constant $\{e\}$ is given by

$$q = a \otimes \{e\} \tag{15}$$

$$a_A = a_0 \oplus a_1, \; a_B = a_2 \oplus a_3$$
$$q_0 = a_1 \oplus a_B$$
$$q_1 = a_A$$
$$q_2 = a_A \oplus a_2$$
$$q_3 = a_A \oplus a_B.$$

Multiplication and inversion in $GF(2^4)$ are the most complex function blocks and rely on Equations 7 and 9.

4   Implementation

Our implementation is based on the architecture described in Sect. 3. It has a maskable affine transformation block, a maskable inverse affine transformation block and a block which calculates the inverse in the finite field $GF(2^8)$. Most of the functionality can be implemented with XOR-gates. Additionally, inverters, AND-gates, and 2-to-1 multiplexers are required, but they can be neglected for performance analysis purposes.

Table 2. Complexity of the SBox

block	d_{XOR}	XORs	instances	sum XORs
aff_trans	3	16	1	16
aff_trans^{-1}	2	12	1	12
map	2	11	1	11
map^{-1}	3	15	1	15
$\oplus$	1	4	3	12
$\wedge^2$	1	2	2	4
$\otimes\{e\}$	2	5	1	5
$\otimes$	2	12	3	36
$\wedge^{-1}$	3	12	1	12
	Max 15			Sum 123

Table 3. Pipelining of the SBox

stages	flipflops	frequency	area
0	0	100%	100%
1	12	178%	111%
3	28	205%	151%

Table 2 lists the resources of all blocks – measured in number of XORs. The overall amount of gates are 123 XOR-gates with two inputs, 16 2-to-1 multiplexers and a dozen of inverters and AND-gates. If XOR-gates with three inputs are available, the number of gates can be reduced. The delay of the blocks is measured in numbers of XOR-gates in series (d_{XOR}). The critical path for encryption is composed of 15 XOR-gates in series. For decryption it is 14 because the inverse of the affine transformation has lower complexity. XOR-gates with three inputs will shorten the critical path and improve performance.

A feature that can be exploited to gain higher throughput is pipelining. Pipelining is a technique which subdivides the critical path by insertion of storing elements (flipflops). Subdividing the SBox functionality into a number of stages

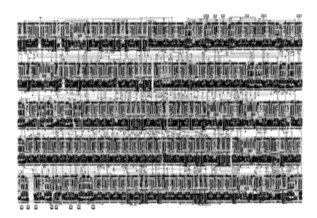

Fig. 3. Layout of the AES-SBox

is easy to accomplish since flipflops can be inserted nearly anywhere when SBoxes are implemented with combinational logic. Pipelining introduces latency but the additional clock cycles are made up by an increased clock frequency as shown in Table 3. This technique will offer best results if the number of SBox instances used for an AES implementation is kept low (e.g. 4), otherwise latency will consume more time than it is saved by shortening the critical path.

Our implementation of the AES-SBox which combines the SubByte-function and InvSubByte-function from the AES algorithm is a standard-cell circuit on a 0.6 μm CMOS process from AMS using two metal layers. It has an area of 0.108 mm^2 and contains 1624 transistors (406 NAND equivalents) when no pipelining is used. A layout is shown in Fig. 3. Layout simulation with typical mean parameters considering parasitics yields a delay below 14.2 ns which equals a maximum clock frequency of 70 MHz for both encryption and decryption. A pipelined version with one stage is slightly bigger (0.120 mm^2), contains nearly 2000 transistors (500 NAND equivalents) and yields a delay below 8 ns (125 MHz). Further increasing the number of pipeline stages will not improve throughput that strong and has a worse ratio of performance benefit to area penalty. It should be considered when the maximum clock frequency is of utmost interest.

For comparison, a lookup-table based approach requires a 4k-bit ROM to store the 256 8-bit entries of SubByte and InvSubByte. An implementation on the same process technology uses about 0.200 mm^2 and has an estimated maximum frequency of 100 MHz [9].

For an SBox implementation using a full-custom design methodology we suggest to use a differential logic style [8]. The logic functions of an SBox based on combinational logic are dominated by XOR-gates and differential XOR-gates offer good performance and have a moderate transistor count. We assume that a full-custom design could halve the required chip area because an SBbox is a small module and does not require the excessive driving capability offered by standard cells. Output transistors of gates can be dimensioned smaller and this will in turn make it possible to scale all transistors down without deteriorating performance. At least three leaf cells are needed: a XOR-gate, an AND-gate and an inverter. To develop these cells will be an rewarding task if the resulting performance gain and area saving is pictured.

5   Related Work

Several AES-implementations have been presented recently. For comparison with our work, only ASIC circuits [11] or circuits exploiting a more efficient finite field arithmetic are of interest [10]. The approach followed in IBM implementation [10] is also based on a conversion of elements in $GF(2^8)$ into two-term polynomials. In contrast to our approach, they calculate the whole round function in this representation. Therefore, they choose the conversion function $map()$ in a way that minimizes the overall gate count. Our primary focus on choosing $map()$ was to minimize the critical path of the complete SBox and secondarily to keep the gate count low. Our $map()$-function has a shorter critical path compared to the IBM implementation, the critical path of $map^{-1}()$ is identical.

6   Conclusion

This article presented an ASIC implementation of the SBoxes from the Advanced Encryption Standard (AES). It is based on finite field arithmetic rather than using lookup-tables. This approach offers higher flexibility. Area requirements can be traded for the maximum clock frequency. The architecture can be easily implemented within a standard-cell design methodology because it completely relies on combinational logic. It is also well suited for a full-custom implementation since it uses only a few leaf cells. We implemented the AES-SBox on 0.6 μm CMOS process with standard-cells. The most promising configuration of design parameters we found is a single stage pipeline architecture. It has a silicon area of 0.12 mm^2 and a maximum clock frequency of 125 MHz. This configuration has a latency of one clock cycle like a ROM based approach. In comparison to ROMs, it offers better performance on smaller area and can even be improved by exploiting better suited logic styles like differential logic.

References

1. NIST, *Advanced Encryption Standard (AES)*, FIPS PUBS 197, National Institute of Standards and Technology, November 2001.
2. A. Menezes, P. van Oorschot, and S. Vanstone, *Handbook of Applied Cryptography*, CRC Press, New York, 1997.
3. R. Lidl and H. Niederreiter, *Introduction to finite fields and their applications*, Cambridge University Press, Cambridge, 1986.
4. V. Rijmen, *Efficient Implementation of the Rijndael SBox*, http://www.esat.ku-leuven.ac.be/~rijmen/rijndael/ .

5. E. Soljanin, R. Urbanke, *An Efficient Architecture for Implementation of a Multiplier and Inverter in GF(2^8)*, Lucent Technologies.
6. E. D. Mastrovito, *VLSI Architectures for Computations in Galois Fields*, PhD thesis, Linköping University, Linköping, Sweden, 1991.
7. C. Paar, *Efficient VLSI Architectures for Bit Parallel Computation in Galois Fields*, PhD thesis, Universität Essen, 1994.
8. J. B. Kuo, J. H. Lou, *Low-Voltage VLSI Circuits*, John Wiley, New York, Jan. 1999.
9. AMS, *Memory Compiler for Diffusion Programmable ROM in 0.6 μm CMOS*, http://www.amsint.com/databooks/.
10. A. Rudra, P. Dubey, C. Jutla, V. Kumar, J. Rao, P. Rohatgi, *Efficient Rijndael Encryption Implementation with Composite Field Arithmetic*, Proceedings of Workshop on Cryptographic Hardware and Embedded Systems, France, 2001, to be published in Springer LNCS.
11. I. Verbauwhede, H. Kuo, *Architectural Optimization for a 1.82 Gbits/sec VLSI Implementation of the AES Rijndael Algorithm*, Proceedings of Workshop on Cryptographic Hardware and Embedded Systems, France, 2001, to be published in Springer LNCS.

On the Impossibility of Constructing
Non-interactive Statistically-Secret Protocols
from Any Trapdoor One-Way Function

Marc Fischlin

Johann Wolfgang Goethe-University,
Frankfurt am Main, Germany
marc@mi.informatik.uni-frankfurt.de
http://www.mi.informatik.uni-frankfurt.de/

Abstract. We show that non-interactive statistically-secret bit commitment cannot be constructed from arbitrary black-box one-to-one trapdoor functions and thus from general public-key cryptosystems. Reducing
the problems of non-interactive crypto-computing, rerandomizable encryption, and non-interactive statistically-sender-private oblivious transfer and low-communication private information retrieval to such commitment schemes, it follows that these primitives are neither constructible
from one-to-one trapdoor functions and public-key encryption in general. Furthermore, our separation sheds some light on statistical zero-
knowledge proofs. There is an oracle relative to which one-to-one trapdoor functions and one-way permutations exist, while the class of promise
problems with statistical zero-knowledge proofs collapses in $\mathcal{P}$. This indicates that nontrivial problems with statistical zero-knowledge proofs
require more than (trapdoor) one-wayness.

1 Introduction

One of the fundamental questions in cryptography deals with the relationship
of cryptographic primitives: does the existence of primitive $\mathcal{A}$ imply the existence of primitive $\mathcal{B}$? As for positive results, such proofs usually give rise to
an explicit construction of primitive $\mathcal{B}$ given an arbitrary instance of primitive $\mathcal{A}$. For instance, given any one-way function we can effectively specify a
secure signature scheme [33,38]. We also know that one-way functions, pseudorandom generators, pseudorandom functions, private-key encryption, signature schemes and computationally-secret bit commitment are all equivalent in
this sense [21,25,33,38,30,24]. Similarly, trapdoor permutations are sufficient for
oblivious transfer and public-key encryption, and for key agreement [13,22,15].

Concerning separations of primitives, Impagliazzo and Rudich [26] have
shown that basing key agreement (and thus trapdoor permutations and oblivious transfer) on any "black-box" one-way permutation is at least as hard as
proving $\mathcal{P} \neq \mathcal{NP}$. The terminology "black-box" refers to the fact that nothing
beyond the structure of a primitive is assumed except for fundamental properties guaranteed by the definition. For instance, in [26] the abstract of a one-way

B. Preneel (Ed.): CT-RSA 2002, LNCS 2271, pp. 79–95, 2002.
© Springer-Verlag Berlin Heidelberg 2002

permutation is a one-way permutation oracle, and the efficient evaluation algorithm of the one-way permutation corresponds to single oracle step that returns the function value. Since reductions where the starting primitive is treated as a black box are common throughout complexity-based cryptography, the result of Impagliazzo and Rudich suggests that showing the equivalence of key agreement and one-way functions is infeasible. Therefore, in a sense, secure key agreement needs more than one-wayness.

Although most known reductions obey the black-box approach, there is at least one example of a reduction which is not black-box, i.e., requires that the description of the evaluation algorithm is explicit. See [26] for a discussion. Thus, oracle-based separations do not completely rule out the possibility that reductions exist. There might still be effective constructions which are not black-box. Yet, as mentioned before, the black-box design is widely used in complexity-based cryptography.

For more impossibility results we refer the reader to [39,45,29,28,17,18]. In particular, the work by Simon [45] separates collision-intractability and one-wayness by defining an oracle relative to which one-way permutations exist, but collision-intractable hash functions do not. Here, relying on the techniques developed in [45], we extend Simon's result. In the first step we present an oracle relative to which non-interactive statistically-secret bit commitment is impossible, yet one-way permutations exists (throughout the paper we refer to non-interactive protocols as schemes where both parties consecutively send a single message only). Relative to our oracle a very weak form of non-interactive statistically-secret bit commitment does not exist. That is, secrecy is only guaranteed with respect to honestly behaving receivers, and a commitment merely binds with very small, yet noticeable probability.

We stress that it is not known whether any kind of bit commitment yields collision-intractable hash functions in general or not. Thus, our extension from collision-intractable hash functions to commitments is not known to be implied by Simon's result directly. We remark that, conversely, collision-intractable hash functions are sufficient for non-interactive statistically-secret commitment [33,10,23]. Furthermore, one can construct perfectly-secret bit commitment from any one-way permutation with linear many rounds in the security parameter [31]. To best of our knowledge, nothing has been reported about improvements concerning either the assumption or the round complexity[1]. Our result provides some evidence that accomplishing non-interactive statistically-secret commitment based on one-wayness alone is impossible. In contrast, non-interactive computationally-secret commitment can be based on one-way functions [24,30].

In addition to showing that non-interactive statistically-secret commitments are impossible in the presence of general one-way functions, in the second extension step we prove that this impossibility result transfers to the case that one adds the "power" of trapdoors to the one-way function. Such (one-to-one) trap-

---

[1] We always refer to the classical Turing machine model in this work. In the Quantum computing model there are indeed results that one-wayness is sufficient for constant-round statistically-secret commitments [14,9].

door functions are a relaxed version of trapdoor permutations. We only demand that the former are one-to-one in order to support unique inversion.

Bellare et al. [5] prove that many-to-one trapdoor functions with super-polynomial preimage size can be derived from any one-way function. Trapdoor functions with polynomially bounded preimage size, among which are one-to-one trapdoor functions, yield public-key cryptosystems, though. In light of [26] they cannot be derived from one-way functions in general, and therefore, in a sense, our result is a strict extension of Simon's separation.

In summery, we broaden Simon's separation in both directions. On one side, we show that relative to an oracle non-interactive weakly-binding honest-receiver statistically-secret commitments schemes do not exist. Such commitment schemes include collision-intractable hash functions, but are not known to imply the existence of such hash functions. On the other side, the negative result holds in the presence of one-way permutation *and* of one-to-one trapdoor functions. The latter functions are presumably not derivable from general one-way permutations.

The relationship of statistical secrecy and one-wayness enables us to obtain a new result about the class $\mathcal{SZK}$ of promise problems with statistical zero-knowledge proofs. We prove that relative to a one-way permutation oracle and to a one-to-one trapdoor function oracle, respectively, the class $\mathcal{SZK}$ breaks down to $\mathcal{P}$. In contrast to our impossibility result, one-way functions suffice to lift the class $\mathcal{CZK}$ of promise problems with computational zero-knowledge proofs to $\mathcal{IP} = \mathcal{PSPACE}$ [43,27,7]. This gives us another, oracle-based separation of $\mathcal{CZK}$ and $\mathcal{SZK}$ in addition to the one implied by the presumably strictness of the polynomial hierarchy: $\mathcal{SZK}$ belongs to $\mathcal{AM} \cap \text{co-}\mathcal{AM}$ [16,1], and likewise lies much lower in the polynomial hierarchy than $\mathcal{CZK}$ (which equals $\mathcal{PSPACE}$ under the assumption that one-way functions exist). From a cryptographer's point of view, our result says that while one-wayness is sufficient and necessary [36] for nontrivial problems in $\mathcal{CZK}$, hard problems in $\mathcal{SZK}$ seem to require more than general one-way permutations and one-to-one trapdoor functions.

Finally, we consider implications to other cryptographic protocols. By constructing non-interactive weakly-binding honest-receiver statistically-secret commitment schemes from other non-interactive statistically-secret cryptographic protocols, we conclude that such protocols cannot be derived from general black-box one-to-one trapdoor functions. Specifically, we prove that this holds for non-interactive crypto-computing, rerandomizable encryption, and non-interactive statistically-sender-private protocols for oblivious transfer and, using a result of Beimel et al. [4], for private information retrieval with low communication complexity.

The paper is organized as follows. We start with basic definitions in Section 2. Then, in Section 3 we introduce the class $\mathcal{SZK}$ of statistical zero-knowledge proofs for motivating our oracle separation constructions in Section 4. In Section 5 we then apply the separation to $\mathcal{SZK}$, and we discuss implications to other cryptographic protocols in the final part.

2   Definitions

We occasionally view probabilistic algorithms as deterministic ones by providing the random coins explicitly. That is, let A be a deterministic algorithm taking two inputs x and r. Then we denote by $A(x, r)$ the output of A for input x, r and by $A(x)$ the random variable that describes the output for fixed x and uniformly chosen r. It will be clear from the context which part of the input is considered as the random coins. Additionally, we denote by $[A(x)]$ the support of $A(x)$, i.e., $a \in [A(x)]$ if and only if there exists r with $a = A(x, r)$. When passing a function as argument to, say, an oracle, it is understood that we pass a circuit description of the function.

A function $\delta(n)$ is *negligible* if it is eventually less than any polynomial fraction, i.e., $\delta(n) < 1/p(n)$ for any positive polynomial $p(n)$ and all sufficiently large n's. A function $\delta(n)$ is *noticeable* if it is not negligible; it is *overwhelming* if $1 - \delta(n)$ is negligible. Two sequences $X = (X_n)_{n \in \mathbb{N}}$ and $Y = (Y_n)_{n \in \mathbb{N}}$ of random variables are called *statistically close*, $X \stackrel{s}{=} Y$, if the statistical difference

$$\mathrm{StatDiff}(X_n, Y_n) = \tfrac{1}{2} \cdot \sum_{s \in [X_n] \cup [Y_n]} |\mathrm{Prob}\,[X_n = s] - \mathrm{Prob}\,[Y_n = s]|$$

is negligible.

2.1   Commitment Schemes

A commitment scheme consists of two phases. In the commitment phase the sender puts a secret bit b into a box and sends the locked box to the receiver. In the decommitment phase the sender assists in opening the box, say, by transmitting the key. Then, on one hand, even a malicious sender $\mathcal{S}^*$ cannot change his mind once the box has been given to the receiver (binding property). The receiver, on the other hand, does not learn anything about the bit b till the decommitment step is carried out (secrecy).

We exclusively present the definition of non-interactive honest-receiver statistically-secret bit commitment schemes. Our definition captures only a very weak binding property, namely, that there is no collision-finder that nearly always succeeds in finding ambiguous decommitments. Usually, the binding property demands that any collision-finder fails with very high probability.

Definition 1. *The tuple* (Gen, Com, Decom, Vf) *of probabilistic polynomial-time algorithms is a non-interactive weakly-binding honest-receiver statistically-secret bit commitment scheme if*

- *generation: on input 1^n algorithm* Gen *outputs a description k_n (wlog. k_n contains 1^n).*
- *meaningfulness: for every $k_n \in [\mathrm{Gen}(1^n)]$ and every $c = \mathrm{Com}_{k_n}(b, r)$ and $d = \mathrm{Decom}_{k_n}(b, r)$ we have $\mathrm{Vf}_{k_n}(b, c, d) = 1$.*
- *honest-receiver statistical secrecy: for every sequence $(k_n)_{n \in \mathbb{N}}$ with $k_n \in [\mathrm{Gen}(1^n)]$ we have $\mathrm{Com}_{k_n}(0) \stackrel{s}{=} \mathrm{Com}_{k_n}(1)$.*

- *weakly binding:* for any probabilistic polynomial-time algorithm $\mathcal{S}^*$ the probability that $\mathcal{S}^*$ on input k_n outputs c and d, d' such that $\mathrm{Vf}_{k_n}(0, c, d) = \mathrm{Vf}_{k_n}(1, c, d') = 1$ is not overwhelming (where the probability is taken over the choice of k_n and the coin tosses of $\mathcal{S}^*$).

Instead of using the notation above, we sometimes adopt the viewpoint of a protocol between a sender $\mathcal{S}$ and a receiver $\mathcal{R}$, in which the honest $\mathcal{R}$ transmits k_n obtained by running $\mathrm{Gen}(1^n)$ and $\mathcal{S}$ (with input b) answers with a sample $\mathrm{Com}_{k_n}(b, r)$ by choosing r at random. Later, in the decommitment phase, the receiver then applies the verification algorithm Vf_{k_n} to check the validity of the decommitment $d = \mathrm{Decom}_{k_n}(b, r)$. Note that, in order to generate the decommitment d, algorithm Decom gets the same random string r as Com.

Wlog. assume that the input length of the commitment function Com_{k_n} is at least as large as the security parameter n. Also, let the output size of $\mathrm{Com}_{k_n}(b)$ be at most the size of the randomness portion of the input. This can always be achieved by padding the input with redundant random bits. Then the domain of Com_{k_n} is at least twice as large as its range.

2.2   Black-Box (Trapdoor) One-Way Functions

We rigidly formalize the concept of black-box (trapdoor) one-way functions as discussed in the introduction.

Definition 2. *A* black-box one-way function *is an oracle* $\mathcal{F} : \{0, 1\}^* \rightarrow \{0, 1\}^*$ *such that for any uniform polynomial-size circuit family* $C = (C_n)_{n \in \mathbb{N}}$ *the inversion probability*

$$\mathrm{Prob}\left[C_n^{\mathcal{F}}(\mathcal{F}(x)) \in \mathcal{F}^{-1}(\mathcal{F}(x))\right]$$

is negligible, where the probability is taken over the random choice of $x \in_R \{0, 1\}^n$ *and the internal coin tosses of* C_n. *If additionally* $\mathcal{F}(\{0, 1\}^n) = \{0, 1\}^n$ *for all* $n \in \mathbb{N}$ *then we say that* $\mathcal{F}$ *is a* black-box one-way permutation.

We remark that C_n is granted oracle access to $\mathcal{F}$. This enables C_n to evaluate $\mathcal{F}$ at values of its choice. Also, we demand that the family C of circuits is uniform. This is necessary to derandomize the probabilistic oracle construction as described in [44,45].

Our definition imitates the one of a one-way function with infinite domain. Instead, one sometimes uses collections of one-way function, where each function of this collection is indexed. Yet, we omit further discussions since the notion of a collection of black-box one-way functions is implicit in the definition of trapdoor one-way functions below, and can be easily inferred. From an existential point of view both notions of one-way functions are equivalent, even in the black-box case.

Definition 3. *A* black-box one-to-one trapdoor function *is an oracle* $\mathcal{T}$ *with three query states* generate, evaluate, invert:

- *generation:* $\mathcal{T}(\mathsf{generate}, \omega)$ *for* $\omega \in \{0, 1\}^n$ *outputs a pair* (t, i). *Wlog. let* 1^n *be recoverable from index* i *and trapdoor* t. *Furthermore, assume that* i *uniquely determines* t *and vice versa.*

- *evaluation: given $x \in \{0,1\}^n$ and an index i with $(t, i) = \mathcal{T}(\text{generate}, \omega)$ for some $\omega \in \{0,1\}^n$, the oracle $\mathcal{T}(\text{evaluate}, i, x)$ returns $y \in \{0,1\}^{poly(n)}$. Also, let $\mathcal{T}(\text{evaluate}, i, \cdot)$ be one-to-one for any index i.*
- *inversion: given $y \in \{0,1\}^{poly(n)}$ and t, the answer $\mathcal{T}(\text{invert}, t, y)$ is some x such that $\mathcal{T}(\text{evaluate}, i, x) = y$ if such an x exists (where i is the uniquely determined index to t), and an undefined symbol otherwise.*

Additionally, $\mathcal{T}$ satisfies the following one-wayness property: for any uniform polynomial-size circuit family $C = (C_n)_{n \in \mathbb{N}}$ the inversion probability

$$\text{Prob}\left[C_n^{\mathcal{T}}(i, \mathcal{T}(\text{evaluate}, i, x)) = x\right]$$

is negligible, where the probability is taken over the choice of i according to $\mathcal{T}(\text{generate}, \omega)$ for a random $\omega \in_R \{0,1\}^n$, over $x \in_R \{0,1\}^n$, and over the randomness of C_n.

The generation step of our definition says that one must externally supply the deterministic oracle $\mathcal{T}$ with randomness ω to get a random function description (t, i). For simplicity and since our construction achieves this, we presume that $\mathcal{T}$ only takes n random bits to produce a random description of complexity n. Additionally, we demand a bijective relationship of trapdoors and indices. More generally, we could allow several matching indices i, i' to a single trapdoor t. Again, as our construction supports this uniqueness property, we do not include this in our definition.

3   Statistical Zero-Knowledge

When talking about complexity classes, we always refer to classes of *promise problems*. A promise problem $\Pi = (\Pi_{\text{YES}}, \Pi_{\text{NO}})$ is a pair of disjoint sets of yes-instances $\Pi_{\text{YES}} \subseteq \{0,1\}^*$ and no-instances $\Pi_{\text{NO}} \subseteq \{0,1\}^*$. The notion of promise problems generalizes the language-based approach: an algorithm putatively deciding membership for some input $x \in \{0,1\}^*$ gets a promise that $x \in \Pi_{\text{YES}} \cup \Pi_{\text{NO}}$.

We briefly introduce the zero-knowledge-based classes we deal with. As we do not use any definitional properties beyond some basic facts about their relationships, we omit formal definitions of these classes and refer the reader to [46] for details. The class $\mathcal{NISZK}$ consists of the promise problems having a non-interactive statistical zero-knowledge proof. The class $\mathcal{SZK}$ is the class of problems having general, possibly interactive statistical zero-knowledge proofs; this is clearly a subset of the class of problems where statistical zero-knowledge holds with respect to honest verifiers, denoted by $\mathcal{HVSZK}$. By [12,20,19] we have $\mathcal{P} \subseteq \mathcal{BPP} \subseteq \mathcal{NISZK} \subseteq \mathcal{SZK} = \mathcal{HVSZK}$.

Sahai and Vadhan [40] introduced the $\mathcal{SZK}$-complete problem statistical difference. Using the completeness of this problem we show the collapse of $\mathcal{SZK}$. To a circuit $X : \{0,1\}^m \to \{0,1\}^n$ (more precisely, to its description) we associate a random variable over $\{0,1\}^n$ by choosing the input uniformly from $\{0,1\}^m$.

Definition 4. *The promise problem statistical difference,* $SD = (SD_{YES}, SD_{NO})$, *is defined by*

$$SD_{YES} = \{(X_0, X_1) \mid \text{StatDiff}(X_0, X_1) \geq 2/3\}$$
$$SD_{NO} = \{(X_0, X_1) \mid \text{StatDiff}(X_0, X_1) \leq 1/3\}$$

To prove that SD is complete for $\mathcal{SZK}$, Sahai and Vadhan [40,41] established the polarization lemma. Basically, this lemma says that one can turn an instance (X_0, X_1) of SD into a pair (Y_0, Y_1) of circuits such that the distributions of Y_0, Y_1 are almost disjoint if $(X_0, X_1) \in SD_{YES}$ and nearly equal if $(X_0, X_1) \in SD_{NO}$. Additionally, the transformation involves an error parameter ℓ that determines how far and close, respectively, the derived distributions are. This parameter ℓ may be independent of X_0, X_1.

Fact 1 (Polarization Lemma [40,41]) *There is a polynomial-time algorithm* Polarize *that on input* $(X_0, X_1, 1^\ell)$ *outputs* (Y_0, Y_1) *such that*

$$(X_0, X_1) \in SD_{YES} \quad \Rightarrow \quad \text{StatDiff}(Y_0, Y_1) \geq 1 - 2^{-\ell}$$
$$(X_0, X_1) \in SD_{NO} \quad \Rightarrow \quad \text{StatDiff}(Y_0, Y_1) \leq 2^{-\ell}$$

Set Polarize(X_0, X_1) = Polarize $\left(X_0, X_1, 1^{|(X_0, X_1)|}\right)$.

Intuitively, one can think of a polarized pair (Y_0, Y_1) as a description of a non-interactive commitment function. The sender splits the bit b into n random pieces $b_1, \ldots, b_n$ such that $b = b_1 \oplus \cdots \oplus b_n$. Given n random instances $(Y_{i,0}, Y_{i,1})$ —among which there will be a no-instance with high probability— the sender commits to each piece b_i individually by sampling according to Y_{i,b_i} and handing this sample to the receiver. If $(Y_{i,0}, Y_{i,1})$ is a (polarized) no-instance then the distribution hides b_i and therefore b statistically; if it corresponds to a yes-instance then the sample determines b_i with very high probability (as long as the sender does not bias the sample too much).

For an ambiguous decommitment to $b' = b \oplus 1$ the sender has to flip at least one piece b_i. Put differently, the sender has to find a random string r' such that $Y_{i,b_i \oplus 1}$ maps this string to the previously given sample $Y_{1,b_i}(r)$. But if $(Y_{i,0}, Y_{i,1})$ is a yes-instance then the distributions are almost disjoint and this is quasi impossible. On the other hand, for a no-instance this is indeed possible. Hence, an ambiguous decommitment tells us the status of (at least) one of the instances. This is basically the reason why deciding membership for SD becomes tractable relative to our oracle: ambiguous decommitments can be found easily given access to the oracle. However, we remark that for a correct membership decision on *each* instance of SD, the oracle must never err. This motivates the investigation of weakly-binding commitment schemes, where the oracle must (nearly) always return ambiguous decommiments to disprove their existence. Still, a very small error for the binding property is acceptable to ensure a perfect oracle.

4   Extensions of Simon's Result

In this section we apply Simon's result [45] to obtain an oracle separation of black-box one-way permutations and trapdoor one-way functions from non-interactive weakly-binding honest-receiver statistically-secret bit commitment.

4.1   Extension to Commitment Schemes

We briefly describe the oracle construction in [45]. One starts with a random oracle Π which contains a random permutation f and a special query state collision. Basically, the random permutation f constitutes a one-way function, and the query state collision enables to find collisions in hash functions. Once proven that this random oracle is one-way, one can then derandomize the construction.

Construction 1 *Let $f : \{0,1\}^* \to \{0,1\}^*$ be a random permutation, i.e., a random function with the constraint $f(\{0,1\}^n) = \{0,1\}^n$ for all $n \in \mathbb{N}$. Define oracle Π to contain a random permutation f, and a special query state collision that takes a circuit description of a many-to-one hash function h and outputs a random element x together with a uniformly chosen value x' from $\{y \mid h(x) = h(y)\}$ (and, besides, repeats the description of the hash function and any oracle queries and answers obtained within the computation of the hash values for x and x').*

In the rest of the paper, we call this way of generating collisions x, x' the *basic sampling procedure*. In [45] it was shown that the collision-finding portion of Π does not help to invert f significantly. Note that the description of the hash function might also include f- and recursive collision-queries and that we let Π append these queries and answers to the output for collision-questions, too. Using an appropriate encoding for collision-queries, e.g., substituting values in f by mapping inputs of the form $(1 \cdots 1, h, \ldots, h)$ to $(1 \cdots 1, h, x, x', \text{queries \& answers})$ and vice versa for a sufficient number of 1's and h-repetitions, the special query state collision can be eliminated and it can be achieved that Π is also a permutation over $\{0,1\}^*$. See [45] for details. In the sequel, we sometimes switch between both approaches for sake of convenience.

We would like to extend the negative result to non-interactive weakly-binding honest-receiver statistically-secret bit commitment schemes. To this end, we change oracle Π to an oracle Σ which allows to open such commitment schemes ambiguously and to contradict the weak binding property. That is, Σ should return valid decommitments for different bits for statistically-secret commitment schemes for any sufficiently large security parameter. For instance, this can be accomplished by letting the oracle always output a non-trivial collision (i.e., with $b \neq b'$) if the statistical difference is, say, less than some bound B, and by reducing one-wayness of this oracle to the one-wayness of Π by querying Π a sufficient number of times in order to simulate the new oracle. However, for this we have to ensure that the oracle's answers to queries with statistical difference more than B are answered consistently compared to the simulation. We will overcome this problem by letting our new oracle Σ generate random ambiguous decommitments in a way that already mimics the simulator's behavior asking several questions to Π:

Construction 2 *Let Π be defined as in Construction 1. Modify Π by replacing the* collision-*query state as follows: if the probability that the basic sampling procedure outputs a random collision $(b, r), (b', r')$ with $b \neq b'$ for the many-to-one function Com_{k_m} is at least $1/6$, then uniformly select some (b, r) from the set of pairs (c, s) for which*

$$C_{c,s} = \{(c \oplus 1, s') \mid \mathrm{Com}_{k_m}(c, s) = \mathrm{Com}_{k_m}(c \oplus 1, s')\}$$

is not empty, together with a uniformly chosen value (b', r') from $C_{b,r}$. Else, for input length ℓ of Com_{k_m}, generate ℓ random collisions $(b, r), (b', r')$ with the basic sampling procedure; if there is some collision with $b \neq b'$ then output the first one that appears among these samples, otherwise return the first of the ℓ samples (which is then of the form $(b, r), (b, r')$, of course). Furthermore, the oracle appends the description of Com_{k_m} and all oracle queries to compute $\mathrm{Com}_{k_m}(b, r)$ and $\mathrm{Com}_{k_m}(b', r')$. Denote this oracle by Σ.

We claim that setting the bound to $1/6$ guarantees that oracle Σ always returns ambiguous decommitments for statistically-secret bit commitments (for sufficiently large security parameter). To see this, let Com_{k_m} be a statistically-secret bit commitment function. Call an input (b, r) good if the number of random strings s that map to the same commitment $\mathrm{Com}_{k_m}(b, r) = \mathrm{Com}_{k_m}(b, s)$ is at most twice the number of random strings s' which map to the same commitment $\mathrm{Com}_{k_m}(b, r) = \mathrm{Com}_{k_m}(b \oplus 1, s')$ for the inverse bit $b \oplus 1$. With probability at least $1/2$ a random value (b, r) is good (otherwise the statistical difference of $\mathrm{Com}_{k_m}(0)$ and $\mathrm{Com}_{k_m}(1)$ would be at least $1/4$ which would contradict the negligible deviation for large security parameters)[2]. In this case, for a uniformly chosen colliding input (b', r') to some good (b, r) it holds that $b' \neq b$ with probability at least $1/3$. Hence, for statistically-secret commitment, with probability at least $1/6$ a random collision represents valid decommitments for distinct bits. See [41] for a tighter bound depending on the actual statistical difference.

Note that if we eliminate the collision-state from Π by encoding such queries in $\{0, 1\}^*$, then Σ inherits this property.

Lemma 1. *Oracle Σ in Construction 2 is a black-box one-way permutation.*

Formally, Σ is a *random* oracle. Hence, we would better say "Picking Σ as in Construction 2 one obtains a black-box one-way permutation." We neglect this as we will later derandomize the construction anyway.

Proof (Sketch). Clearly, the permutation property is not affected by the modification, even if we encode collision-queries by bit strings. It thus suffices to prove

---

[2] By assumption the input length of commitment functions is at least as large as the security parameter. This allows us to use the same bound $1/6$ when considering the input length as replacement for the security parameter, as done in Construction 2.

one-wayness. Assume that there exists a (uniform) polynomial-size circuit family $D = (D_n)_{n \in \mathbb{N}}$ that takes advantage of the modification in Construction 2. That is, D is able to invert Σ with noticeable probability. Let the polynomial $q(n)$ bound the size of D, and let D_n invert a random image under Σ with probability at least $1/p(n)$ for a polynomial $p(n)$ and infinitely many $n \in \mathbb{N}$. Note that the total number of oracle queries, including the recursive ones in collision-queries, is bounded above by the size $q(n)$ of D_n. From D we construct a polynomial-size circuit family $C = (C_n)_{n \in \mathbb{N}}$ interacting with a random oracle Π according to Construction 1.

Basically, C_n gets an image y as input and simulates D_n on y. Each f-query of D_n is answered by asking the f-oracle of Σ. Every time D_n submits a collision-query for some Com_{k_m} with input length ℓ, then circuit C_n essentially (details below) asks Π altogether ℓ collision-queries by padding the description of Com_{k_m} with redundant bits in each query (after all, Π is a random *function* and always returns the same answer to the same question again; padding the commitment function description thus yields independent random collisions). Then C_n selects an adequate collision and hands it to D_n.

We explain in detail how C_n finds an appropriate collision. Assume for the moment that C_n picks a collision by querying Π as described above ℓ times. If the commitment function Com_{k_m} is above the limit $1/6$, then circuit C_n would find a proper collision with probability at least $1 - (5/6)^{\ell}$; if ℓ is large this is very close to 1 and almost identical to the answer of Σ. For Com_{k_m}-functions below the bound $1/6$, circuit C_n would give identically distributed answers to collision-queries in comparison to Σ. A problem occurs if ℓ is too small. Then C_n's output would differ noticeably from Σ's answer for commitment function above the bound $1/6$. To overcome this, we let C_n search for the right answers for commitment functions Com_{k_m} with small input length. Then the simulation error will still leave enough mass for C_n's success probability. We use the limit $L(n) = 4\log_2(2p(n)q(n)) \geq \log_{6/5}(2p(n)q(n))$ to identify a small input length.

Recall that the description of many-to-one functions in collision-queries may also include recursive collision-request. We assert that the same solution as before applies. Either the input length is "very short", or using enough samples yields a sufficiently good approximation. More formally, we can first modify the commitment function description by adding an "if-then-else"-check for recursive collision-queries. This check simply imitates C_n's strategy, i.e., compares the input length to $L(n)$ and proceeds accordingly.

It is not hard to show that by the choice of the parameters the simulation error for a single collision-query, either one on top level or a recursive one, is at most $1/(2p(n)q(n))$. Since D_n puts at most $q(n)$ oracle queries, by the union bound the simulation therefore fails with probability at most $1/(2p(n))$ for any sufficiently large n. Thus, at most half of the cases of D_n's success are covered by the simulation error, and C_n successfully inverts Π with probability at least $1/(2p(n))$ infinitely often. But this contradicts the result in [45].   □

Derandomizing the oracle construction by taking an appropriate oracle which works for all of the countable many uniform circuits [44,45], we obtain a one-way permutation oracle relative to which there cannot exist non-interactive weakly-binding honest-receiver statistically-secret bit commitment schemes, even such ones that use oracle queries.

Theorem 1. *Relative to an oracle there exist black-box one-way functions and permutations, but no non-interactive weakly-binding honest-receiver statistically-secret bit commitment schemes.*

4.2   Extension to Black-Box Trapdoor Functions

The essence of our construction of a black-box trapdoor function utilizes the idea of the construction of signature schemes from one-way functions [33,38]: the public and the private key of the signature scheme are the value of a one-way function and its preimage. Here, the index i of the trapdoor function is the value of Σ at the trapdoor t. Incorporating i into the evaluation process by setting the function to $\Sigma(i, \cdot)$ gives the desired trapdoor one-way function. To invert some y in the range of $\Sigma(i, \cdot)$ one has to provide the matching trapdoor t to i to the inversion oracle.

Construction 3 *Let Σ (over $\{0,1\}^*$) be as in Construction 2. Define $\mathcal{T}$ as follows:*

- *generation: on input $\omega \in \{0,1\}^n$ oracle $\mathcal{T}$ outputs $t = \omega$ and $i = \Sigma(\omega)$.*
- *evaluation: on input $i, x \in \{0,1\}^n$ the evaluation algorithm of $\mathcal{T}$ returns $\Sigma(i, x) \in \{0,1\}^{2n}$*
- *inversion: given $y \in \{0,1\}^{2n}$ and $t \in \{0,1\}^n$ the oracle $\mathcal{T}$ first checks that $\Sigma(t)$ equals the left half of $(i, x) = \Sigma^{-1}(y)$. If so, it outputs x, else some undefined symbol.*

Some remarks are in place. Apparently, our function is one-to-one but not a permutation. Hence, iteration techniques for trapdoor permutations, like feeding the output into the function again, are impossible. Nevertheless, we can apply a tree construction of logarithmic depth by iterating the function on each output half. This may replace the permutation in some settings. Similarly, it may suffice to iterate the function on, say, the left half of the result and output the right half "in clear".

Also, observe how our construction circumvents the problem of claws. A pair of claw-free functions is pair of functions with identical range, but such that finding inputs for each function that both map to the same output is infeasible. Any impossibility result about the construction of non-interactive statistically-secret commitment schemes based on any trapdoor function implies that the trapdoor functions do not yield claw-free functions. In our case, any distinct trapdoor functions $(t, i) \neq (t', i')$ have disjoint ranges (because $\Sigma(i, x) \neq \Sigma(i', x')$ for all x, x' for the permutation Σ).

The proof that $\mathcal{T}$ is a trapdoor one-way function follows by reduction to the one-wayness of Σ. While generation and evaluation queries for $\mathcal{T}$ can be easily emulated given access to Σ, we have to ensure that inversion queries do not lend significant power to an adversary. Indeed, for a given index $i \in \{0,1\}^n$ the range of $\Sigma(i, \cdot)$ forms a sparse subset of $\{0,1\}^{2n}$ of size 2^n. Therefore, any algorithm that tries to invert an image y by guessing a trapdoor t' and asking $\mathcal{T}$ to invert a large y with respect to t' almost certainly gets the undefined symbol as reply. In other words, inversion queries for large images essentially lead to reasonable answers only if the corresponding image has been computed previously by querying the evaluation oracle of $\mathcal{T}$. But then the preimage is already known and gives no additional information. For short images, a preimage can be computed efficiently by searching the domain, and thus inversion queries do not give any advantage in this case either.

Lemma 2. *Oracle $\mathcal{T}$ in Construction 3 is a black-box one-to-one trapdoor one-way function.*

Basically, the proof follows the one of Lemma 1 of the one-wayness of oracle Σ, regarding that inversion queries do not help significantly according to the previous discussion. It is omitted for space reasons. Derandomizing Construction 3 we conclude:

Theorem 2. *Relative to an oracle there are black-box one-to-one trapdoor functions and black-box one-way functions and permutations, but no non-interactive weakly-binding honest-receiver statistically-secret bit commitment schemes.*

In analogy to [45] we can turn $\mathcal{T}$ into a single oracle that operates on bit strings. A description of this will be given in the full version.

5   Nontrivial Statistical Zero-Knowledge Requires More Than Black-Box One-Wayness

In this section we prove the collapse of $\mathcal{SZK}$ relative to an appropriate one-way permutation oracle and to a one-to-one black-box trapdoor function. It is known that hard-to-predict problems in $\mathcal{SZK}$ imply one-way functions [34]. The premise of this implication was later relaxed to $\mathcal{CZK} \neq$ average-case-$\mathcal{BPP}$ [36]. Our result presents some evidence that nontrivial problems in $\mathcal{SZK}$ actually need more than one-wayness. Furthermore, we supplement the result that $\mathcal{SZK} \neq \mathcal{BPP}$ relative to an oracle [2] by showing that $\mathcal{SZK} = \mathcal{BPP} = \mathcal{P}$ relative to a one-way permutation oracle. Note that the existence of (black-box) one-way functions implies that $\mathcal{NP} \nsubseteq \mathcal{BPP}$.

The construction of our oracle Γ, relative to which SD is easy, is a slight modification of Σ in Construction 3. In order to preserve the interpretation that one-wayness does not suffice for nontrivial problems in $\mathcal{SZK}$, we allow instances of SD to include query gates for the oracle Γ. We remark that this extended version of SD is complete for this relativized class of $\mathcal{SZK}$; this shows for example in the completeness proof given in [46]. Hence, if we show the tractability of SD relative to Γ it follows that the whole relativized class of $\mathcal{SZK}$ collapses.

In the construction of Γ we presume wlog. that the input size of circuits Y_0 and Y_1 of a polarized instance (for any complexity parameter) is at least the output length, and that both circuit have the same input size. Otherwise algorithm Polarize pads the input length with a minimal number of bits. Then we can view an input (X_0, X_1) as a description of a commitment function $\text{Com}_{(X_0, X_1)}(b, r) = Y_b(r)$, where Y_0, Y_1 are derived by applying the polarization lemma (together with the length convention).

Construction 4 *Let Σ be as in Construction 3. Alter Σ to Γ by modifying the* collision-*state as follows: Γ only accepts pairs (X_0, X_1) of circuits as arguments to* collision-*queries. Then Γ polarizes (X_0, X_1) with parameter $\ell = |(X_0, X_1)|$ to obtain (Y_0, Y_1). If (X_0, X_1) is a yes-instance for* SD *then return a pair $(b, r), (b, r')$ such that $Y_b(r) = Y_b(r')$ (generated by the basic sampling procedure with the restriction that the second value is chosen uniformly among the collisions with the same leftmost bit b); if $(X_0, X_1) \in$* SD_{NO} *then sample a random collision $(b, r), (b \oplus 1, r')$ accordingly. Otherwise, if $(X_0, X_1) \notin$* $\text{SD}_{\text{YES}} \cup \text{SD}_{\text{NO}}$*, compute ℓ random collisions with the basic procedure, output the first one with $b \neq b'$, if such a collision exists, otherwise return the first sample. Each time, also append (X_0, X_1) and all oracle queries made to compute the circuits' outputs.*

Clearly, for *any* polarized no-instance (Y_0, Y_1) with probability at least $1/6$ the basic sampling procedure returns a collision $(b, r), (b', r')$ with $b \neq b'$. See the discussion in Section 4.1. Next we show that for yes-instances (with statistical difference close to 1) this rarely happens; the proof is deferred from this version:

Lemma 3. *Let $(X_0, X_1) \in$* SD_{YES}*. Then the probability that the basic sampling procedure for $(Y_0, Y_1) = \text{Polarize}(X_0, X_1)$ yields a collision $(b, r), (b, r')$ is at most $2^{-\ell/2+1}$ for $\ell = |(X_0, X_1)|$.*

We omit a formal proof that we can reduce a circuit D inverting Γ to a circuit C finding preimages for Π. The argument is almost identical to the one of Lemma 1, taking into account that the basic sampling procedure almost never yields collisions for the same bit b for yes-instances according to the previous lemma. Additionally, replacing Σ by Γ in Construction 3 of $\mathcal{T}$, we obtain a one-to-one trapdoor function granting access to Γ.

Theorem 3. *There exists an oracle relative to which $\mathcal{P} = \mathcal{BPP} = \mathcal{NISZK} = \mathcal{SZK} = \mathcal{HVSZK}$ but relative to which one-to-one trapdoor functions and one-way permutations exist.*

Proof. Obviously, relative to our (derandomized) oracles Γ and $\mathcal{T}$ we have $\mathcal{SZK} \subseteq \mathcal{P} \subseteq \mathcal{BPP}$, because if we simply query the oracle about the input instance (X_0, X_1) and output 1 (respectively, 0) if and only if we are given a collision $(b, r), (b, r')$ (respectively, $(b, r), (b \oplus 1, r')$), then we correctly decide membership for SD in polynomial time. Furthermore, the proofs [20,19] that $\mathcal{NISZK} \subseteq \mathcal{HVSZK} \subseteq \mathcal{SZK}$ relativize, and together with $\mathcal{BPP} \subseteq \mathcal{NISZK}$ the assertion follows.   □

6   Implications to Other Cryptographic Protocols

We show that various problems imply non-interactive weakly-binding honest-receiver statistically-secret commitment schemes. It follows that our oracle separation transfers to these cases as well. Because of lack of space, a formal treatment is deferred from this abstract, and we give a rather informal description here.

The problem of non-interactive crypto-computing [42] deals with computing on encrypted data. The server, possessing a secret circuit C, receives an encryption of some input x from the client. Then the server inattentively evaluates $C(x)$ and returns some value to the client upon which the client can extract the value $C(x)$ but learns nothing more about C in a statistically sense.

It is straightforward to devise non-interactive weakly-binding honest-receiver statistically-secret bit commitment schemes from such crypto-computing protocols. Namely, the committing party splits bit $b = b_1 \oplus \cdots \oplus b_n$ into n random pieces, then the honest receiver in this commitment protocol sends n encryptions of random bits $a_1, \ldots, a_n$, and the sender replies with the inattentive circuit evaluations of $\mathsf{OR}_{b_i}(a_i) = b_i \vee a_i$ for $i = 1, \ldots, n$. If the honest receiver sends some a_i with $a_i = 1$ (which happens with probability at least $1 - 2^{-n}$) then he does not learn anything about b_i and thus about b. On the other hand, to open a commitment ambiguously, a malicious sender needs to distinguish with noticeable advantage between 0-encryptions —for which b_j is pinned down because the receiver learns b_j— and 1-encryptions when the receiver does not gain any information about b_j. But this would contradict the security of the encryption scheme.

Sander et al. [42] construct non-interactive crypto-computing protocols from any semantically-secure *rerandomizable* bit encryption scheme; a rerandomizable bit encryption system allows to renew the distribution of an encrypted bit without knowing the secret key, i.e., from the ciphertext and the public key alone. Hence, we can also construct a statistically-secret bit commitment protocol from a rerandomizable encryption scheme.

With a one-out-of-two oblivious transfer (OT) protocol [37,15] a party transfers one of two bits (the choice is made at random) to a receiver such that the receiver does not learn anything about the other bit, and such that the sender does not know which bit has been sent. In order to derive a commitment protocol we apply the same splitting technique as in the case of crypto-computers. Specifically, the sender splits b into two random pieces and obliviously transfers one to the honest receiver. Therefore, if the oblivious transfer protocol is non-interactive and provides statistically-sender-privacy with respect to honest receivers we obtain an appropriate commitment scheme. Additionally, this approach works with other kinds of oblivous transfers, like chosen-one-out-of-two protocols. Details are omitted.

Until recently, non-interactive oblivious transfer protocols were only known in a public-key infrastructure setting [6,11]. But lately, under the decisional Diffie-Hellman assumption, Naor and Pinkas [32] and Aiello et al. [3] devised statistically-sender-private chosen-one-out-of-two OT protocols which require both parties to send a single message only and without any setup assumptions. Hence, such oblivious transfers may be impossible using general public-key cryptosystems but they are constructible from specific intractability assumptions.

A private information retrieval (PIR) scheme [8] is a special oblivous transfer protocol in which one out of n bits is transferred. Nothing is guaranteed about the sender's privacy, though, i.e., even the honest receiver might learn more than a single bit. In contrast to oblivious transfer the communication complexity must not exceed n bits in PIR schemes. In [4] it has been shown that non-interactive *low-comunication* PIR schemes, i.e., where less than $n/2$ bits are communicated, imply non-interactive statistically-secret bit commitment. In summery,

Corollary 1. *There is an oracle relative to which black-box one-to-one trapdoor functions and black-box one-way permutations exist, but relative to which non-interactive honest-client statistically-server-private crypto-computing for orgates, rerandomizable bit encryption, non-interactive honest-receiver statistically-sender-private oblivious transfer and non-interactive low-communication private information retrieval are impossible.*

Acknowledgements

We are grateful to Dan Simon for helpful discussions about his paper. We would also like to thank all anonymous reviewers for their comments.

References

1. W.AIELLO, J.HÅSTAD: Statistical Zero-Knowledge Languages can be Recognized in Two Rounds, *Journal of Computer and System Science, Vol. 42, pp. 327–345*, 1991.
2. W.AIELLO, J.HÅSTAD: Relativized Perfect Zero-Knowledge is not BPP, *Information and Computation, Vol. 93, pp. 223–240*, 1991.
3. W.AIELLO, Y.ISHAI, O.REINGOLD: Priced Oblivious Transfer: How to Sell Digital Goods, *Eurocrypt 2001, Lecture Notes in Computer Science, Vol. 2045, Springer-Verlag*, 2001.
4. A.BEIMEL, Y.ISHAI, E.KUSHILEVITZ, T.MALKIN: One-Way Functions are Essential for Single-Server Private Information Retrieval, *Proceedings of the 31st Annual ACM Symposium on the Theory of Computing (STOC), pp. 89–98*, 1999.
5. M.BELLARE, S.HALEVI, A.SAHAI, S.VADHAN: Many-To-One Trapdoor Functions and Their Relation to Public-Key Cryptosystems, *Crypto '98, Lecture Notes in Computer Science, Vol. 1462, Springer-Verlag, pp. 283–298*, 1998.
6. M.BELLARE, S.MICALI: Non-Interactive Oblivious Transfer and Applications, *Crypto '89, Lecture Notes in Computer Science, Vol. 435, Springer-Verlag, pp. 547–559*, 1990.
7. M.BEN-OR, O.GOLDREICH, S.GOLDWASSER, J.HÅSTAD, J.KILLIAN, S.MICALI, P.ROGAWAY: Everything Provable is Provable in Zero-Knowledge, *Crypto '88, Lecture Notes in Computer Science, Vol. 403, Springer-Verlag, pp. 37–56*, 1990.
8. B.CHOR, O.GOLDREICH, E.KUSHILEVITZ, M.SUDAN: Private Information Retrieval, *Journal of ACM, vol. 45, pp. 965–981*, 1998.

9. C.CRÉPEAU, F.LÉGARÉ, L.SAVAIL: How to Convert a Flavor of Quantum Bit Commitment, *Eurocrypt 2001, Lecture Notes in Computer Science, Vol. 2045, Springer-Verlag,* 2001.

10. I.DAMGÅRD, T.PEDERSEN, B.PFITZMANN: On the Existence of Statistically Hiding Bit Commitment Schemes and Fail-Stop Signatures, *Crypto '93, Lecture Notes in Computer Science, Vol. 773, Springer-Verlag, pp. 250–255,* 1993.

11. A.DE SANTIS, G.DI CRESCENZO, G.PERSIANO: Public-Key Cryptography and Zero-Knowledge Arguments, *Information and Computation, Vol. 121, No. 1, pp. 23–40,* 1995.

12. G.DI CRESCENZO, T.OKAMOTO, M.YUNG: Keeping the SZK-Verifier Honest Unconditionally, *Crypto '97, Lecture Notes in Computer Science, Vol. 1294, Springer-Verlag, pp. 31–45,* 1997.

13. W.DIFFIE, M.HELLMAN: New Directions in Cryptography, *IEEE Transaction on Information Theory, Vol. 22, pp. 644–654,* 1976.

14. P.DUMAIS, D.MAYERS, L.SALVAIL: Perfectly Concealing Quantum Bit Commitment from Any One-Way Permutation, *Eurocrypt 2000, Lecture Notes in Computer Science, Vol. 1807, Springer-Verlag, pp. 300–315,* 2000.

15. S.EVEN, O.GOLDREICH, A.LEMPEL: A Randomized Protocol for Signing Contracts, *Communication of the ACM, vol. 28, pp. 637–647,* 1985.

16. L.FORTNOW: The Complexity of Perfect Zero-Knowledge, *Proceedings of the 19th Annual ACM Symposium on the Theory of Computing (STOC), pp. 204–209,* 1987.

17. R.GENNARO, L.TREVISAN: Lower Bounds on the Efficiency of Generic Cryptographic Constructions, *Proceedings of the 41st IEEE Symposium on Foundations of Computer Science (FOCS),* 2000.

18. Y.GERTNER, S.KANNAN, T.MALKIN, O.REINGOLD, M.VISWANATHAN: The Relationship Between Public Key Encryption and Oblivious Transfer, *Proceedings of the 41st IEEE Symposium on Foundations of Computer Science (FOCS),* 2000.

19. O.GOLDREICH, A.SAHAI, S.VADHAN: Can Statistical Zero-Knowledge be made Non-Interactive? or On the Relationship of SZK and NISZK, *Crypto '99, Lecture Notes in Computer Science, Springer-Verlag,* 1999.

20. O.GOLDREICH, A.SAHAI, S.VADHAN: Honest-Verifier Statistical Zero-Knowledge Equals General Statistical Zero-Knowledge, *Proceedings of the 30th Annual ACM Symposium on Theory of Computing (STOC), ACM Press, pp. 399–408,* 1998.

21. S.GOLDWASSER, O.GOLDREICH, S.MICALI: How to Construct Random Functions, *Journal of ACM, vol. 33, pp. 792–807,* 1986.

22. S.GOLDWASSER, S.MICALI: Probabilistic Encryption, *Journal of Computer and System Science, Vol. 28, pp. 270–299,* 1984.

23. S.HALEVI, S.MICALI: Practical and Provably-Secure Commitment Schemes from Collision-Free Hashing, *Crypto '96, Lecture Notes in Computer Science, Vol. 1109, Springer-Verlag, pp. 201–215,* 1996.

24. J.HÅSTAD, R.IMPAGLIAZZO, L.LEVIN, M.LUBY: A Pseudorandom Generator from any One-way Function, *SIAM Journal on Computing, vol. 28(4), pp. 1364–1396,* 1999.

25. R.IMPAGLIAZZO, M.LUBY: One-Way Functions are Essential for Complexity Based Cryptography, *Proceedings of the 30th IEEE Symposium on Foundations of Computer Science (FOCS), pp. 230–235,* 1989.

26. R.IMPAGLIAZZO, S.RUDICH: Limits on the Provable Consequences of One-Way Permutations, *Proceedings of the 21st Annual ACM Symposium on the Theory of Computing (STOC), pp. 44–61,* 1989.

27. R.IMPAGLIAZZO, M.YUNG: Direct Minimum-Knowledge Computations, *Crypto '87, Lecture Notes in Computer Science, Vol. 293, Springer-Verlag, pp. 40–51*, 1987.

28. J.KAHN, M.SAKS, C.SMYTH: A Dual Version of Reimer's Inequality and a Proof of Rudich's Conjecture, *Proceedings of 15th IEEE Conference on Computational Complexity*, 2000.

29. J.KIM, D.SIMON, P.TETALI: Limits on the Efficiency of One-Way Permutation-Based Hash Functions, *Proceedings of the 40th IEEE Symposium on Foundations of Computer Science (FOCS)*, 1999.

30. M.NAOR: Bit Commitment Using Pseudo-Randomness, *Journal of Cryptology, vol. 4, pp. 151–158*, 1991.

31. M.NAOR, R.OSTROVSKY, R.VENKATESAN, M.YUNG: Perfect Zero-Knowledge Arguments for NP Using Any One-Way Permutation, *Journal of Cryptology, vol. 11, pp. 87–108*, 1998.

32. M.NAOR, B.PINKAS: Efficient Oblivious Transfer Protocols, *Twelfth Annual ACM-SIAM Symposium on Discrete Algorithms*, 2001.

33. M.NAOR, M.YUNG: Universal One-Way Hash Functions and Their Cryptographic Applications, *Proceedings of the 21st Annual ACM Symposium on the Theory of Computing (STOC), pp. 33–43*, 1989.

34. R.OSTROVSKY: One-Way Functions, Hard on Average Problems, and Statistical Zero-Knowledge Proofs, *IEEE Conference on Structure in Complexity Theory, pp. 133–138*, 1991.

35. R.OSTROVSKY, R.VENKATESAN, M.YUNG: Fair Games Against an All-Powerful Adversary, *AMS DIMACS Series in Discrete Mathematics and Theoretical Computer Science, Vol. 13, pp. 155–169*, 1993.

36. R.OSTROVSKY, A.WIGDERSON: One-Way Functions are Essential for Non-Trivial Zero-Knowledge, *Proceedings of the Second Israel Symposium on Theory of Computing and Systems*, 1993.

37. M.RABIN: How to Exchange Secrets by Oblivious Transfer, *Technical Report TR-81, Harvard*, 1981.

38. J.ROMPEL: One-Way Functions are Necessary and Sufficient for Secure Signatures, *Proceedings of the 22nd Annual ACM Symposium on the Theory of Computing (STOC), pp. 387–394*, 1990.

39. S.RUDICH: The Use of Interaction in Public Cryptosystems, *Crypto '91, Lecture Notes in Computer Science, Vol. 576, Springer-Verlag, pp. 242–251*, 1992.

40. A.SAHAI, S.VADHAN: A Complete Promise Problem for Statistical Zero-Knowledge, *Proceedings of the 38th IEEE Symposium on Foundations of Computer Science (FOCS), pp. 448–457*, 1997.

41. A.SAHAI, S.VADHAN: Manipulating Statistical Difference, *AMS DIMACS Series in Discrete Mathematics and Theoretical Computer Science, Vol. 43, pp. 251–270*, 1999.

42. T.SANDER, A.YOUNG, M.YUNG: Non-Interactive Crypto-Computing for NC^1, *Proceedings of the 40th IEEE Symposium on Foundations of Computer Science (FOCS)*, 1999.

43. A.SHAMIR: IP=PSPACE, *Proceedings of the 31st IEEE Symposium on Foundations of Computer Science (FOCS)*, 1990.

44. D.SIMON: On the Power of Quantum Computation, *Proceedings of the 35th IEEE Symposium on Foundations of Computer Science (FOCS), pp. 124–134*, 1994.

45. D.SIMON: Finding Collisions on a One-Way Street: Can Secure Hash Functions be Based on General Assumptions?, *Eurocrypt '98, Lecture Notes in Computer Science, Vol. 1403, Springer-Verlag, pp. 334–345*, 1998.

46. S.VADHAN: A Study of Statistical Zero-Knowledge Proofs, *Ph.D. thesis, MIT, available at* http://theory.lcs.mit.edu/~salil/, September 1999.

The Representation Problem Based on Factoring

Marc Fischlin and Roger Fischlin

Johann Wolfgang Goethe-University
Frankfurt am Main, Germany
{marc,fischlin}@mi.informatik.uni-frankfurt.de
http://www.mi.informatik.uni-frankfurt.de/

Abstract. We review the representation problem based on factoring and show that this problem gives rise to alternative solutions to a lot of cryptographic protocols in the literature. And, while the solutions so far usually either rely on the RSA problem or the intractability of factoring integers of a special form (e.g., Blum integers), the solutions here work with the most general factoring assumption. Protocols we discuss include identification schemes secure against parallel attacks, secure signatures, blind signatures and (non-malleable) commitments.

1 Introduction

The RSA representation problem deals with the problem of finding a decomposition of a value into an RSA-like representation. Specifically, given a modulus $N = pq$ of two secret primes p, q, an exponent e relatively prime to Euler's totient function $\varphi(N)$ and a value $g \in \mathbb{Z}_N^*$, find to $X \in \mathbb{Z}_N^*$ a representation $x \in \mathbb{Z}_e$ and $r \in \mathbb{Z}_N^*$ with $X = g^x r^e \bmod N$. It is well-known that given N, e, g coming up with some X and distinct representations $(x_1, r_1), (x_2, r_2)$ is as hard as the RSA problem [Ok92].

The RSA representation problem has a vast number of applications: for instance, Okamoto [Ok92] constructs an identification protocol secure against (parallel) active attacks which Pointcheval and Stern [PS00] subsequently turn into a secure signature scheme and a blind signature scheme. Fischlin and Fischlin [FF00] as well as Di Crescenzo et al. [CKOS01] use the RSA representation problem to derive efficient non-malleable commitment schemes based on RSA. Brands [B97] shows how to prove linear relations on committed values with an extended version of the RSA representation problem.

Interestingly, there is a seemingly less popular analogue to the RSA representation problem relying on the assumed hardness of factoring integers. In this case, a representation of X with respect to N, g and a number t is a pair $x \in \mathbb{Z}_{2^t}$ and $r \in \mathbb{Z}_N^*$ with $X = g^x r^{2^t} \bmod N$. Brassard et al. [BCC88] introduce this representation type for the special case $t = 1$. Damgård [D95] generalizes this to arbitrary $t \geq 1$ for Blum integers N where $p, q = 3 \bmod 4$. In order to advance to general moduli we introduce an "adjustment" parameter τ which depends on the prime factorization of N (and which equals 0 for Blum integers, for example), and we define a representation of X with respect to N, τ, g and t to

B. Preneel (Ed.): CT-RSA 2002, LNCS 2271, pp. 96–113, 2002.
© Springer-Verlag Berlin Heidelberg 2002

be a pair $x \in \mathbb{Z}_{2^t}$ and $r \in \mathbb{Z}_N^*$ such that $X = g^x r^{2^{\tau+t}} \bmod N$. As we will elaborate, for appropriate choices of τ, g the task of finding a value X and different representations becomes equivalent to the factoring problem for *arbitrary moduli*.

One reason for the unpopularity of the factoring representation problem may stem from the fact that Okamoto's previously proposed identification scheme based on this problem is flawed. It is sufficient to solve the RSA problem to pass the identification scheme with constant probability, without necessarily being able to factor the modulus. We review this shortcoming in Appendix A. Fortunately, the bug in Okamoto's scheme is fixable, and we can indeed devise a secure identification scheme using the factoring representation problem. We show that for suitable parameters the protocol becomes provably secure under the factoring assumption.

Among other identification schemes provably secure as factoring, the presumably most popular are the Feige-Fiat-Shamir protocol [FFS88] and its variation due to Ong-Schnorr [OS90,S96] as well as Shoup's system [Sh99]. For these schemes there is a trade-off between the key size and security against parallel attacks. While the Feige-Fiat-Shamir protocol provides security against such parallel attacks, and therefore forms a fundament for secure resettable identification [BFGM01] and blind signatures with parallel signature generation [PS97,PS00], it also requires large secret and public keys. The Shoup and the Ong-Schnorr system, on the other hand, admit short keys but are conceivably not secure against parallel attacks[1].

Our protocol fills the gap and achieves security against parallel attacks and requires only short keys. With the techniques introduced in [PS00] we therefore obtain a secure signature scheme and a secure blind signature scheme withstanding up to poly-logarithmically many concurrent signature request, both in the random oracle model. Furthermore, we derive a secure resettable identification protocol by the general transformation presented in [BFGM01].

As for further applications, our result generalizes the result by Halevi [H99] that two-round commitment schemes does not only work with William integers but rather with any moduli. Also, plugging our result into the constructions of [FF00,CKOS01], we conclude that efficient non-malleable commitment schemes can be constructed under the assumption that factoring is hard. In fact, our variation of the protocols in [CKOS01] does not only base the security on a milder assumption, but also simplifies and improves the scheme concerning computational effort and communication complexity.

The paper is structured as follows. In Section 2 we formally state the representation problem based on factoring and prove equivalence to the intractability

---

[1] Schnorr [S96,S97] claims that the Ong-Schnorr protocol with short keys is secure against parallel attacks for very special system parameters where a large power 2^m divides $p - 1$ or $q - 1$ (e.g., $m \geq 25$ for reasonable choices). Such primes form only a small subspace of all primes and may be much harder to find. Moreover, although we are not aware of any factoring method today taking advantage of this property, such moduli are in principle more vulnerable to improved factoring procedures.

of factoring large numbers. Section 3 discusses applications of the representation problem to identification and (blind) signatures. In Section 4 we deal with commitments and show how to construct efficient non-malleable commitment schemes based on the factoring representation problem.

2   Representation Problem

We state the RSA and factoring representation problems formally in Sections 2.1 and 2.2, respectively. In Section 2.3 we prove the equivalence of the factoring representation problem to the factoring problem.

2.1   RSA Representation Problem

An RSA modulus $N = pq$ is the product of two distinct primes p, q. A corresponding RSA exponent $e \neq \pm 1 \bmod \varphi(N)$ is relatively prime to Euler's totient function $\varphi(N) = (p - 1)(q - 1)$ [RSA78]. We say that N is an n-bit modulus if n bits are sufficient and necessary for the binary representation of N, that is, if $2^{n-1} \leq N < 2^n$.

We presume that there is an efficient index generator RSAIndex for the representation problem which, on input 1^n, returns an n-bit RSA modulus N, a corresponding RSA exponent e and a random element $g \in_R \mathbb{Z}_N^*$. Let $(N, e, g) \leftarrow \mathsf{RSAIndex}(1^n)$ denote the sampling process. An *RSA representation* for a value $X \in \mathbb{Z}_N^*$ with respect to a tuple (N, e, g) is a pair (x, r) with $x \in \mathbb{Z}_e$ and $r \in \mathbb{Z}_N^*$ such that

$$X = g^x r^e \bmod N.$$

Every $X \in \mathbb{Z}_N^*$ has exactly e representations with respect to (N, e, g), because for each $x \in \mathbb{Z}_e$ there is a unique $r \in \mathbb{Z}_N^*$ such that $r^e = Xg^{-x} \bmod N$. We usually omit mentioning the reference to (N, e, g) if it is clear from the context, and simply say that (x, r) is a representation of X.

Definition 1 (RSA Representation Problem). *Given* $(N, e, g) \leftarrow$ RSAIndex(1^n) *return some* $X \in \mathbb{Z}_N^*$ *as well as two different representations* $(x_1, r_1), (x_2, r_2) \in \mathbb{Z}_e \times \mathbb{Z}_N^*$ *of* X.

In contrast, the ordinary RSA problem asks to compute the e-th root $g^{1/e} \bmod N$ given $(N, e, g) \leftarrow \mathsf{RSAIndex}(1^n)$. This task is widely assumed to be intractable, i.e., no polynomial-time algorithm solves the RSA problem with more than negligible success probability. This implies that factoring N, too, is believed to be intractable. Yet, it is an open problem if RSA is indeed equally hard as factoring (see also [BV98] for a discussion).

Provided one can solve the RSA problem, then the RSA representation problem becomes tractable, e.g., for any $r \in \mathbb{Z}_N^*$ both $(0, r)$ and $(1, rg^{-1/e} \bmod N)$ are representations of $X = r^e \bmod N$. The converse holds as well [Ok92], and the equivalence reveals that both problems can be solved with the same success/running time characteristics, neglecting minor extra computations (in the sequel we keep on disregarding the effort for such additional minor computations).

2.2 Factoring Representation Problem

We next address the factoring representation problem. We replace the RSA exponent e by some power of 2. Namely, we substitute e by $2^{\tau+t}$ where t describes the bit length of $x \in \mathbb{Z}_{2^t}$ and the integer τ depends on the prime factorization of the modulus N; we will explain the choice and role of this adjustment parameter τ later. Then a representation for $X \in \mathbb{Z}_N^*$ with respect to $N = pq$, $g \in \mathbb{Z}_N^*$ and $\tau \geq 0$, $t \geq 1$ is a pair $(x, r) \in \mathbb{Z}_{2^t} \times \mathbb{Z}_N^*$ such that

$$X = g^x r^{2^{\tau+t}} \bmod N.$$

Apparently, given the factorization of N one can easily come up with two different representations. The converse does not hold in general: for example (x, r) and $(x, -r)$ represent the same X. Since we are mainly interested in finding distinct x-components we therefore call representations (x_1, r_1) and (x_2, r_2) *different* or *distinct* if and only if $x_1 \neq x_2$. Observe that this subsumes the RSA case where distinct x-components imply different r's and vice versa.

Basically, the RSA and the factoring representation problem diverge concerning the equivalence to the underlying number-theoretic assumption because of the number of preimages of r^e and $r^{2^{\tau+t}}$, respectively. For RSA parameters the mapping $r \mapsto r^e \bmod N$ constitutes a permutation on $\mathbb{Z}_N^*$. Squaring on $\mathbb{Z}_N^*$, however, is a 4:1 mapping for $N = pq$. Restricting the modulus to a Blum integer where $p, q = 3 \bmod 4$ squaring becomes a permutation on the subgroup of quadratic residues QR_N. More generally, for any odd modulus N with prime factorization $N = \prod_{i=1}^r p_i^{e_i}$ where $p_1, p_2, \ldots, p_r$ are distinct odd primes and $e_1, e_2, \ldots, e_r \geq 1$, let η denote the smallest integer such that $2^{\eta+1}$ does not divide any $\varphi(p_i^{e_i})$. Then squaring is a permutation on the subgroup

$$\mathrm{HQR}_N := \{x^{2^\eta} \mid x \in \mathbb{Z}_N^*\} = \{x \in \mathbb{Z}_N^* \mid \mathrm{ord}_N(x) \text{ is odd}\}$$

of the "highest quadratic" residues, namely the 2^η-th powers (see, for example, [S96,H99]):

Proposition 1. *For any odd modulus N squaring is a permutation on* HQR_N.

Squaring permutes the 2^k-th powers for any $k \geq \eta$ for any odd n-bit modulus N. In other words, as long as $k \geq \eta$, the set of the 2^k-th powers of the elements in $\mathbb{Z}_N^*$ is the subgroup of elements with odd order. Since $\eta \leq n$, even without knowledge of η the set HQR_N is efficiently samplable by taking a random element from $\mathbb{Z}_N^*$ and raising it to its 2^n-th power.

With the similarity of Blum integers and QR_N to general moduli and HQR_N we are ready to state the factoring representation problem turning out to be equivalent to the factoring problem. But before, some words of clarification about the parameter τ follow. Recall that a representation for X is a pair (x, r) with

$X = g^x r^{2^{\tau+t}} \bmod N$. In the following we demand that $\tau \geq \eta - 1$ and thus τ may reveal some information about the factors of N. But because $1 \leq \eta \leq n$ we can easily guess this information with probability $\frac{1}{n}$, or, in case of Blum moduli for instance, the fact $\eta = 1$ is publicly known anyway. In particular, for Blum integers we may set $\tau = 0$ and the representation problem in this case equals the one stated by Damgård [D95].

Let FactIndex denote an efficient index generator that outputs an n-bit RSA modulus N, $\tau \geq \eta - 1$, $t \geq 1$ and an independently chosen element $g \in_R \mathrm{HQR}_N$ for input 1^n, and write $(N, \tau, t, g) \leftarrow \mathsf{FactIndex}(1^n)$ for the sampling process.

Definition 2 (Factoring Representation Problem). *Given $(N, \tau, t, g) \leftarrow$ FactIndex(1^n) return some $X \in \mathbb{Z}_N^*$ as well as two different representations $(x_1, r_1), (x_2, r_2) \in \mathbb{Z}_{2^t} \times \mathbb{Z}_N^*$ of X, i.e., with $x_1 \neq x_2$.*

An important observation for our identification and commitment protocols is that each $X \in \mathrm{HQR}_N$ has exactly 2^t different representations. It follows that for a random representation (x, r) the value $X := g^x r^{2^{\tau+t}} \bmod N$ does not reveal anything about the specific x.

2.3   Factoring Representation Problem and Factoring

Given the factorization of N it is easy to compute a $2^{\tau+t}$-th root of $g \in \mathrm{HQR}_N$ and the corresponding representation problem becomes tractable. On the other hand, by solving the representation problem one efficiently determines the prime factors of N. Before we prove this we present a technical lemma:

Lemma 1. *If a probabilistic algorithm solves the factoring representation problem $(N, \tau, t, g) \leftarrow$ FactIndex(1^n), then a $2^{\tau+1}$-th root $b \in \mathbb{Z}_N^*$ of g can be computed within the same time bound and same success probability.*

Proof. Given two different representations (x_1, r_1) and (x_2, r_2) of some $X \in \mathbb{Z}_N^*$, let $\Delta x := x_1 - x_2$ and $r := r_2 r_1^{-1} \bmod N$ where $0 < |\Delta x| < 2^t$. Then

$$g^{\Delta x} = g^{x_1 - x_2} = r_2^{2^{\tau+t}} r_1^{-2^{\tau+t}} = r^{2^{\tau+t}} \bmod N. \tag{1}$$

Notice that the exponents Δx and $2^{\tau+t}$ may not be relatively prime. So suppose $2^k = \gcd(\Delta x, 2^{\tau+t})$ where $0 \leq k < t$. Computing $u, v \in \mathbb{Z}$ subject to $u\Delta x + v2^{\tau+t} = 2^k$ by applying the extended Euclidean algorithm we derive

$$g^{2^k} = g^{u\Delta x + v2^{\tau+t}} = (g^{\Delta x})^u \cdot (g^v)^{2^{\tau+t}} = (r^u g^v)^{2^{\tau+t}} \bmod N.$$

Since $g \in \mathrm{HQR}_N$ and squaring permutes HQR_N, the value $b := (r^u g^v)^{2^{t-k-1}}$ is a $2^{\tau+1}$-th root of g modulo N.   □

We next prove that factoring is reducible to the factoring representation problem:

Theorem 1. *If a probabilistic algorithm solves the factoring representation problem $(N, \tau, t, g) \leftarrow$ FactIndex(1^n) with probability ϵ, then N can be factored within the same time bound and success probability at least $\frac{1}{2}\epsilon$.*

Proof. Pick a random $a \in_R \mathbb{Z}_N^*$ and set $g := a^{2^{\tau+1}} \bmod N$ which is uniformly distributed in HQR_N. Based on Lemma 1 compute some $2^{\tau+1}$-th root b of g. Let $c := ab^{-1} \bmod N$. Then:

$$c^{2^{\tau+1}} = 1 = c^{2^\eta} \bmod N.$$

The second equation follows from $\tau + 1 \geq \eta$ and since squaring is a permutation on HQR_N. We next consider the equation modulo the prime factors p, q of N. Suppose $p - 1 = 2^{\eta_p} p'$ and $q - 1 = 2^{\eta_q} q'$ for odd p', q', and therefore $\eta = \max\{\eta_p, \eta_q\}$. Wlog. let $\eta = \eta_p$. Because of $\eta_q \leq \eta_p = \eta$ we have

$$c^{2^\eta} = 1 \bmod p \qquad\qquad c^{2^{\eta-1}} = \sigma_p \bmod p$$
$$c^{2^\eta} = 1 \bmod q \qquad\qquad c^{2^{\eta-1}} = \sigma_q \bmod q$$

for some $\sigma_p, \sigma_q \in \{\pm 1\}$ [2]. To complete the proof we show that $\sigma_p \sigma_q = -1$ holds with probability $\frac{1}{2}$, because in that case one of the GCD computations $\gcd(c^{2^{\eta-1}} \pm 1, N)$ yields the factorization of N.

Note that c is uniformly distributed among the 2^η-th roots of 1, because g does not reveal any information about the random root a we have actually chosen, thus the element b determined by the representation finder's output is independent of a. Hence, $c \bmod p$ and $c \bmod q$ are independently and uniformly distributed among the 2^η-th roots of 1 modulo p and modulo q. Consequently, σ_p and σ_q are independent.

For half of the 2^η-th roots w of 1 modulo p we have $w^{2^{\eta-1}} = 1 \bmod p$ and otherwise $w^{2^{\eta-1}} = -1 \bmod p$. Since $c \bmod p$ is a random 2^η-th root of 1 modulo p, the value σ_p is uniformly distributed in $\{\pm 1\}$. As σ_p does not depend on σ_q we have $\sigma_p \sigma_q = -1$ with probability $\frac{1}{2}$. $\qquad\square$

Figure 1 illustrates the proof idea of Theorem 1. The root of each tree is labeled with $+1$. Descending from one node to the successors corresponds to taking a square root modulo the prime p or q, e.g., in the left tree the tree's root $+1$ has the successors $+1$ and -1 as squaring is still a 2:1-mapping modulo p, whereas in the right tree $\eta_q < \eta_p$ and squaring permutes HQR_q, implying that the tree's root $+1$ only has the square root $+1$. Hence, the leaves in each tree represent all 2^{η_p} and 2^{η_q} many 2^η-th roots of 1 modulo p and q, respectively.

The path to the leftmost leaf in each tree represents the 2^η-th root 1 of 1 modulo p and modulo q (1-*path*). In the proof we use the representation finder to derive a random 2^η-th root c of 1 in $\mathbb{Z}_N^*$. Thus, the values $c \bmod p$ and $c \bmod q$ each describe a random path to some leaf in the corresponding tree (*c-path*), and each path is independent of the other one. We are able to find the prime factors of N if and only if at some level k one of the c-paths branches from the 1-path whilst the other one still follows the 1-path. For example, in Figure 1 this happens in the marked nodes for $k = \eta_p - 1$: there we have $c^{2^{k-1}} = 1 \bmod p$

---

[2] While both values for σ_p may occur, if $\eta_q < \eta$ then we always have $\sigma_q = +1$ as squaring is one-to-one on HQR_q and $+1$ is the unique square root.

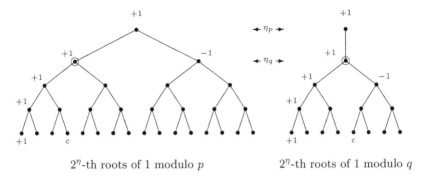

2^η-th roots of 1 modulo p 2^η-th roots of 1 modulo q

Fig. 1. Factoring via 2^η-th root of 1

but $c^{2^{k-1}} = -1 \bmod q$ and a GCD computation yields the prime factors of N. In fact, in the proof of Theorem 1 we only check the divergence of the paths for $k = \eta_p = \eta$. Therefore, except for Blum integers, the probability is actually higher then $\frac{1}{2}$.

Theorem 1 even holds for fixed $g \in \mathrm{HQR}_N$, given that $\tau \geq \eta$ and some $2^{\tau+1}$-th root $a \notin \mathrm{QR}_N$ of g with $-a \notin \mathrm{QR}_N$ is publicly known. Besides that variant the factoring representation problem gives rise to other modifications and generalizations:

1. One may substitute the RSA modulus by an arbitrary odd integer N. Then the algorithm of Theorem 1 retrieves a non-trivial factor of N.
2. The problem can be relaxed such t is not given as part of the output of FactIndex, but the representation finder rather gets the freedom to select an arbitrary $t \geq 1$ on its own after seeing (N, τ, g). Given two representations (x_1, r_1, t_1) and (x_2, r_2, t_2) where wlog. $t_1 \geq t_2$ use $r := r_2 r_1^{-2^{t_1-t_2}}$ for the proof of Lemma 1.
3. One may replace $2^{\tau+t}$ by $e^{\tau+t}$ for some $e = \mathcal{O}(\log n)$. In this case, η denotes the smallest integer such that $e^{\eta+1}$ neither divides $p - 1$ nor $q - 1$ and use $\{x^{e^\eta} \mid x \in \mathbb{Z}_N^*\}$ instead of HQR_N. The hardness is also based on factoring as Ohto and Okamoto [OO88] have shown that taking e-th roots in this case is equivalent to factoring N.

3   Identification and Signature Schemes

In this section we show how to repair Okamoto's identification protocol [Ok92] obtaining a provably secure identification scheme withstanding parallel active attacks. Exploiting the relationship to signature schemes via the Fiat-Shamir heuristic [FS86], we then show that this identification protocol can be used for ordinary as well as blind signatures.

3.1   Identification Scheme

Our identification protocol in Figure 2 follows the framework of Okamoto [Ok92] for the RSA setting, which in turn is an extension of the Guillou-Quisquater setup [GQ88]. The values $(N, \tau, t, g) \leftarrow \mathsf{FactIndex}(1^n)$ are public parameters. The public key of a user is $X \in_R \mathrm{HQR}_N$ and the corresponding secret key is a random representation $(x, r) \in \mathbb{Z}_{2^t} \times \mathbb{Z}_N^*$ of X. The user is not required to be aware of the factorization of N and several users may share the same public parameters N, τ, g (even with different t's). The prover $\mathcal{P}$ tries to convince the verifier $\mathcal{V}$ that

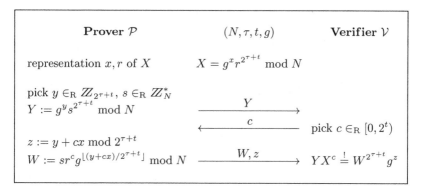

Fig. 2. Identification Scheme using Factoring Representation Problem

$\mathcal{P}$ knows a representation of X with respect to (N, τ, t, g). In the first step $\mathcal{P}$ sends an initial commitment Y to $\mathcal{V}$ who answers with a challenge $c \in_R [0, 2^t)$, and $\mathcal{P}$ finally hands the response W, z to $\mathcal{V}$ which determines acceptance or rejection.

Obviously, this protocol is complete in the sense that the honest prover $\mathcal{P}$ always passes the examination of honest verifiers $\mathcal{V}$. We show that this identification scheme is secure against active attacks, i.e., where the adversary $\mathcal{A}$ may run executions with the honest prover before trying to impersonate. But first we start with passive attacks in which the adversary tries to intrude given the public key only:

Lemma 2. *If a passive adversary $\mathcal{A}$ passes the identification scheme in Figure 2 with time bound T and success probability ϵ, then the modulus N can be factored in expected time $\mathcal{O}(T)$ with probability at least $\frac{1}{4}(\epsilon - 2^{-t})$.*

Proof. The proofs follows the one in [Ok92] for RSA. Let $N = pq$, τ, t and a random $g \in_R \mathrm{HQR}_N$ be given, i.e., $(N, \tau, t, g) \leftarrow \mathsf{FactIndex}(1^n)$. We show how to compute with probability $\frac{1}{2}(\epsilon - 2^{-t})$ some $2^{\tau+1}$-th root of g with the help of $\mathcal{A}$. As a result, the claim is a consequence of Theorem 1.

Pick $x \in_R \mathbb{Z}_{2^t}$ and $r \in_R \mathbb{Z}_N^*$. Next, simulate an attack of $\mathcal{A}$ for $N, t, g, X := g^x r^{2^{\tau+t}} \bmod N$. After $\mathcal{A}$ has sent W, z rewind to the situation where $\mathcal{A}$ faces the challenge. By this, we obtain in expected time $\mathcal{O}(T)$ with probability $\epsilon - 2^{-t}$ two successful intrusion attempts in which $\mathcal{A}$ has sent the same Y but has answered with W, z and W', z' to different challenges c, c'. Then

$$X^{-c} W^{2^{\tau+t}} g^z = Y = X^{-c'} (W')^{2^{\tau+t}} g^{z'} \bmod N$$

or, rewritten,

$$g^{z-z'+x(c'-c)} = (r^{c-c'} W^{-1} W')^{2^{\tau+t}} \bmod N \qquad (2)$$

Let $\Delta z := z - z'$ and $\Delta c := c' - c$. Now we have an equation similar to Equation (1) in the proof of Lemma 1. If $\gcd(\Delta z + x\Delta c, 2^{\tau+t}) = 2^k$ for some $k < t$, then we are able to retrieve some $2^{\tau+1}$-th root of g. To complete the proof it thus suffices to give an upper bound $\frac{1}{2}$ for the probability that the GCD exceeds 2^{t-1}. Obviously,

$$\gcd(\Delta z + x\Delta c, 2^{\tau+t}) \geq 2^t \quad \Longleftrightarrow \quad x \cdot \Delta c = -\Delta z \bmod 2^t.$$

Whenever this modular equation is solvable, then for fixed $\Delta c, \Delta z$ the number of solutions for x equals $2^j := \gcd(\Delta c, 2^t)$ where $0 \leq j < t$ because $0 < |\Delta c| < 2^t$. Observe that in the actual protocol execution the selection of the parameters $\Delta c, \Delta z$ for the equation is done *after* the variable x has been chosen. But Δc is distributed independently of x because the challenges are simply picked at random, and the distribution of the adversary's choice z, z' for Δz does not depend on x either since the public key X does not reveal anything about the specific choice of x. Therefore, we can view the process as first fixing $\Delta c, \Delta z$ and then picking $x \in \mathbb{Z}_{2^t}$ at random. But then the probability that the random x matches the equation is bounded above by 2^{j-t}. From $j < t$ it follows that this probability is at most $\frac{1}{2}$. $\qquad\square$

Note that this approach factors N but unlike the corresponding RSA based scheme it does not extract a representation of the prover. Hence, once more we have a secure identification protocol which does not constitute a proof of knowledge in the sense of Bellare and Goldreich [BG92]. See [OS90,S96,Sh99] for other examples.

In order to prove security against active adversaries, we follow the approach in [Ok92] and show that even executions with the prover before the intrusion attempt do not disclose any information about x (called witness-indistinguishability [FS90]):

Lemma 3. *The protocol in Figure 2 is perfectly witness-indistinguishable.*

Proof. We have to justify that even in the case of a dishonest verifier $\mathcal{V}$ the view (i.e., the distribution of the communication Y, c, W, z) is independent of the representation actually known by the prover $\mathcal{P}$. We show that for any communication (Y, c, W, z) of an execution of $\mathcal{V}$ with $\mathcal{P}$, another prover $\mathcal{P}'$ knowing another representation (x', r') of X generates this communication with the same probability in an execution with $\mathcal{V}$.

Let $\Delta x := x' - x$ and $\Delta r = r/r' \bmod N$. Since $r^{2^{\tau+t}} = Xg^{-x}$ we have $\Delta r^{2^{\tau+t}} = g^{\Delta x}$. Assume that (Y, c, W, z) is a transcript of a communication with $\mathcal{P}$ having chosen y, s at the outset. The probability that $\mathcal{P}'$ picks $y' := y - c \cdot \Delta x \bmod 2^{\tau+t}$ and $s' := s \cdot \Delta r^c \bmod N$ in the first step is exactly the same as for $\mathcal{P}$ choosing y, s; both times the values are uniformly distributed. For this choice of y', z' we have $Y' = Y$, and therefore $\mathcal{V}$ returns the challenge $c' = c$ with equal probability in both executions. Now, W', z' and W, z are deterministically determined by the secret key, the challenge and the random values from the first step, and it is easily shown that $(W', z') = (W, z)$ here. Hence, the probability that a run with $\mathcal{P}'$ generates (Y, c, W, z) equals the one for $\mathcal{P}$. This completes the proof. $\qquad\square$

It follows that the identification scheme is also secure against active attacks:

Theorem 2. *If an active adversary $\mathcal{A}$ passes the identification scheme in Figure 2 with time bound T and success probability ϵ, then the modulus N can be factored in expected time $\mathcal{O}(T)$ with probability at least $\frac{1}{4}(\epsilon - 2^{-t})$.*

Proof. Given N, t, g pick a random secret key (x, r) and simulate an attack $\mathcal{A}$ on N, t, g and the public key $X := g^x r^{2^{\tau+t}} \bmod N$. This includes several interactions of $\mathcal{A}$ with the prover before trying to fool the verifier. But we can easily run these prover-adversary executions as we know the secret key. Due to the witness-indistinguishability, these executions still hide x perfectly, and the argument of Lemma 2 applies. $\qquad\square$

The proposition even holds if the adversary is allowed to run concurrent executions with the prover. Hence, the scheme can be turned into one secure against reset attacks under the factoring assumption; for details see [BFGM01].

3.2   Signature Schemes

The identification scheme in Figure 2 gives rise to a signature scheme secure against chosen-message attacks [GMR88] using the Fiat-Shamir heuristic. The challenge is generated by applying a hash function H to the message and the initial commitment of the prover. More specifically, publish (N, τ, t, g) as public parameter, X as public key and use the representation (x, r) as the secret key of the signer $\mathcal{S}$. In order to sign a message m pick $y \in_R \mathbb{Z}_{2^{\tau+t}}$ and $s \in_R \mathbb{Z}_N^*$ at random, calculate $Y := g^y s^{2^{\tau+t}} \bmod N$, $c := H(Y, m)$ and compute z, W as in the case of the identification scheme. The signature to m becomes $\sigma(m) := (Y, W, z)$. Verification is straightforward.

Provided the hash function H behaves like a random function, then even with the help of an signature oracle any adversary fails to come up with a valid signature for a new message of his own choice [PS00, Sec. 3.2]:

Proposition 2. *In the Random Oracle Model the signature scheme based on the factoring representation problem is secure against existentially forgery under adaptive chosen-message attacks relative to the hardness of factoring.*

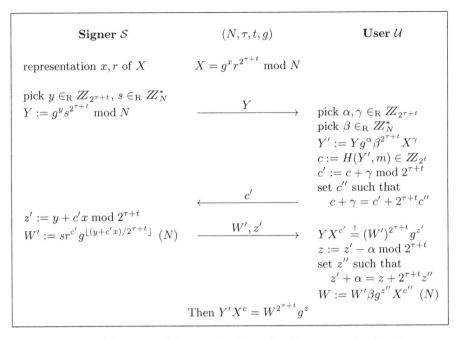

Fig. 3. Blind Signature Scheme using Factoring Representation Problem

An important variant of signatures schemes are blind signatures. In this case, the user $\mathcal{U}$ blinds the actual message m and requests a signature from the signer $\mathcal{S}$, which $\mathcal{U}$ later turns into a valid signature for the message m while the signer $\mathcal{S}$ cannot infer something about m. In a "one-more" forgery the adversary $\mathcal{A}$ tries to generate one more signed message than $\mathcal{A}$ originally requested from the signer $\mathcal{S}$ [PS00]. For example, in the ecash setting where messages signed by the bank represent anonymous digital coins, $\mathcal{U}$ cannot spent more money than $\mathcal{U}$ has actually withdrawn from the bank.

The blind signature scheme based on the factoring problem is given in Figure 3; it is heavily influenced by the discrete-log and RSA protocols of Pointcheval and Stern [PS00]. In the first step, the signer $\mathcal{S}$ commits to Y. Then the user $\mathcal{U}$ blinds Y by multiplying with $g^{\alpha}\beta^{2^{\tau+t}}X^{\gamma}$ for random values α, β, γ. The actual challenge $c \in \mathbb{Z}_{2^t}$ is hidden by $c' := c + \gamma \in \mathbb{Z}_{2^{\tau+t}}$. $\mathcal{S}$ replies the challenge by sending W', z' subject to $YX^{c'} = (W')^{2^{\tau+t}}g^{z'}$. Now, $\mathcal{U}$ undoes the blinding and finally retrieves the signature $\sigma(m) := (Y', W, z)$:

$$
\begin{aligned}
W^{2^{\tau+t}}g^z &= (W'\beta g^{z''}X^{c''})^{2^{\tau+t}}g^{z'}g^{z-z'} \\
&= YX^{c'}(\beta g^{z''}X^{c''})^{2^{\tau+t}}g^{z-z'} = YX^{c+\gamma}\beta^{2^{\tau+t}}g^{\alpha} = Y'X^c.
\end{aligned}
$$

The scheme is perfectly blind as (Y, c', W', z') and (Y', c, W, z) are independently distributed. Security follows as in [PS00]:

Theorem 3. *In the Random Oracle Model the blind signature scheme based on the factoring representation problem is secure against a "one-more" forgery under a parallel attack (where up to poly-logarithmic signature generations are executed concurrently) relative to the hardness of factoring.*

Note that this scheme is provable secure against interleaving attacks meanwhile the one based on the Ong-Schnorr identification is only known to be secure against sequential attacks [PS97].

4   Commitment Schemes

A commitment scheme is a protocol of three stages (initialization, commitment and decommitment) between to parties called the sender S and the receiver R. In the commitment stage S binds himself to a message m by sending a commitment meanwhile the receiver R cannot deduce any information about m. Later, the sender S reveals m and R checks whether this message indeed matches the commitment.

4.1   Non-interactive Commitment Scheme

In this section we set up a commitment scheme based on the factoring representation problem following the well-known scheme derived from the RSA representation problem and generalizing Halevi's scheme [H99].

Assume for the moment that a trusted third party selects a valid instance $(N, \tau, t, g) \leftarrow \mathsf{FactIndex}(1^n)$ for the factoring representation problem and publishes it; we afterwards discuss how to delegate this task to the receiver. In any case, S must not know the factorization of N. To commit to a message $m \in \mathbb{Z}_{2^t}$, the sender S picks a random $r \in_R \mathbb{Z}_N^*$ and sends

$$\mathrm{com}(m, r) := g^m r^{2^{\tau+t}} \bmod N \tag{3}$$

to the receiver R. For the decommitment, S reveals the commited message m and the random value r. The receiver R verifies that (m, r) is indeed a representation of $\mathrm{com}(m, r)$.

If we let the receiver instead choose $(N, \tau, t, g) \leftarrow \mathsf{FactIndex}(1^n)$ and send it to S in the first step, then there is no guarantee that a malicious receiver does not select inproper values like $g \notin \mathrm{HQR}_N$ or $\tau < \eta - 1$. To prevent this we take $\tau := n$ and use a method suggested in [H99] to make sure that g really is an element from HQR_N, even if N is not the product of two primes. Namely, let the sender verify that N is odd and raise g to the 2^n-th power first. S then transmits

$$\mathrm{com}(m, r) := \left(g^{2^n}\right)^m \left(r^{2^n}\right)^{2^{t+n}} = g^{m2^n} r^{2^{2n+t}} \bmod N \tag{4}$$

Given factoring N is intractable, then Theorem 1 implies that S cannot come up the a different representation of $\mathrm{com}(m, r)$, in either case (3) or (4). Hence, a commitment is computationally binding and S cannot ambiguously open the commitment. On the other hand, the distribution of $\text{com}(m, r) \in \text{HQR}_N$ is independent of the message m, that is, even a computationally unbounded malicious receiver $\mathcal{R}$ is unable to deduce any information about m given only the commitment. To summarize:

Proposition 3. *The factoring representation commitment scheme* (3) *respectively* (4) *has the following properties:*

1. *Computational unambiguity relative to the hardness of factoring.*
2. *Perfect privacy.*

We compare this commitment scheme with the one introduced by Halevi [H99]. To commit to a message $m \in \mathbb{Z}_{2^t}$ with a trusted setup mechanism providing a correct N pick at random $r \in_R \mathbb{Z}_N^*$ and publish

$$\text{com}(m, r) := 4^m r^{2^{t+1}} \bmod N \tag{5}$$

for a William integer N, i.e., an RSA modulus $N = pq$ with $p = 3 \bmod 4$ and $q = 7 \bmod 8$. The binding property relative to the hardness of factoring N can be proven in a direct way [H99]. Alternatively one may apply Theorem 1. We have $\eta = 1$, $\tau := \eta$ and $4 \in \text{HQR}_N$ because its square root $(+2, -2) \in \text{QR}_p \times \text{QR}_q$ is a square, too. As $\pm 2 \notin \text{QR}_N$, the adversary has to compute some other square root of 4 yielding the factorization of N.

4.2   Non-malleable Commitment Scheme

Roughly speaking a commitment scheme is non-malleable if for any adversary $\mathcal{A}$ seeing the commitment of an honest sender $\mathcal{S}$ to an unknown message m it is infeasible to commit to a related (but different) message m^*. Depending on the level of security, the adversary may also be obliged to provide a valid decommitment after having learned the decommitment of $\mathcal{S}$ (called non-malleability with respect to opening). See [DDN00,FF00] for details.

Fischlin and Fischlin [FF00] and Di Crescenzo et al. [CKOS01] present efficient non-malleable commitment schemes based either on the discrete-log or the RSA assumption. All protocols work in the public parameter model, where public data like an RSA modulus N and a value $g \in \mathbb{Z}_N^*$ are published by a trusted party. Also, both solutions apply so-called trapdoor or equivocable commitments: knowledge of a secret information, the trapdoor, enables to open a given commitment with any message later on. For instance, for RSA an e-th root of g allows to fake commitments. Here, in case of the factoring representation commitment scheme (3), a $2^{\tau+t}$-th root h of $g \in \text{HQR}_N$ provides a trapdoor, because a commitment $g^m r^{2^{\tau+t}}$ can be opened for m' by transmitting m' and $r' := h^{m-m'} r \bmod N$.

We discuss how to modify the non-malleable commitments schemes based on the RSA representation problem [FF00,CKOS01] to derive non-malleable commitments schemes as secure as factoring. Fischlin and Fischlin [FF00] present interactive schemes that work with the RSA representation problem, one time

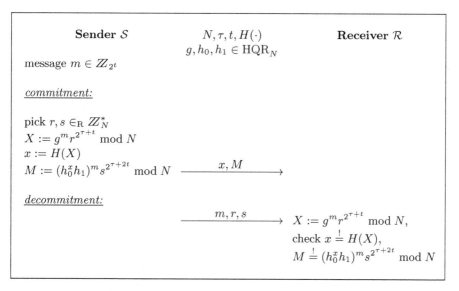

Fig. 4. Non-Interactive Non-Malleable Commitment Scheme

using standard proofs of knowledge, the other time with a more sophisticated variant based on the Chinese Remainder Theorem. We cannot plug in the factoring representation problem into the proof-of-knowledge-based approach as the protocol given in Figure 2 does not constitute a proof of knowledge. However, we can use the Chinese Remainder Theorem protocol with the factoring representation problem instead of the RSA problem. By this, we obtain a non-malleable commitment scheme with statistical privacy. Details are omitted.

The non-interactive scheme in [CKOS01] achieves a weaker notion of non-malleability than the one in [FF00], where the adversary does not have any side information about the message m of $\mathcal{S}$. In Figure 4 we present a modification of their RSA protocol which is based on the factoring representation problem. Surprisingly, although this modification works under a potentially weaker assumption than RSA, it is even more efficient than the RSA protocol in [CKOS01]. In fact, transferring our protocol to the RSA or discrete-log representation problem setting also improves these protocols in [CKOS01] with respect to computational and communication complexity.

Basically, we let the sender commit twofold to the message m: one time by X with the standard factoring representation problem, the other time by M with a base where the hash value of the former commitment enters (and for which we use $\tau + 2t$ rather than $\tau + t$, see below). For this, let $H : \mathrm{HQR}_N \to \mathbb{Z}_{2^t}$ be some universal one-way hash function [NY90] with which we hash down $X \in \mathbb{Z}_N^*$ to $x \in \mathbb{Z}_{2^t}$. In case of $t \geq n$ one may eliminate the hash function by using X as exponent x.

Theorem 4. *There exists (efficient) commitments schemes with the following properties relative to the hardness of factoring:*

1. *Non-malleable with respect to opening.*
2. *Computationally binding.*
3. *Statistical privacy (and perfect privacy for the scheme in Figure 4).*

We outline the non-malleability proof for the non-interactive commitment scheme given in Figure 4. The definition of non-malleability essentially requires that for any adversary that is given a commitment of the sender and generates another commitment for which it is also able to adapt the sender's opening to one of a related message, there is a simulator that is almost as successful but without interacting with the sender at all.

We briefly recall the proof method in [CKOS01]. There, the simulator prepares a commitment on behalf the original sender which includes a trapdoor. The simulator submits it to the adversary who answers with its commitment. Then the simulator samples a sufficient number of random messages and sequentially opens the trapdoor commitment (by adapting the decommitment with the trapdoor accordingly). By this, the adversary reveals with sufficiently high probability a valid opening for its commitment to *some* message. The probability that the adversary finds different valid openings is negligible under the discrete-log or RSA assumption, hence, the simulator extracts *the* message of the adversary that is related to the original message of the sender.

In our case, given a commitment (x, M) for some unknown message m, the adversary $\mathcal{A}$ tries to commit to a related but different message m^* by sending (x^*, M^*). We first condition on the event that the adversary selects $x^* \neq x$. Assume towards contradiction that the adversary succeeds in an actual attack by sending $x^* = x$ with noticeable probability. For simplicity, we presume that the sender's value $X := g^m r^{2^{\tau+t}} \bmod N$ equals the adversary's choice $X^* := g^{m^*}(r^*)^{2^{\tau+t}} \bmod N$; otherwise we find a collision for the universal one-way hash function H. But then the decommitment step yields distinct representations of X and this allows to efficiently solve the factoring representation problem with noticeable success. We may thus consider only the adversary's success on values $x^* \neq x$ without sacrificing more than a negligible probability.

It remains to describe the trapdoor in our scheme to apply the technique of [CKOS01]. Given $(N, \tau, t, h_0) \leftarrow \mathsf{FactIndex}(1^n)$ select a universal one-way hash function H and define $M := s^{2^{\tau+t}}$, $X := r^{2^{\tau+t}}$, $g := u^{2^{\tau+t}}$, $h_1 := h_0^{-x} v^{2^{\tau+2t}}$ for random $r, s, u, v \in_R \mathbb{Z}_N^*$ and $x := H(X)$. Take $(N, \tau, t, H, g, h_0, h_1)$ as public parameters and send (x, M) on behalf of the honest sender.

For the data in the simulation we know a $2^{\tau+2t}$-root v of $h_0^x h_1 = v^{2^{\tau+2t}}$. This, together with the $2^{\tau+t}$-th root of X, enables us to correctly open the commitment (x, M) with any message later. In contrast, even if we know the trapdoor, the adversary will not be able to find distinct openings for its commitment since, by assumption, $x^* \neq x$. The reason for this is that any valid decommitments of the adversary including (m_1^*, r_1^*), (m_2^*, r_2^*) for M^* imply that

$$(h_0^{x^*} h_1)^{m_1^*}(r_1^*)^{2^{\tau+2t}} = M^* = (h_0^{x^*} h_1)^{m_2^*}(r_2^*)^{2^{\tau+2t}}$$

and, substituting $h_1 = h_0^{-x} v^{2^{\tau+2t}}$,

$$h_0^{(x^*-x)m_1^*}(v^{m_1^*}r_j)^{2^{\tau+2t}} = M^* = h_0^{(x^*-x)m_2^*}(v^{m_2^*}r_2^*)^{2^{\tau+2t}}.$$

Since $x^*-x \neq 0$ and both products with $m_1^* \neq m_2^*$ are less than 2^{2t} this results in different representations with x-components $(x^*-x)m_1^*$, $(x^*-x)m_2^* \in \mathbb{Z}_{2^{2t}}$ for M^*. The probability that this happens is therefore negligible under the factoring assumption. With these preliminaries the rest of the proof is the same as in [CKOS01].

Acknowledgments

This work has been stimulated by discussions with Stefan Brands at Crypto 2000 about the factoring based representation problem mentioned in his Eurocrypt '97 paper. We also thank the anonymous reviewers for their comments.

References

BFGM01. M. BELLARE, M. FISCHLIN, S. GOLDWASSER and S. MICALI: *Identification Protocols Secure Against Reset Attacks*, Eurocrypt 2001, Lecture Notes in Computer Science, vol. 2045, pp. 495–511, Springer Verlag, 2001.

BG92. M. BELLARE and O. GOLDREICH: *On Defining Proofs of Knowledge*, Crypto '92, Lecture Notes in Computer Science, vol. 740, pp. 390–420, Springer Verlag, 1993.

BR93. M. BELLARE and P. ROGAWAY: *Random Oracles are Practical: a Paradigm for Designing Efficient Protocols*, First ACM Conference on Computer and Communication Security, ACM Press, pp. 62–73, 1993.

B97. S. BRANDS: *Rapid Demonstration of Linear Relations Connected by Boolean Operators*, Eurocrypt '97, Lecture Notes in Computer Science, vol. 1233, pp. 318–333, Springer-Verlag, 1997.

BCC88. G. BRASSARD, D. CHAUM and C. CRÉPEAU: *Minimum Disclosure Proofs of Knowledge*, Journal Computing System Science, vol. 37(2), pp. 156–189, 1988.

BV98. D. BONEH and R. VENKATESAN: *Breaking RSA may Not be Equivalent to Factoring*, Eurocrypt '98, Lecture Notes in Computer Science, vol. 1403, pp. 59–71, Springer Verlag, 1998.

CKOS01. G. DI CRESCENZO, J. KATZ, R. OSTROVSKY and A. SMITH: *Efficient And Non-Interactive Non-Malleable Commitment*, Eurocrypt 2001, Lecture Notes in Computer Science, vol. 2045, pp. 40–59, Springer Verlag, 2001.

D95. I. DAMGÅRD: *Practical and Provable Secure Release of a Secret and Exchange of Signature*, Journal of Cryptology, vol. 8, pp. 201–222, 1995.

DDN00. D. DOLEV, C. DWORK and M. NAOR: *Nonmalleable Cryptography*, SIAM Journal on Computing, vol. 30(2), pp. 391-437, 2000.

FFS88. U. FEIGE,A. FIAT and A. SHAMIR: *Zero-Knowledge Proofs of Identity*, Journal of Cryptology, vol. 1(2), pp. 77–94, 1988.

FS86. A. FIAT and A. SHAMIR: *How to Prove Yourself: Practical Solutions to Identification and Signature Schemes*, Crypto '86, Lecture Notes in Computer Science, vol. 263, Springer-Verlag, pp. 186–194, 1986.

FS90.   A. FIAT and A. SHAMIR: *Witness Indistinguishable and Witness Hiding Protocols*, Proceedings of the 22nd Annual ACM Symposium on the Theory of Computing (STOC), pp. 416–426, ACM Press, 1990.

FF00.   M. FISCHLIN and R. FISCHLIN: *Efficient Non-Malleable Commitment Schemes*, Crypto 2000, Lecture Notes in Computer Science, vol. 1880, pp. 414–432, Springer Verlag, 2000.

GQ88.   L.C. GUILLOU and J.-J. QUISQUATER: *A Practical Zero-Knowledge Protocol Fitted to Security Microprocessors Minimizing Both Transmission and Memory*, Eurocrypt '88, Lecture Notes in Computer Science, vol. 330, pp. 123–129, Springer Verlag, 1988.

GMR88.   S. GOLDWASSER, S. MICALI and R.L. RIVEST: *A Digital Signature Scheme Secure Against Adaptive Chosen-Message Attacks*, SIAM Journal of Computing, vol. 17(2), pp. 281–308, 1988.

H99.   S. HALEVI: *Efficient Commitment Schemes with Bounded Sender and Unbounded Receiver*, Journal of Cryptology, vol. 12(2), pp. 77–90, 1999.

NY90.   M. NAOR and M. YUNG: *Universal Oneway Hash Functions and Their Cryptographic Applications*, Proceedings of the 21st Annual ACM Symposium on the Theory of Computing (STOC), pp. 33–43, ACM Press, 1989.

OS90.   H. ONG and C.P. SCHNORR: *Fast Signature Generation with as Fiat-Shamir-Like Scheme*, Eurocrypt '90, Lecture Notes in Computer Science, vol. 473, pp. 432–440, Springer Verlag, 1991.

OO88.   K. OHTA and T. OKAMOTO: *A Modification of the Fiat-Shamir Scheme*, Crypto '88, Lecture Notes in Computer Science, vol. 403, pp. 232–243, Springer Verlag, 1989.

Ok92.   T. OKAMOTO: *Provable Secure and Practical Identification Schemes and Corresponding Signature Schemes*, Crypto '92, Lecture Notes in Computer Science, vol. 740, pp. 31–53, Springer Verlag, 1993.

PS97.   D. POINTCHEVAL and J. STERN: *New Blind Signatures Equivalent to Factorization*, Proceedings of the 4th ACM Conference on Computer and Communications Security (CCS) '97, pp. 92-99, ACM Press, 1997.

PS00.   D. POINTCHEVAL and J. STERN: *Security Arguments for Digital Signatures and Blind Signatures*, Journal of Cryptology, vol. 13(3), pp. 361–396, 2000.

RSA78.   R.L. RIVEST, A. SHAMIR and L. ADLEMAN: *A Method for Obtaining Digital Signatures and Public-Key Cryptosystems*, Communications of the ACM, vol. 21, pp. 120–126, 1978.

S96.   C.P. SCHNORR: *Security of 2^t-Root Identification and Signatures*, Crypto '96, Lecture Notes in Computer Science, vol. 1109, pp. 143–156, Springer Verlag, 1996.

S97.   C.P. SCHNORR: *Erratum: Security of 2^t-Root Identification and Signatures*, in Crypto '97, Lecture Notes in Computer Science, vol 1294, page 540, Springer Verlag, 1997.

Sh99.   V. SHOUP: *On the Security of a Practical Identification Scheme*, Journal of Cryptology, vol. 12, pp. 247–260, 1999.

A   On Okamoto's Identification Scheme

Okamoto [Ok92] presents witness-indistinguishable identification schemes based on the hardness of discrete log and RSA. In the same paper he also suggests the modified RSA scheme given in Figure 5. Compared to the RSA based scheme the prime RSA exponent is replaced by $2e$ for some prime e.

Prover $\mathcal{P}$	N, e, g	**Verifier** $\mathcal{V}$
representation x, r of X	$X = g^x r^{2e} \bmod N$	
pick $y \in \mathbb{Z}_{2e}$, $s \in \mathbb{Z}_N^*$		
$Y := g^y s^{2e} \bmod N$	$\xrightarrow{\quad Y \quad}$	
	$\xleftarrow{\quad c \quad}$	pick $c \in_R [0, 2e)$
$z := y + cx \bmod (2e)$		
$W := s r^c g^{\lfloor (y+cx)/2e \rfloor} \bmod N$	$\xrightarrow{\quad W, z \quad}$	$Y X^c \overset{!}{=} W^{2e} g^z$

Fig. 5. Okamoto's Identification Scheme

Okamoto claims that the security is based on the hardness of factoring the modulus N. However, we show that the security de facto relies on the RSA problem rather than on the factoring problem. Namely, we show that computing e-th roots enables an adversary to pass the protocol with probability $\frac{1}{2}$.

Suppose we are given $g \in \mathrm{QR}_N$ and the public key X. Hence, $X \in \mathrm{QR}_N$. Now, pick $x \in_R \mathbb{Z}_{2e}$ and compute $r^2 = (Xg^{-x})^{\frac{1}{e}}$ by solving the RSA problem. Apparently, (x, r) is a representation for $X = g^x r^{2e}$ but we are just aware of x and the square r^2. Nevertheless, we are able to compute $z := y + cx \bmod 2e$ and whenever the challenge c is even then knowledge of r^2 suffices to determine W:

$$W = s r^c g^{\lfloor (y+cx)/2e \rfloor} = s (r^2)^{\frac{c}{2}} g^{\lfloor (y+cx)/2e \rfloor}.$$

Thus, solving the RSA problem allows to pass the protocol with probability $\frac{1}{2}$ since the challenge is even with this probability. Whenever $g \notin \mathrm{QR}_N$, one computes $r^4 = (Xg^{-x})^{\frac{2}{e}}$ and succeeds if the challenge satisfies $c = 0 \bmod 4$.

It is tempting to restrict the challenge c to odd values. Still, we are not aware of any proof in this case (e.g., we were unable to modify the proof of Lemma 2 about security against passive adversaries to work in this case).

Ciphers with Arbitrary Finite Domains

John Black[1] and Phillip Rogaway[2]

[1] Dept. of Computer Science, University of Nevada, Reno NV 89557, USA
jrb@cs.unr.edu, http://www.cs.unr.edu/~jrb
[2] Dept. of Computer Science, University of California at Davis,
Davis, CA 95616, USA
rogaway@cs.ucdavis.edu, http://www.cs.ucdavis.edu/~rogaway

Abstract. We explore the problem of enciphering members of a finite set $\mathcal{M}$ where $k = |\mathcal{M}|$ is arbitrary (in particular, it need not be a power of two). We want to achieve this goal starting from a block cipher (which requires a message space of size $N = 2^n$, for some n). We look at a few solutions to this problem, focusing on the case when $\mathcal{M} = [0, k-1]$. We see ciphers with arbitrary domains as a worthwhile primitive in its own right, and as a potentially useful one for making higher-level protocols.

Keywords: Ciphers, Modes of Operation, Provable security, Symmetric Encryption.

1 Introduction

A MOTIVATING EXAMPLE. Consider the following problem: a company wishes to generate distinct and unpredictable ten-digit credit-card numbers. One way to accomplish this involves keeping a history of all previously-issued numbers. But the company wishes to avoid storing a large amount of sensitive information. Another approach is to use some block cipher E under a randomly-selected key K and then issue credit-card numbers $E_K(0), E_K(1), \cdots$. But the domains of contemporary block ciphers are inconvenient for this problem: this company needs distinct numbers in $[0, 10^{10} - 1]$ but block cipher have a domain $[0, 2^n - 1]$ for some n such as 64 or 128. Is there an elegant solution to this problem?

ENCIPHERING WITH ARBITRARY DOMAINS. More generally now, we have good tools—block ciphers—to encipher points when the message space $\mathcal{M}$ is strings of some particular length, $\mathcal{M} = \{0, 1\}^n$. But what if you want to encipher a number between one and a million? Or a point in Z_N or Z_N^*, where N is a 1024-bit number? Or a point from some elliptic-curve group? This paper looks at the question of how to construct ciphers whose domain is *not* $\{0, 1\}^n$.

That is, we are interested in how to make a cipher which has some desired but "weird" domain: $F \colon \mathcal{K} \times \mathcal{M} \to \mathcal{M}$ where $\mathcal{K}$ is the key space and $\mathcal{M}$ is the finite message space that we have in mind. A tool from which we may start our construction is a block cipher: a map $E \colon \mathcal{K}' \times \{0, 1\}^n \to \{0, 1\}^n$ where $\mathcal{K}'$ is the key space and n is the block length. A solution to this problem immediately solves the credit-card problem: for a block cipher $F \colon \mathcal{K} \times [0, 10^{10} - 1] \to [0, 10^{10} - 1]$, the

B. Preneel (Ed.): CT-RSA 2002, LNCS 2271, pp. 114–130, 2002.
© Springer-Verlag Berlin Heidelberg 2002

company chooses a random $K \in \mathcal{K}$ and issues the (distinct) credit-card numbers $F_K(0), F_K(1), F_K(2), \ldots, F_K(i)$, and has only to remember the last i value used.

MEASURING SUCCESS. We would like to make clear right away what is the security goal that we are after. Let's do this by way of an example. Suppose that you want to encipher numbers between one and a million: $\mathcal{M} = [1, 10^6]$. Following [7,2], we imagine two games. In the first game one chooses a random key K from $\mathcal{K}$ and hands to an adversary an oracle $E_K(\cdot)$. In the second game one chooses a random permutation π on $[1, 10^6]$ and hands the adversary an oracle for $\pi(\cdot)$. The adversary should be unable to distinguish these two types of oracles without spending a huge amount of time. Note that the domain is so small that the adversary might well ask for the value of the oracle $f(\cdot) \in \{E_K(\cdot), \pi(\cdot)\}$ at *every* point in the domain. This shouldn't help the adversary win. So, for example, if the adversary asks the value of $E_K(\cdot)$ at all points except 1 and 2 (a total of $10^6 - 2$ points), then the adversary will know what are the two "missing" numbers, c_1 and c_2, but the adversary won't be able to ascertain if $E_K(1) = c_1$ and $E_K(2) = c_2$, or if $E_K(1) = c_2$ and $E_K(2) = c_1$, instead.

OUR CONTRIBUTIONS. Though the problem of enciphering on an arbitrary domain has been considered before [13], here we draw attention to this problem and give the first rigorous treatment, providing a few solutions together with their analyses. Our solutions focus on the case in which the message space is $\mathcal{M} = [0, k-1]$, though we sketch extensions to some other message spaces, like Z_{pq}^* and common elliptic-curve groups.

Our first method assumes that we have a block cipher E that acts on $N = 2^n$ points, where $N \geq k$. To encipher $\mathcal{M} = [0, k-1]$ one just enciphers these points with block cipher E and uses the ordering of $E_K(0)$, $E_K(1)$, up to $E_K(k-1)$ to name the desired permutation on $[0, k-1]$. This method is computationally reasonable only for small k, such as $k < 2^{30}$.

A second method, similar to known techniques used in other settings, enciphers a message $m \in \mathcal{M}$ by repeatedly applying the block cipher, starting at m, until one gets back to a point in $\mathcal{M}$. (Assume once again that $N \geq k$.) This method is good if $\mathcal{M}$ is "dense" in the domain of the block cipher, $\{0, 1\}^n$. So, for example, one can use this method to encipher a string in Z_N, where N is a 1024-bit number, using a block cipher with block length of 1024 bits. (A block cipher with a long block length, like this, can be constructed from a "standard" block cipher by following works like [9,11,3].) This construction has been suggested before [13]; our main contribution here is the analysis of the construction.

A final method which we look at chooses an a, b where $ab \geq k$ and performs a Feistel construction on the message m, but uses a left-hand side in Z_a and a right-hand side in Z_b. Our analysis of this is an adaptation of Luby and Rackoff's [9]. This method can be quite efficient, though the proven bounds are weak when the message space is small (eg, $k < 2^{128}$).

With each of our ciphers we provide a deciphering algorithm, though this may not be required in all domains (eg, in our credit-card example above).

Note that the three methods above solve our problem for small and large domains, but there is a gap which remains: intermediate-sized values where our first method requires too much space and time, and our second method requires too many block-cipher invocations, and our third method may work but the bound is too weak. This gap occurs roughly from $k = 2^{30}$ up to about $k = 2^{60}$, depending on your point of view. Our credit-card example ($k = 10^{10} \approx 2^{33.2}$) falls into this gap. This problem remains open.

WHY CIPHERS ON NON-STANDARD SETS? Popular books on cryptography speak of enciphering the points in the message space $\mathcal{M}$, whatever that message space may be, but few seem to have thought much about how to actually do this when the message space is something other than a set of bit strings, often of one particular length. This omission is no doubt due to the fact that it is usually fine to embed the desired message space into a larger one, using some padding method, and then apply a standard construction to encipher in the larger space. For example, suppose you want to encipher a random number m between one and a million. Your tool is a 128-bit block cipher E. You could encode m as a 128-bit string M by writing m using 20 bits, prepending 108 zero-bits, and computing $C = E_K(M)$. Ignoring the fact that the ciphertext C "wastes" 108 bits, this method is usually fine. But not always.

One problem with the method above is that it allows one to tell if a candidate key K' might have been used to produce C. To illustrate the issue, suppose that the key space is small, say $|\mathcal{K}| = 2^{30}$. Suppose the adversary sees a point $C = E_K(M)$. Then the adversary has everything she needs to decrypt ciphertext $C = E_K(M)$: she just tries all keys $K' \in \mathcal{K}$ until she finds one for which $E_{K'}^{-1}(C)$ begins with 108 zeros. This is almost certainly the right key. The objection that "we shouldn't have used a small key space" is not a productive one if the point of our efforts was to make due with a small key space.

If we had used a cipher with message space $\mathcal{M} = [1, 10^6]$ we would not have had this problem. Every ciphertext C, under every possible key K, would correspond to a valid message M. The ciphertext would reveal nothing about which key had been used.

Of course there are several other solutions to the problem we have described, but many of them have difficulties of their own. Suppose, for example, that one pads with random bits instead of zero bits. This is better, but still not perfect: in particular, an adversary can tell that a candidate key K' could not have been used to encipher M if decrypting C under K' yields a final 20 bits whose decimal value exceeds 1,000,000. If one had 1,000 ciphertexts of random plaintexts enciphered in the manner we have described, the adversary could, once again, usually determine the correct key.

As a more realistic example related to that above, consider the Bellovin-Merritt "EKE" protocol [4]. This entity-authentication protocol is designed to defeat password-guessing attacks. The protocol involves encrypting, under a possibly weak password K, a string $g^x \bmod p$, where p is a large prime number and g is a generator of Z_p^*. In this context it is crucial that from the resulting ciphertext C one can not ascertain if a candidate password K' could possibly have produced the ciphertext C. This can be easily and efficiently done by enciphering with message space $\mathcal{M} = Z_p^*$. Ordinary encryption methods won't work.

Another problem with ciphertext-expansion occurs when we are constrained by an existing record format: suppose we wish to encrypt a set of fields in a database, but the cost of changing the record size is prohibitive. Using a cipher whose domain is the set of values for the existing fields allows some measure of added security without requiring a complete restructuring of the database. And if the data have additional restrictions beyond size (eg, the fields must contain printable characters), we can further restrict the domain as needed.

In addition to these (modest) applications, the question is interesting from a theoretical standpoint: how can we construct new ciphers from existing ones? In particular, can we construct ciphers with arbitrary domains without resorting to creating new ciphers from scratch? It certainly "feels" like there should be a good way to construct a block cipher on 32 bits given a block cipher on 64 bits, but, even for this case, no one knows how to do this in a practical manner with good security bounds.

RELATED WORK. We assume that one has in hand a good block cipher for any desired block length. Since "standard" block ciphers come only in "convenient" block lengths, such as $n = 128$, here are some ways that one might create a block cipher for some non-standard block length. First, one could construct the block cipher from scratch. But it is probably better to start with a well-studied primitive like SHA-1 or AES. These could then be used within a balanced Feistel network [14], which creates a block cipher for any (even) block length $2n$, starting with something that behaves as a pseudorandom function (PRF) from n bits to n bits. Luby and Rackoff [9] give quantitative bounds on the efficacy of this construction (when using three and four rounds), and their work has spawned much related analysis, too. Naor and Reingold [11] provide a different construction which extends a block cipher on n bits to a block cipher on $2ni$ bits, for any $i \geq 1$. A variation on their construction due to Patel, Ramzan and Sundaram [12] yields a cipher on ni bits for any $i \geq 1$. Lucks [10] generalizes Luby-Rackoff to consider a three-round unbalanced Feistel network, using hash functions for round functions. This yields a block cipher for any given length N starting with a PRF from r bits to ℓ bits and another from ℓ bits to r bits, where $\ell + r = N$. Starting from an n-bit block cipher, Bellare and Rogaway [3] construct and analyze a length-preserving cipher with domain $\{0,1\}^{\geq n}$. This is something more than making a block cipher on arbitrary $N \geq n$ bits. Anderson and Biham [1] provide two constructions for a block cipher (BEAR and LION) which use a hash function and a stream cipher. This again uses an unbalanced Feistel network.

It is unclear how to make any of the constructions above apply to message spaces which are not sets of strings. Probably several of the constructions can modified, and in multiple ways, to deal with a message space $\mathcal{M} = [0, k - 1]$, or with other message spaces.

The Hasty Pudding Cipher of Schroeppel and Orman [13] is a block cipher which works on any domain $[0, k-1]$. They use what is essentially "Method 2," internally iterating the cipher until a proper domain point is reached. Schroeppel believes that the idea underlying this method dates back to the rotor machines used in the early 1900's.

Our notion of a pseudorandom function is due to Goldreich, Goldwasser and Micali [6]. Pseudorandom permutations are defined and constructed by Luby and Rackoff [9]. We use the adaptation of these notions to deal with finite objects, which first appears in Bellare, Kilian and Rogaway [2].

2 Preliminaries

NOTATION. If A and B are sets then $\text{Rand}(A, B)$ is the set of all functions from A to B. If A or B is a positive number, n, then the corresponding set is $[0, n-1]$. We write $\text{Perm}(A)$ to denote the set of all permutations on the set A and if n is a positive number then the set is assumed to be $[0, n-1]$. By $x \xleftarrow{R} A$ we denote the experiment of choosing a random element from A.

A function family is a multiset $F = \{f : A \to B\}$, where $A, B \subseteq \{0, 1\}^*$. Each element $f \in F$ has a name K, where $K \in \text{Key}$. So, equivalently, a function family F is a function $F : \text{Key} \times A \to B$. We call A the domain of F and B the range of F. The first argument to F will be written as a subscript. A cipher is a function family $F : \text{Key} \times A \to A$ where $F_K(\cdot)$ is always a permutation; a block cipher is a function family $F : \text{Key} \times \{0, 1\}^n \to \{0, 1\}^n$ where $F_K(\cdot)$ is always a permutation. An ideal block cipher is a block cipher in which each permutation on $\{0, 1\}^n$ is realized by exactly one $K \in \text{Key}$.

An adversary is an algorithm with an oracle. The oracle computes some function. We write $A^{f(\cdot)}$ to indicate an adversary A with oracle $f(\cdot)$. Adversaries are assumed to never ask a query outside the domain of the oracle, and to never repeat a query.

Let $F : \text{Key} \times A \to B$ be a function family and let $\mathcal{A}$ be an adversary. In this paper, we measure security as the maximum advantage obtainable by some adversary; we use the following statistical measures:

$$\mathbf{Adv}_F^{\text{prf}}(\mathcal{A}) \stackrel{\text{def}}{=} \Pr[f \xleftarrow{R} F : \mathcal{A}^{f(\cdot)} = 1] - \Pr[R \xleftarrow{R} \text{Rand}(A, B) : \mathcal{A}^{R(\cdot)} = 1] \,,$$

and when $A = B$

$$\mathbf{Adv}_F^{\text{prp}}(\mathcal{A}) \stackrel{\text{def}}{=} \Pr[f \xleftarrow{R} F : \mathcal{A}^{f(\cdot)} = 1] - \Pr[\pi \xleftarrow{R} \text{Perm}(A) : \mathcal{A}^{\pi(\cdot)} = 1] \,.$$

USEFUL FACTS. It is often convenient to replace random permutations with random functions, or vice versa. The following proposition lets us easily do this. For a proof see Proposition 2.5 in [2].

Lemma 1. [PRF/PRP Switching] *Fix $n \geq 1$. Let $\mathcal{A}$ be an adversary that asks at most p queries. Then*

$$\left| \Pr[\pi \xleftarrow{R} \text{Perm}(n) : \mathcal{A}^{\pi(\cdot)} = 1] - \Pr[\rho \xleftarrow{R} \text{Rand}(n, n) : \mathcal{A}^{\rho(\cdot)} = 1] \right| \leq p^2 / 2^{n+1}.$$

Algorithm Init_Px$_K$
  for $j \leftarrow 0$ **to** $k - 1$ **do** $I_j \leftarrow E_K(j)$
  for $j \leftarrow 0$ **to** $k - 1$ **do** $J_j \leftarrow \mathrm{Ord}(I_j, \{I_j\}_{j \in [0, k-1]})$
  for $j \leftarrow 0$ **to** $k - 1$ **do** $L_{J_j} \leftarrow j$

Algorithm Px$_K(m)$
  return J_m

Algorithm Px$_K^{-1}(m)$
  return L_m

Fig. 1. Algorithms for the Prefix Cipher. First the initialization algorithm Init_Px$_K$ is run. Then encipher with Px$_K(m)$ and decipher with Px$_K^{-1}(m)$.

3   Method 1: Prefix Cipher

Fix some integer k and let $\mathcal{M}$ be the set $[0, k - 1]$. Our goal is to build a cipher with domain $\mathcal{M}$.

Our first approach is a simple, practical method for small values of k. We name this cipher Px. Our cipher will use some existing block cipher E with keyspace $\mathcal{K}$ and whose domain is a superset of $\mathcal{M}$. The key space for Px will also be $\mathcal{K}$. To compute Px$_K(m)$ for some $m \in \mathcal{M}$ and $K \in \mathcal{K}$ we first compute the tuple

$$I = (E_K(0) \ E_K(1) \ \cdots \ E_K(k - 1)).$$

Since each element of I is a distinct string, we may replace each element in I with its ordinal position (starting from zero) to produce tuple J. And now to encipher any $m \in \mathcal{M}$ we compute Px$_K(m)$ as simply the m-th component of J (again counting from zero). The enciphering and deciphering algorithms are given in Figure 1.

EXAMPLE. Suppose we wish to encipher $\mathcal{M} = \{0, 1, 2, 3, 4\}$. We choose some random key K for some block cipher E. Let's assume E is an 8-bit ideal block cipher; therefore E_K is a uniformly chosen random permutation on $[0, 255]$. Next we encipher each element of $\mathcal{M}$. Let's say $E_K(0) = 166$, $E_K(1) = 6$, $E_K(2) = 130$, $E_K(3) = 201$, and $E_K(4) = 78$. So our tuple I is (166 6 130 201 78) and J is (3 0 2 4 1). We are now ready to encipher any $m \in \mathcal{M}$: we return the m-th element from J, counting from zero. For example we encipher 0 as 3, and 1 as 0, etc..

ANALYSIS. Under the assumption that our underlying block cipher E is ideal, I is equally likely to be any of the permutations on $\mathcal{M}$. The proof of this fact is trivial and is omitted. The method remains good when E is secure in the sense of a PRP. The argument is standard and is omitted.

PRACTICAL CONSIDERATIONS. Enciphering and deciphering are constant-time operations. The cost here is $O(k)$ time and space used in the initialization step. This clearly means that this method is practical only for small values of k. A further practical consideration is that, although this initialization is a one-time cost, it results in a table of sensitive data which must be stored somewhere.

$$\begin{array}{l|l}
\textbf{Algorithm Cy}_K(m) & \textbf{Algorithm Cy}_K^{-1}(m) \\
\quad c \leftarrow E_K(m) & \quad c \leftarrow E_K^{-1}(m) \\
\quad \textbf{if } c \in \mathcal{M} \textbf{ return } c & \quad \textbf{if } c \in \mathcal{M} \textbf{ return } c \\
\quad \textbf{else return } \mathrm{Cy}_K(c) & \quad \textbf{else return } \mathrm{Cy}_K^{-1}(c)
\end{array}$$

Fig. 2. Algorithms for the Cycle-Walking Cipher. We encipher with $\mathrm{Cy}_K(\cdot)$ and decipher with $\mathrm{Cy}_K^{-1}(\cdot)$.

4   Method 2: Cycle-Walking Cipher

This next method uses a block cipher whose domain is larger than $\mathcal{M}$, and then handles those cases where a point is out of range. Again we fix an integer k, let $\mathcal{M}$ be the set $[0, k-1]$, and devise a method to encipher $\mathcal{M}$.

Let N be the smallest power of 2 larger or equal to k, let n be $\lg N$, and let $E_K(\cdot)$ be an n-bit block cipher. We construct the block cipher Cy_K on the set $\mathcal{M}$ by computing $t = E_K(m)$ and iterating if $c \notin \mathcal{M}$. The enciphering and deciphering algorithms are shown in Figure 2.

EXAMPLE. Let $\mathcal{M} = [0, 10^6]$. Then $N = 2^{20}$ and so $n = 20$. We use some known method to build a 20-bit block cipher $E_K(\cdot)$ on the set $\mathcal{T} = [0, 2^{20} - 1]$. Now suppose we wish to encipher the point $m = 314159$; we compute $c_1 = E_K(314159)$ which yields some number in $\mathcal{T}$, say 1040401. Since $c_1 \notin \mathcal{M}$, we iterate by computing $c_2 = E_K(1040401)$ which is, say, 1729. Since $c_2 \in \mathcal{M}$, we output 1729 as $\mathrm{Cy}_K(314159)$. Decipherment is simply the reverse of this procedure.

ANALYSIS. Let's view the permutation $E_K(\cdot)$ as a family of cycles: any point $m \in \mathcal{M}$ lies on some cycle and repeated applications of $E_K(\cdot)$ can be viewed as a particle walking along the cycle, starting at m. In fact, we can now think of our construction as follows: to encipher any point $m \in \mathcal{M}$ walk along the cycle containing m until you encounter some point $c \in \mathcal{M}$. Then $c = \mathrm{Cy}_K(m)$. Of course this method assumes that one can efficiently test for membership in $\mathcal{M}$. This is trivial for our case when $\mathcal{M} = [0, k-1]$, but might not be for other sets.

Now we may easily see that $\mathrm{Cy}_K(\cdot)$ is well-defined: given any point $m \in \mathcal{M}$ if we apply $E_K(\cdot)$ enough times, we will arrive at a point in $\mathcal{M}$. This is because walking on m's cycle must eventually arrive back at *some* point in $\mathcal{M}$, even if that point is m itself. We can also see that $\mathrm{Cy}_K(\cdot)$ is invertible since inverting $\mathrm{Cy}_K(m)$ is equivalent to walking *backwards* on m's cycle until finding some element in $\mathcal{M}$. Therefore, we know $\mathrm{Cy}_K(\cdot)$ is a permutation on $\mathcal{M}$. However the question arises, "how much security do we lose in deriving this permutation?" The fortunate answer is, "nothing."

Theorem 1. [Security of Cycle-Walking Cipher] *Fix $k \geq 1$ and let $\mathcal{M} = [0, k-1]$. Let $E_K(\cdot)$ be an ideal block cipher on the set $\mathcal{T}$ where $\mathcal{M} \subseteq \mathcal{T}$. Choose a key K uniformly at random and then construct $\mathrm{Cy}_K(\cdot)$ using $E_K(\cdot)$. Then $\mathrm{Cy}_K(\cdot)$ is a uniform random permutation on $\mathcal{M}$.*

Proof. Fix some permutation π on the set $\mathcal{M}$. We will show that an equal number of keys K will give rise to π; this will imply the theorem.

We proceed by induction, showing that the number of permutations on $\{0, \ldots, k-1, x\}$ which give rise under our construction to π is constant. Since $\mathcal{M} \subseteq \mathcal{T}$ we can repeatedly add all elements $x \in \mathcal{T} - \mathcal{M}$ while maintaining that the number of permutations which give rise to π is constant.

Decompose π into r cycles of lengths $l_1, l_2, \cdots, l_r$. We count the number of ways to insert the new element x. There are l_i ways to insert x into the ith orbit corresponding to the ith cycle, and one way to insert x into a new orbit of its own (ie, the permutation which fixes x). Therefore there are $\sum_{i=1}^{r} l_i + 1 = k$ ways to add element x to π yielding a permutation which will give rise to π by repeated iterations. This holds no matter what π we choose.

Let $|\mathcal{T}| = t$. Then by induction we see that there are exactly $\prod_{i=k}^{t} i$ keys K under which our construction reduces $E_K(\cdot)$ to π.

Similar to the Prefix Cipher, our construction has retained all of the security of the underlying block cipher.

Theorem 1 is an information-theoretic result. Passing to the corresponding complexity-theoretic result is standard. Because no security is lost in the information-theoretic setting, and because we apply E an expected two times (or fewer), an adversary's maximal advantage to distinguish $E_K(\cdot)$ from a random permutation of Z_{2^n} in expected time $2t$ approximately upper bounds an adversary's maximal advantage to distinguish $\text{Cy}_K(\cdot)$ from a random permutation on $\mathcal{M}$ in time t.

5   Method 3: Generalized-Feistel Cipher

Our final method works as follows: we decompose all the numbers in $\mathcal{M}$ into pairs of "similarly sized" numbers and then apply the well-known Feistel construction [14] to produce a cipher. Again we fix an integer k, let $\mathcal{M}$ be the set $[0, k-1]$, and devise a method to encipher $\mathcal{M}$.

We call our cipher $\text{Fe}[r, a, b]$ where r is the number of rounds we use in our Feistel network and a and b are positive numbers such that $ab \geq k$. We use a and b to decompose any $m \in \mathcal{M}$ into two numbers for use as the inputs into the network. Within the network we use r random functions $F_1, \ldots, F_r$ whose ranges contain $\mathcal{M}$. The algorithms to encipher and decipher are given in Figure 3. Notice that if using the Feistel construction results in a number not in $\mathcal{M}$, we iterate just as we did for the Cycle-Walking Cipher.

EXAMPLE. In order to specify some particular $\text{Fe}[r, a, b]_K(\cdot)$ we must specify the numbers a and b, the number of Feistel rounds r, and the choice of underlying functions $F_1, \cdots, F_r$ we will use.

As a concrete example, let's take $k = 2^{35}$, $r = 3$, and $a = 185360$ and $b = 185368$ (methods for finding a and b will be discussed later). Note that $ab \geq k$ as required. Since ab is 74112 larger than k, our Feistel construction will be on the set $\mathcal{M}' = [0, (2^{35} - 1) + 74112]$, meaning there are 74112 values

---

Algorithm $\text{Fe}[r, a, b]_K(m)$
$\quad c \leftarrow \text{fe}[r, a, b]_K(m)$
$\quad$ **if** $c \in \mathcal{M}$ **return** c
$\quad$ **else return** $\text{Fe}[r, a, b]_K(c)$

Algorithm $\text{fe}[r, a, b]_K(m)$
$\quad L \leftarrow m \bmod a; \quad R \leftarrow \lfloor m/a \rfloor$
$\quad$ **for** $j \leftarrow 1$ **to** r **do**
$\quad\quad$ **if** (j is odd) **then** $tmp \leftarrow (L + F_j(R)) \bmod a$
$\quad\quad$ **else** $tmp \leftarrow (L + F_j(R)) \bmod b$
$\quad\quad L \leftarrow R; \quad R \leftarrow tmp$
$\quad$ **if** (r is odd) **then return** $aL + R$
$\quad$ **else return** $aR + L$

Algorithm $\text{Fe}[r, a, b]_K^{-1}(m)$
$\quad c \leftarrow \text{fe}[r, a, b]_K^{-1}(m)$
$\quad$ **if** $c \in \mathcal{M}$ **return** c
$\quad$ **else return** $\text{Fe}[r, a, b]_K^{-1}(c)$

Algorithm $\text{fe}[r, a, b]_K^{-1}(m)$
$\quad$ **if** (r is odd) **then** $R \leftarrow m \bmod a; \quad L \leftarrow \lfloor m/a \rfloor$
$\quad$ **else** $L \leftarrow m \bmod a; \quad R \leftarrow \lfloor m/a \rfloor$
$\quad$ **for** $j \leftarrow r$ **to** 1 **do**
$\quad\quad$ **if** (j is odd) **then** $tmp \leftarrow (R - F_j(L)) \bmod a$
$\quad\quad$ **else** $tmp \leftarrow (R - F_j(L)) \bmod b$
$\quad\quad R \leftarrow L; \quad L \leftarrow tmp$
$\quad$ **return** $aR + L$

Fig. 3. Algorithms for the Generalized-Feistel Cipher. We encipher with $\text{Fe}[r, a, b]_K(\cdot)$ and decipher with $\text{Fe}[r, a, b]_K^{-1}(\cdot)$. Here a and b are the numbers used to bijectively map all $m \in \mathcal{M}$ into L, and R, and r is the number of rounds of Feistel we will apply. The key K is implicitly used to select the r functions $F_1, \ldots, F_r$.

which are in $\mathcal{M}' - \mathcal{M}$ for which we will have to iterate (just as we did for the Cycle-Walking Cipher). Let's use DES with independent keys as our underlying PRFs. DES is a 64-bit cipher which uses a 56-bit key; we will regard the 64-bit strings on which DES operates as integers in the range $[0, 2^{64} - 1]$ in the natural way. We need three PRFs so our key $K = K_1 \parallel K_2 \parallel K_3$ will be $3 \times 56 = 168$ bits. Now to compute $\text{Fe}[3, 185360, 185368](m)$ we compute $L = m \bmod 185360$, and $R = \lfloor m/185360 \rfloor$, and then perform three rounds of Feistel using $\text{DES}_{K_1}(\cdot)$, $\text{DES}_{K_2}(\cdot)$, and $\text{DES}_{K_3}(\cdot)$ as our underlying PRFs. The first round results in $L \leftarrow \lfloor m/185360 \rfloor$ and $R \leftarrow (m \bmod 185360 + DES_{K_1}(\lfloor m/185360 \rfloor)) \bmod 185360$, and so on.

ANALYSIS. First we note that $\text{Fe}[r, a, b](\cdot)$ is a permutation: it is well-known that the Feistel construction produces a permutation, and we showed previously that iterating any permutation is a permutation. We now analyze the how good is this Generalized-Feistel Cipher for the three-round case.

Assuming the underlying functions F_1, F_2, and F_3 used in our construction are *truly random* functions, we will compare how close $Fe[3, a, b](\cdot)$ is to a truly random permutation. Passing to the complexity-theoretic setting is then standard, and therefore omitted.

Theorem 2. [Security of Generalized-Feistel Cipher] *Fix $k \geq 1$ and let $\mathcal{M} = [0, k-1]$. Fix two numbers $a, b > 0$ such that $ab \geq k$. Let $\Delta = ab - k$. Fix an n such that $2^n > a$ and $2^n > b$. Let D be an adversary which asks q queries of her oracle. Then*

$$\mathbf{Adv}_{Fe}^{\mathrm{prf}}(D) = \Pr[F_1, F_2, F_3 \xleftarrow{R} \mathrm{Rand}(2^n, 2^n) : D^{Fe[3,a,b](\cdot)} = 1]$$
$$- \Pr[\rho \xleftarrow{R} \mathrm{Rand}(k, k) : D^{\rho(\cdot)} = 1]$$
$$\leq \frac{(q + \Delta)^2}{2^{n+1}} \left(\lceil 2^n/a \rceil + \lceil 2^n/b \rceil\right).$$

The proof is an adaptation of Luby's analysis from Lecture 13 of [8], which is in-turn based on [9]. It can be found in Appendix A.

Finally, we must adjust this bound to account for the fact that we have compared $Fe[3, a, b]_K(\cdot)$ with a random function instead of a random permutation. We can invoke Lemma 1 which gives us a final bound quantifying the quality of our construction:

$$\mathbf{Adv}_{Fe}^{\mathrm{prp}}(D) = \Pr[F_1, F_2, F_3 \xleftarrow{R} \mathrm{Rand}(2^n, 2^n) : D^{Fe[3,a,b](\cdot)} = 1]$$
$$- \Pr[\pi \xleftarrow{R} \mathrm{Perm}(k) : D^{\pi(\cdot)} = 1]$$
$$\leq \frac{(q + \Delta)^2 + q^2}{2^{n+1}} \left(\lceil 2^n/a \rceil + \lceil 2^n/b \rceil\right).$$

6 Discussion

PREFIX CIPHER. Our first method, the Prefix Cipher, is useful only for suitably small k. Since enciphering one point requires enciphering all k points in $[0, k-1]$, many applications would find this prohibitively expensive for all but fairly small values of k.

CYCLE-WALKING CIPHER. Our second method, the Cycle-Walking Cipher, can be quite practical. If k is just smaller than some power of 2, the number of points we have to "walk through" during any given encipherment is correspondingly small. In the worst case, however, k is one larger than a power of 2, and (with extremely bad luck) might require k calls to the underlying block cipher to encipher just one point. But if the underlying block cipher is good we require, in the worst case, an expected two calls to it in order to encipher and decipher any point.

GENERALIZED-FEISTEL CIPHER. To get the best bound we should select a and b such that these numbers are somewhat close together and such that $\Delta = ab - k$

is small. One obvious technique is to try numbers near $\sqrt{k}$; for example, taking $a = b = \lceil\sqrt{k}\rceil$ means that $ab - k$ will never be more than $2\sqrt{k} + 1$. But often one can do better.

Another way to improve the bound is to ensure n is suitably large. The "tail effects" spoken of in the proof are diminished as n grows (because as 2^n gets larger $\lceil 2^n/a\rceil/2^n$ gets closer to $1/a$).

THE ONE-OFF CONSTRUCTION. Another method, not mentioned above, works well for domains which are one element larger than a domain we can accommodate efficiently. Say we have a cipher E with domain $[0, k - 1]$ and we wish to construct a cipher E' with domain $[0, k]$. We choose a key $K' = \{K, r\}$ for E' by choosing a key K for E and a random number $r \in [0, k]$. We then compute $E'_{K'}(X)$ as follows:

$$E'_{K'}(X) = \begin{cases} r & \text{if } X = k \\ k & \text{if } X = E_K^{-1}(r) \\ E_K(X) & \text{otherwise} \end{cases}$$

The security of this construction is tightly related to the security of E and the method for selecting r. The analysis is omitted.

Of course we can use this method to repeatedly extend the domain of any cipher to the size of choice, but for most settings it is impractical to do this more than a few times. A typical method for generating r would be to take $r = E_{K^*}(0) \bmod (k + 1)$ where K^* is a new randomly-selected key. The "tail effect" here is not too bad, but will cause a rapid deterioration of the security bound when used too often. Also, the scheme begins to become quite inefficient when we extend the domain in this way too many times.

OTHER DOMAINS. Though we have spoken in terms of the domain $[0, k - 1]$ the same methods work for other domains, too. For example, to encipher in Z_N^*, where $N = pq$ is a 1024-bit product of two primes, one can use either cycle-walking or the generalized-Feistel construction, iterating in the highly unlikely event that a point is in Z_N but not in Z_N^*.

We may also use our methods to encipher points from an elliptic curve group (EC group). There are well-known "compact" representations of the points in EC groups, and these representations form our starting point. For example, one finds in [5] simple algorithms to compress the representation of a point in an EC group. Consider the EC group G over the field F_q where q is either a power of two or a prime. Then any point $(x, y) \in G$ may be represented as a member of F_q together with a single bit. Let's consider first the case where $q = 2^m$ with $m > 0$. The Hasse theorem (see [5], page 8) guarantees at least $d(r) = r + 1 - 2\sqrt{r}$ points in G. Since it is possible to represent any point in G with $m + 1$ bits and it is also possible to efficiently test for membership in G, we could use the cycle-walking construction over a 2^{m+1}-bit cipher. The expected number of invocations of this cipher to encipher a point in G is then $2^{m+1}/d(2^m) \approx 2$.

If q is instead a prime p, we can represent any point in G as a number $x \in [0, p - 1]$ and a single bit y. We may again use any of our methods to encipher these $2p$ points. Here the Hasse theorem ([5], page 7) guarantees at least $d(p)$ points in G and once again an efficient test for membership in G exists. Therefore we may use the cycle-walking construction over some $\lceil \lg 2p \rceil$-bit cipher. However if $2p$ is not close to a power of 2, we may wish to instead use the generalized-Feistel construction.

OPEN PROBLEMS. As mentioned already, we have not provided any construction which works well (and provably so) for intermediate-sized values of k. For example, suppose you are given an ideal block cipher Π on 128-bit strings, and you want to approximate a random permutation π on, say, 40-bit strings. Probably enough rounds of Feistel work, but remember that our security goal is that even if an adversary inquires about all 2^{40} points, still she should be unable to distinguish π from a random permutation on 40 bits. Known bounds are not nearly so strong. Of course the prefix method works, but spending 2^{40} time and space to encipher the first point is not practical.

Acknowledgments

Special thanks to Richard Schroeppel who made many useful comments on an earlier draft. Thanks also to Mihir Bellare, David McGrew, and Silvio Micali for their helpful comments. This paper was written while Rogaway was on leave of absence from UC Davis, visiting the Department of Computer Science, Faculty of Science, Chiang Mai University. This work was supported under NSF CAREER award CCR-9624560, and by a generous gift from Cisco Systems.

References

1. ANDERSON, R., AND BIHAM, E. Two practical and provably secure block ciphers: BEAR and LION. In *Fast Software Encryption* (1996), vol. 1039 of *Lecture Notes in Computer Science*, Springer-Verlag, pp. 114–120.
2. BELLARE, M., KILIAN, J., AND ROGAWAY, P. The security of the cipher block chaining message authentication code. *Journal of Computer and System Sciences 61*, 3 (2000), 362–399. Earlier version in CRYPTO '94. See www.cs.ucdavis.edu/~rogaway.
3. BELLARE, M., AND ROGAWAY, P. On the construction of variable-input-length ciphers. In *Fast Software Encryption* (1999), vol. 1636 of *Lecture Notes in Computer Science*, Springer-Verlag. See www.cs.ucdavis.edu/~rogaway.
4. BELLOVIN, S., AND MERRITT, M. Encrypted key exchange: password-based protocols secure against dictionary attacks. In *1992 IEEE Computer Society Symposium on Research in Security and Privacy* (1992), IEEE Computer Society Press, pp. 72–84.
5. CERTICOM RESEARCH. *Standards for efficient cryptography, SEC1: Elliptic curve cryptography*, version 1, Sept. 2000. Available on-line at www.secg.org.
6. GOLDREICH, O., GOLDWASSER, S., AND MICALI, S. How to construct random functions. *Journal of the ACM 33*, 4 (1986), 210–217.

7. GOLDWASSER, S., MICALI, S., AND RIVEST, R. A digital signature scheme secure against adaptive chosen-message attacks. *SIAM Journal of Computing 17*, 2 (Apr. 1988), 281–308.
8. LUBY, M. *Pseudorandomness and cryptographic applications*. Princeton University Press, Princeton, New Jersey, 1996.
9. LUBY, M., AND RACKOFF, C. How to construct pseudorandom permutations from pseudorandom functions. *SIAM Journal of Computing 17*, 2 (Apr. 1988).
10. LUCKS, S. Faster Luby-Rackoff ciphers. In *Fast Software Encryption* (1996), vol. 1039 of *Lecture Notes in Computer Science*, Springer-Verlag.
11. NAOR, M., AND REINGOLD, O. On the construction of pseudorandom permutations: Luby-Rackoff revisited. *Journal of Cryptology 12*, 1 (1999), 29–66.
12. PATEL, S., RAMZAN, Z., AND SUNDARAM, G. Towards making Luby-Rackoff ciphers optimal and practical. In *Fast Software Encryption* (1999), vol. 1636 of *Lecture Notes in Computer Science*, Springer-Verlag.
13. SCHROEPPEL, R., AND ORMAN, H. Introduction to the hasty pudding cipher. In Proceedings from the First Advanced Encryption Standard Candidate Conference, National Institute of Standards and Technology, Aug. 1998. See http://www.cs.arizona.edu/~rcs/hpc/.
14. SMITH, J. L. The design of Lucifer: A cryptographic device for data communications. Tech. Rep. IBM Research Report RC 3326, IBM T.J. Watson Research Center, Yorktown Heights, N.Y., 10598, U.S.A., Apr. 1971.

A   Proof of Theorem 2

Proof. To simplify the exposition, we will initially assume that $k = ab$. In other words, that no iterating is required to compute $\text{Fe}[3, a, b]_K(\cdot)$. Once we establish the result in this setting, we can make some minor changes to get the general result.

We begin by defining a couple of games. Let us call "Game Fe" the game in which we choose three random functions $F_1, F_2, F_3 \leftarrow \text{Rand}(2^n, 2^n)$ and then answer D's queries according to $\text{Fe}[3, a, b](\cdot)$ using F_1, F_2, and F_3 as our underlying functions. Let us call "Game Rn" the game in which we choose a random function $\rho \in \text{Rand}(k, k)$ and then answer D's queries according to $\rho(\cdot)$. Let's denote by P_{Fe} the probability that D outputs 1 in Game Fe, and denote by P_{Rn} the probability that D outputs 1 in Game Rn. We are trying to show that

$$P_{\text{Fe}} - P_{\text{Rn}} \leq \frac{(q + ab - k)^2}{2^{n+1}} \left(\lceil 2^n/a \rceil + \lceil 2^n/b \rceil \right).$$

Without loss of generality, assume D never repeats a query. We begin by describing a new game called "Game B'". Game B' will look the same to adversary D as Game Fe, but Game B' will be played completely differently. Instead of choosing three random functions F_1, F_2, F_3, we'll choose only some random numbers $x_1, \ldots, x_q, y_1, \ldots, y_q$, and $z_1, \ldots, z_q$. Each of these numbers is in $[0, 2^n - 1]$. The *only* random choices we will make in playing game B' is in the choice of the x_i, y_i, and z_i. We describe Game B' in Figure 4. It is played as follows: first choose random numbers $x_1, \ldots, x_q, y_1, \ldots, y_q$, and $z_1, \ldots, z_q$. Now answer the i-th query with $a\beta_i + \gamma_i$, where β_i and γ_i are described in the figure.

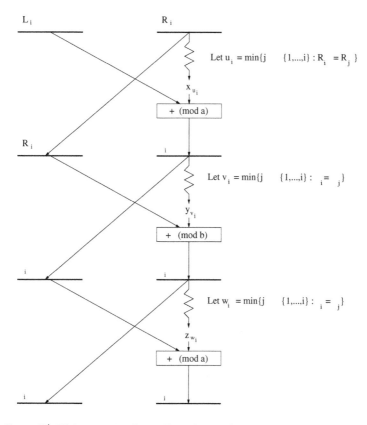

Fig. 4. Game B′. This game is identical, as far as the adversary can tell, to Game Fe. Begin by choosing $x_1, \ldots, x_q$, $y_1, \ldots, y_q$, and $z_1, \ldots, z_q$ at random. Then answer the i-th query, L_i, R_i, by β_i, γ_i, computed as in the figure.

It should be obvious that Game B′ is the same, as far as the adversary can see, to Game Fe. Thus $P_{\text{Fe}} = \Pr[D^{B'} = 1]$.

We now modify Game B′ to a Game B which is identical, from the adversary's point of view, to Game B′ (and therefore to Game Fe). This modification is unusual: we will subtract R_{v_i} from the second sum, and we will subtract α_{w_i} from the final sum. The new game is shown in Figure 5.

The reason that these new addends do not change the adversary's view of the game stems from the fact that the $((y_{v_i} - R_{v_i}) \bmod b, (z_{w_i} - \alpha_{w_i}) \bmod a)$ in Game B retain the same distribution as (y_{v_i}, z_{w_i}) had in game B′.

We now have that $P_{\text{Fe}} = \Pr[D^B = 1]$. The probability is taken over the random q-vectors x, y, and z with coordinates in $[0, 2^n - 1]$.

We now consider one final game, Game C. This game is identical to B except that we output $ay_i + z_i$ (instead of $a\beta_i + \gamma_i$). Obviously $P_{\text{Rn}} = \Pr[D^C = 1]$. Again the probability is over the random vectors x, y, z.

We will now make some observations and calculations about Games B and C which will allow us to conclude with the theorem. The idea is that Games B and

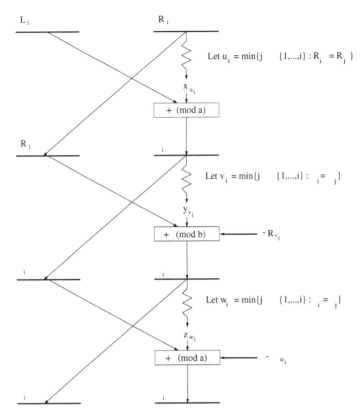

Fig. 5. Game B. We modify B' by adding the quantities indicated by the emboldened arrows. This game is once again identical, from the adversary's perspective, to Game Fe.

C usually coincide. We will manage to bound adversarial advantage by looking at the chance that games B and C do not coincide.

First we define some events. These events are defined in Game C. (It is important that we do this in Game C, not Game B.) Define the event REPEAT_α as true if $\alpha_i = \alpha_j$ for some $i < j$–that is, some α_i arises twice. Define the event REPEAT_β as true if $\beta_i = \beta_j$ for some $i < j$–that is, some β_i arises twice. Define the event REPEAT as the disjunct of α_i and β_i–that is, either an α_i repeats or a β_i repeats. Again, these events are defined in Game C.

Claim. $\Pr[\mathsf{REPEAT}_\alpha] \leq \dfrac{q^2 \lceil 2^n/a \rceil}{2^{n+1}}.$

Look at query i. If R_i itself is a repetition of an earlier R_j, then we know for sure that $\alpha_i \neq \alpha_j$, since all queries are assumed to be distinct. It is possible, however, that α_i could coincide with some α_j where R_j was different from R_i. But we have provided the adversary no information about internal x_i and α_i values. If the cardinality of $[0, 2^n - 1]$ were evenly divisible by a then we would know the chance for any particular α_j to coincide with α_i would be $1/a$. This is because we are taking the sum of L_i with a random member of $[0, 2^n - 1]$ and then taking this (mod a). But of course 2^n may not be divisible by a and this modulus will create an "tail effect" slightly biasing the probability. We can easily measure this, however, as follows: the amount of probability mass on some points will be $\lfloor 2^n/a \rfloor / 2^n$ and on the others it will be $\lceil 2^n/a \rceil / 2^n$. We will simply take the latter as a bound. If R_i is a new, unrepeated value, then x_{u_i} will be a random number in $[0, 2^n - 1]$ and so the chance that α_i will collide with any particular prior α_j is again bounded by $\lceil 2^n/a \rceil / 2^n$. Thus the chance that α_i will collide with an earlier query is at most $(i - 1)\lceil 2^n/a \rceil / 2^n$, and the chance that there will eventually be a collision in α_i-values is at most $\sum_{i=1}^{q}(i-1)\lceil 2^n/a \rceil / 2^n \leq \frac{q^2}{2}\lceil 2^n/a \rceil / 2^n$. $\Diamond$

Claim. $\Pr[\mathsf{REPEAT}_\beta | \overline{\mathsf{REPEAT}_\alpha}] \leq \dfrac{q^2 \lceil 2^n/b \rceil}{2^{n+1}}$.

By assumption, the v_i values are all distinct, so y is being evaluated on distinct points. The chance that two β_i values coincide is determined similar to the case in the previous claim where the R_i values were distinct. So analogously we have $\sum_{i=1}^{q}(i-1)\lceil 2^n/b \rceil / 2^n \leq \frac{q^2}{2}\lceil 2^n/b \rceil / 2^n$. $\Diamond$

Putting this together we have that

Claim. $\Pr[\mathsf{REPEAT}] \leq \dfrac{q^2}{2^{n+1}}\left(\lceil 2^n/a \rceil + \lceil 2^n/b \rceil\right)$.

The reason is that

$$\begin{aligned}
\Pr[\mathsf{REPEAT}] &= \Pr[\mathsf{REPEAT}_\alpha] + \Pr[\mathsf{REPEAT}_\beta \wedge \overline{\mathsf{REPEAT}_\alpha}] \\
&= \Pr[\mathsf{REPEAT}_\alpha] + \Pr[\mathsf{REPEAT}_\beta \mid \overline{\mathsf{REPEAT}_\alpha}] \cdot \Pr[\overline{\mathsf{REPEAT}_\alpha}] \\
&\leq \Pr[\mathsf{REPEAT}_\alpha] + \Pr[\mathsf{REPEAT}_\beta \mid \overline{\mathsf{REPEAT}_\alpha}],
\end{aligned}$$

and we have just bounded each of the above addends. $\Diamond$

Now for the key observation:

Claim. $\Pr[D^B = 1 \mid \overline{\mathsf{REPEAT}}] = \Pr[D^C = 1 \mid \overline{\mathsf{REPEAT}}]$.

Both probabilities are over random choices of x, y, z. On the right-hand we output y_i, z_i in response to the ith query. On the left-hand side, assuming that REPEAT does not hold *in Game* C, once again we output y_i, z_i. This would be clear if we had said "assuming that REPEAT does not hold *in Game* B," and we defined this even in Game B in the obvious manner. But notice that as long as REPEAT does not hold in Game C, Game C and Game B behave identically, always returning y_i, z_i in response to query i. This is easily established by induction. $\Diamond$

We claim that, because of the last claim,

$$\begin{aligned}
P_{\mathrm{Fe}} - P_{\mathrm{Rn}} &= \Pr[D^B = 1] - \Pr[D^C] = 1 \\
&\leq \Pr[\mathsf{REPEAT}]
\end{aligned}$$

Let A, B, C be arbitrary events and assume $\Pr[A \mid \overline{C}] = \Pr[B \mid \overline{C}]$. Now

$$\Pr[A] - \Pr[B] = \Pr[A \mid \overline{C}] \Pr[\overline{C}] + \Pr[A \mid C] \Pr[C]$$
$$- \Pr[B \mid \overline{C}] \Pr[\overline{C}] - \Pr[B \mid C] \Pr[C]$$

and so $\Pr[A \mid \overline{C}] = \Pr[B \mid \overline{C}]$ tells us that first and third addends cancel. Now upperbound the second addend by dropping the $\Pr[A|C]$ (that is, upperbound this by 1) and drop the final addend (which is negative) entirely, thereby getting an upperbound of $\Pr[C]$, as desired.

We now address the case where we iterate the cipher. In other words, what happens when $ab - k > 0$? In this case we may invoke $\text{fe}[3, a, b]_K(\cdot)$ multiple times per encipherment, and we must account for this in the bound. The crucial point in the proof affected by iterating is when we are calculating REPEAT_α. In the worst case, the first encipherment could cause us to compute $\text{fe}[3, a, b](m)$ for *all* $m \in [k, ab-1]$. In this case up to $ab - k$ values of α_i may already have been computed. We therefore include these points in the computation of $\Pr[\mathsf{REPEAT}_\alpha]$. The new bound is therefore $\sum_{i=1}^{q+(ab-k)} (i-1)\lceil 2^n/a \rceil / 2^n \le \frac{(q+ab-k)^2}{2} \lceil 2^n/a \rceil / 2^n$ for $\Pr[\mathsf{REPEAT}_\alpha]$ and similarly for $\Pr[\mathsf{REPEAT}_\beta \mid \overline{\mathsf{REPEAT}_\alpha}]$. So the overall bound is now

$$\Pr[\mathsf{REPEAT}] \le \frac{(q + ab - k)^2}{2^{n+1}} \left(\lceil 2^n/a \rceil + \lceil 2^n/b \rceil \right).$$

And setting $\Delta = ab - k$ we obtain the bound of Theorem 2.

Known Plaintext Correlation Attack against RC5

Atsuko Miyaji[1], Masao Nonaka[1], and Yoshinori Takii[2]

[1] School of Information Science, Japan Advanced Institute of Science and Technology
1-1, Asahidai, Tatsunokuchi, Nomi, 923-1292, Japan
{miyaji,m-nonaka}@jaist.ac.jp
http://grampus.jaist.ac.jp:8080/miyaji-lab/index.html
[2] Japan Air Self Defense Force
5-1, Ichigaya Honmura-cho, Shinjyuku-ku, Tokyo, 162-0845, Japan

Abstract. We investigate a known plaintext attack on RC5 based on correlations. Compared with the best previous known-plaintext attack on RC5-32, a linear cryptanalysis by Borst, Preneel, and Vandewalle, our attack applies to a larger number of rounds. RC5-32 with r rounds can be broken with a success probability of 90% by using $2^{6.14r+2.27}$ plaintexts. Therefore, our attack can break RC5-32 with 10 rounds (20 half-rounds) with $2^{63.67}$ plaintexts with a probability of 90%. With a success probability of 30%, our attack can break RC5-32 with 21 half-rounds by using $2^{63.07}$ plaintexts.

1 Introduction

RC5, designed by Rivest ([11]), is a block cipher which is constructed by only simple arithmetic such as an addition, a bit-wise exclusive-or(XOR), and a data dependent rotation. Therefore, RC5 can be implemented efficiently by software with small amount of memory. RC5-32/r means that two 32-bit-block plaintexts are encrypted by r rounds, where one round consists of two half-rounds. Various attacks against RC5 have been analyzed intensively ([1,2,4,5,6,7]). The best chosen-plaintext attack ([1]), up to the present, breaks RC5-32/12 by using $(2^{44}, 2^{54.5})$ pairs of chosen plaintexts and known plaintexts. However, it requires many stored plaintexts such as $2^{54.5}$. Even in the case of RC5-32/10, it requires $(2^{36}, 2^{50.5})$ pairs of chosen plaintexts and known plaintexts with stored $2^{50.5}$ plaintexts. In a realistic sense, it would be infeasible to employ such an algorithm on a modern computer. On the other hand, a known plaintext attack can work more efficiently and practically, even though it has not been reported far higher round like 12.

The best known-plaintext attack against RC5 is a linear cryptanalysis ([2]). They have reported that RC5-32 with 10 rounds is broken by 2^{64} plaintexts under the heuristic assumption, that is, RC5-32 with r rounds is broken with a success probability of 90% by using 2^{6r+4} plaintexts. However, their assumption seems to be highly optimistic. Table 1 shows both their results and our results. In fact, their experimental results report that RC5-32 with 3 or 4 rounds is broken with a

B. Preneel (Ed.): CT-RSA 2002, LNCS 2271, pp. 131–148, 2002.
© Springer-Verlag Berlin Heidelberg 2002

Table 1. Required plaintexts for attack on RC5. The second column provides a theoretical estimate based on heuristics

	$r-round$ estimate	2 rounds		3 rounds		4 rounds		5 rounds	
		#texts	#keys	#texts	#keys	#texts	#keys	#texts	#keys
[2]	$2^{6.8r+2.4}(2^{6r+4})$	2^{16}	92/100	2^{22}	81/100	2^{28}	82/100	2^{34}	9/10
our attack	$2^{6.14r+2.27}$	2^{15}	100/100	2^{22}	100/100	2^{28}	99/100	2^{33}	90/100
		2^{14}	95/100	2^{21}	95/100	2^{27}	96/100	2^{32}	60/100

success probability of 81% or 82% if we use 2^{22} or 2^{28} plaintexts respectively. This means that their estimation does not hold even in such lower rounds as 3 or 4. On the other hand, they also discussed the theoretical complexity of breaking RC5-32 with r rounds: RC5-32 with r rounds can be broken with a success probability of 90% by using $2^{6.8r+2.4}$ plaintexts. According to their theoretical assumption, it requires $2^{22.8}$ or $2^{29.6}$ plaintexts in order to break RC5-32/3 or RC5-32/4 with a success probability of 90%. Actually, it seems that the theoretical estimate reflects their experimental results. Note that, under the theoretical assumption, their known plaintext attack can break RC5-32/9 but not RC5-32/10 with a success probability of 90%.

In this paper, we investigate a known plaintext attack by improving a correlation attack against RC6 ([7]). RC6 is the next version of RC5, which has almost the same construction as RC5: RC6 consists of a multiplication, an addition, XOR, and a data dependent rotation. While the input of RC5 consists of 2 words such as (L_0, R_0), that of RC6 consists of 4 words. This is why approach of attacks on RC5 is similar to that on RC6, but a slight difference is needed. Correlation attack makes use of correlations between an input and an output, which is measured by the χ^2 test: the specific rotation in both RC5 and RC6 is considered to cause the correlations between the corresponding two 5-bit integer values. In [7], correlation attacks against RC6-32 recover subkeys from the 1st round to the r-th round by handling a plaintext in such a way that the χ^2-test after one round becomes significantly higher value. Their main idea is to choose such a plaintext that the least significant five bits in the first and third words are constant after one-round encryption as follows: 1. the least significant five bits in the first and third words are zero; 2. the fourth word is set to the values that introduce a zero rotation in the 1st round. Their attack controls a plaintext in two parts with 5 bits: 5 bits corresponding to the χ^2-test, and 5 bits in relation to data dependent rotations. We apply their attack to RC5, where a plaintext is represented by 2 words (L_0, R_0). According to their approach, it is necessary to control a plaintext in each block with each 5 bits: 1. the least significant five bits of L_0, $lsb_5(L_0)$, is 0; 2. $lsb_5(R_0)$ introduces a zero rotation in the 1st round. As a result, available plaintexts are reduced by 2^{10}. Compared with RC6, available plaintexts to attack RC5 are extremely fewer because the block size of RC5 is just half of RC6. Therefore, it is critical to reduce available plaintexts of RC5 by 2^{10} in order to break RC5 with a higher round. This is why their attack does not work well on RC5 directly. In fact, they also report that their attacks do not work well on RC5 compared with the existing attack ([1]). In [3], a correlation attack is also applied to RC5. Their algorithm searches subkeys from the final round to the 1st round by fixing both $lsb_5(R_0)$ and $lsb_5(L_0)$ to be 0. Therefore, their attack also suffers from the same problem of fewer available plaintexts. The important factor to target at RC5 is how to increase the available plaintexts.

We investigate how an output after h half-rounds, L_{h+1}, depends on a chosen plaintext, and find experimentally the following features of RC5:

1. The χ^2-values for the least significant five bits on L_{h+1} become significantly high by simply setting such R_0 that fixes a rotation amount in the 1st half-round. Note that any rotation amount, even large one, outputs the higher χ^2-values.
2. Any consecutive five bits on L_{h+1} outputs similarly high χ^2-values by simply setting such R_0 that fixes a rotation amount in the 1st half-round.

Usually, we know that output of RC5 is highly unlikely to be uniformly distributed if a plaintext introduces small rotation amounts such as a zero in the 1st half-round ([7]). However, from Feature 1, output of RC5 is also highly unlikely to be uniformly distributed if only a rotation in the 1st half-round is fixed. Apparently, a rotation in the 1st half-round is fixed if and only if the least significant five bits of R_0 is fixed. This means that any given plaintext can be used for correlation attack by classifying it in the same least significant five bits. In this way, we can extend a chosen plaintext correlation attack to a known plaintext attack without any cost. From Feature 2, any consecutive five bits on L_{h+1} can be used to compute the χ^2-values in the similar success probability.

We improve a correlation attack as a known plaintext attack by taking full advantage of the above features. The main points of our attack on RC5 are as follows:

1. Use any plaintext by classifying it into the same least significant five bits;
2. Determine the parts, on which the χ^2-statistic is measured, according to the ciphertexts.

We also present two algorithms to recover 31 bits of the final half-round key: one recovers each 4 bits in serial, and the other recovers each 4 bits in parallel. By employing our correlation attack, RC5-32 with r rounds(h half-rounds) can be broken with a success probability of 90% by using $2^{6.14r+2.27}(2^{3.07h+2.27})$ plaintexts. As a result, our attack can break RC5-32/10 with $2^{63.67}$ plaintexts in a probability of 90%. In the case of success probability 30%, our attack can break RC5-32 with r rounds(h half-rounds) by using $2^{5.90r+1.12}(2^{2.95h+1.12})$ plaintexts. Therefore, our attack can break RC5-32 with 21 half-rounds by using $2^{63.07}$ plaintexts in a probability of 30%.

This paper is organized as follows. Section 2 summarizes some notations and definitions in this paper. Section 3 applies Knudsen-Meier's correlation attack to RC5, and discusses the differences between RC5 and RC6. Section 4 describes some experimental results including the above features of RC5. Section 5 presents the chosen plaintext algorithm, Main algorithm. Section 6 discusses how to extend Main algorithm to the known plaintext algorithm, Extended algorithm. Section 7 applies Extended algorithm to 31-bit key recovery in the final round.

2 Preliminary

This section denotes some notations, definitions, and experimental remarks. First we describe RC5 algorithm after defining the following notations.

$\boxplus$ ($\boxminus$): an addition (subtraction) mod 2^{32};

$\oplus$: a bit-wise exclusive OR;

r : the number of rounds;

h : the number of half-rounds ($h = 2r$);

$a \lll b (a \ggg b)$: a cyclic rotation of a to the left (right) by b bits;

(L_i, R_i): an input of the i-th half-round, and (L_0, R_0) is a plaintext; We

S_i : the i-th subkey(S_{h+1} is a subkey of the h-th half-round);

$lsb_n(X)$: the least significant n bits of X;

X^i : denotes the i-th bit of X;

$X^{[i,j]}$: denotes from the i-th bit to the j-th bit of X ($i > j$);

$\overline{X}$: a bit-wise inversion of X.

denote the least significant bit(LSB) to the 1st bit, and the most significant bit(MSB) as the 32-th bit for any 32-bit element. RC5 encryption is defined as follows: a plaintext (L_0, R_0) is encrypted to (L_{h+1}, R_{h+1}) by h half-rounds iterations of a main loop, which is called one half-round. Two consecutive half-rounds correspond to one round of RC5.

Algorithm 1 (Encryption with RC5)
1. $L_1 = L_0 + S_0$; $R_1 = R_0 + S_1$;
2. for $i = 1$ to h do: $L_{i+1} = R_i$; $R_{i+1} = ((L_i \oplus R_i) \lll R_i) + S_{i+1}$.

We make use of the χ^2-tests for distinguishing a random sequence from non-random sequence ([5,7,8]). Let $X = X_0, ..., X_{n-1}$ be a sequence with $\forall X_i \in \{a_0, \cdots, a_{m-1}\}$. Let $N_{a_j}(X)$ be the number of X_i which equals a_j. The χ^2-statistic of X, $\chi^2(X)$, estimates the difference between X and the uniform distribution as follows: $\chi^2(X) = \frac{m}{n} \sum_{i=0}^{m-1} \left(N_{a_i}(X) - \frac{n}{m}\right)^2$. We use the threshold for 31 degrees of freedom in Table 2. For example, (level, χ^2)=(0.95, 44.99) in Table 2 means that the value of χ^2-statistic exceeds 44.99 in the probability of 5% if the observation X is uniform. Here, we set the level to 0.95 in order to distinguish a sequence X from a random sequence.

Table 2. χ^2-distribution with 31 degree of freedom

Level	0.50	0.60	0.70	0.80	0.90	0.95	0.99	0.999	0.9999
χ^2	30.34	32.35	34.60	37.36	41.42	44.99	52.19	61.10	69.11

In our experiments, all plaintexts are generated by using m-sequence ([9]). For example, Main or Extended algorithm uses 59-bit or 64-bit random number generated by m-sequence, respectively. The platforms are IBM RS/6000 SP (PPC 604e/332MHz $\times$ 256) with memory of 32 GB.

3   Applying Knudsen-Meier's Correlation Attack to RC5

Knudsen and Meier ([7]) proposed a key-recovery attack to RC6, which estimates a subkey from the 1st round to the r-th round by handling a plaintext. Their main idea is to choose such a plaintext that the least significant five bits in the first and third words are constant after one-round encryption. Therefore plaintexts in RC6 are chosen as follows: 1. the least significant five bits in the first and third words are zero; 2. the fourth word is set to the values that introduce a zero rotation in the 1st round. To sum up, their attack controls a plaintext in two parts with 5 bits: 5 bits corresponding to the χ^2-test and 5 bits in relation to data dependent rotations. Let us apply their idea to RC5 directly.

Algorithm 2 (Knudsen-Meier's attack to RC5)
This algorithm recovers $lsb_5(S_1)$. Set $s = lsb_5(S_1)$, and $lsb_5(R_0) = x$.
1. For each $s(s = 0, 1, \cdots, 31)$, set such an x that leads to a zero rotation in the 1-st half-round, that is, set $x + s = 0 \pmod{32}$.
2. Choose plaintexts (L_0, R_0) with $(lsb_5(L_0), lsb_5(R_0)) = (0, x)$, and set $y = lsb_5(L_{h+1})$
3. For each (L_0, R_0), update each array by incrementing $count[s][y]$.
4. For each s, compute the χ^2-value $\chi^2[s]$, and output s with the highest value $\chi^2[s]$ as $lsb_5(S_1)$.

Table 3. Success probability of Algorithm 2 (in 100 trials)

4 half-rounds			6 half-rounds			8 half-rounds		
#texts	#keys	χ^2-value*	#texts	#keys	χ^2-value*	#texts	#keys	χ^2-value*
2^{10}	25	530.12	2^{13}	20	139.59	2^{17}	13	69.56
2^{19}	28	256034.12	2^{19}	31	6639.60	2^{21}	26	443.06
2^{23}	28	4097164.86	2^{23}	32	105660.92	2^{26}	29	13303.08

* The average of the maximum χ^2-value in each trial.

Table 3 shows the experimental results of Algorithm 2. From Table 3, we see that a correct key can not be efficiently recovered, even though the maximum χ^2-value is enough high. Apparently, plaintexts with a zero rotation in the 1st round outputs the high χ^2-value, but a small-absolute-value one such as $\pm 1, \pm 2$, etc., also outputs the high χ^2-value. Since Algorithm 2 uses only plaintexts with a zero rotation in the 1st round, it suffers from other high-χ^2-value plaintexts, and thus cannot recover keys efficiently. Algorithm 2 is distinguishable, but unrecoverable. Furthermore, Algorithm 2 can use only 2^{54} plaintexts. In general, the number of plaintexts on RC5 is not so large as RC6. Another problem may occur in recovering other bits of S_1 because the rotation in the 1st half-round is determined only by $lsb_5(S_1)$. In RC6, the rotation in the 1st half-round is determined by all bits of S_1, and thus their algorithm can work well to recover all bits of S_1. This is why it is not efficient to apply their attack to RC5.

4 χ^2-Statistic of RC5

In this section, we investigate how to reduce the constraint of plaintexts in order to increase available plaintexts. In RC5, $lsb_5(R_0)$ determines the 1st half-round data dependent rotation, so it would be desirable to handle $lsb_5(R_0)$ in some way. On the other hand, the effect of $lsb_5(L_0) = 0$ deeply depends on $lsb_5(R_0)$ as follows: 1. if $lsb_5(R_0)$ is fixed to a value that leads a zero rotation in the 1st half-round like Algorithm 2, then $lsb_5(L_0) = 0$ can fix $lsb_5(R_2)$, that is, fix the rotation amount of the 2nd half-round, and can also fix $lsb_5(L_3)$ for any available plaintext; 2. if $lsb_5(R_0)$ is fixed to just 0 ([3]), then $lsb_5(L_0) = 0$ can not fix $lsb_5(R_2)$ (i.e. $lsb_5(L_3)$) for any available plaintext. We experimentally compare the effect of $lsb_5(L_0) = 0$, $lsb_5(R_0) = 0$, or both. We also investigate which parts output the higher χ^2-statistics. To observe these, we conduct the following five experiments in each h half-round.

Test 1: χ^2-test on $lsb_5(L_{h+1})$ with $lsb_5(R_0) = lsb_5(L_0) = 0$.
Test 2: χ^2-test on $lsb_5(L_{h+1})$ with $lsb_5(R_0) = 0$.
Test 3: χ^2-test on $lsb_5(L_{h+1})$ with $lsb_5(L_0) = 0$.
Test 4: χ^2-test on $lsb_5(L_{h+1})$ with $lsb_5(R_0) = x$ $(x = 0, 1, ..., 31)$.
Test 5: χ^2-test on any consecutive 5bits of L_{h+1} with $lsb_5(R_0) = 0$.

4.1 Test 1, 2, and 3

Here we show the experimental results of Test 1, 2, and 3 after discussing the differences among these Tests. The condition of $lsb_5(R_0) = 0$ means that the rotation amount of the 1st half-round is fixed. The purpose of Tests 1 and 2 is to observe the effect of handling a plaintext in the part corresponding to the χ^2-test: Test 1 handles it; and Test 2 does not handle it. On the other hand, Test 3 sets only $lsb_5(L_0) = 0$, so it cannot control the rotation amount of the 1st half-round at all.

Table 4 shows the experimental results of Test 1, 2, and 3 which represent the number of plaintexts required for χ^2-value exceeding 44.99. These tests are computed to the second decimal place, and the χ^2-value is computed on the average of 100 different keys. From Table 4, we see that the χ^2-value in Test 3 is much lower than that in Test 1, and also lower than that in Test 2. Test 3 requires about $2^5(2^3)$ times as many plaintexts as Test 1(Test 2) in order to get the same effect as Test 1(Test 2). As a result, we see that fixing the rotation amount of the 1st half-round, that is, $lsb_5(R_0) = 0$, causes highly nonuniform distribution, and that Test 3 has no advantage to both Tests 1 and 2.

Next we focus on the effect of $lsb_5(L_0) = 0$ under $lsb_5(R_0) = 0$. From Table 4, we see that in each half-round, the χ^2-value in Test 1 is higher than that in Test 2, but that almost the same effect of Test 1 is expected in Test 2 if we use about 2^2 times plaintexts as many as Test 1. The number of plaintexts required for χ^2-value exceeding 44.99 on h half-rounds, $log_2(\#text)$, is estimated

$$log_2(\#text) = 3.03h - 3.21 \text{ (Test 1)}, \quad log_2(\#text) = 2.93h - 0.73 \text{ (Test 2)}$$

Table 4. #texts required for χ^2-value > 44.99 in Test 1, 2 and 3 (on the average of 100 keys)

#half-rounds	4	5	6	7	8	9	10
Test 1($\log_2(\#texts)$)	9.08	11.77	14.92	18.05	20.93	24.36	26.98
Test 2($\log_2(\#texts)$)	10.94	14.05	16.83	19.84	22.79	25.74	28.57
Test 3($\log_2(\#texts)$)	14.26	17.26	19.62	23.14	25.75	—	—

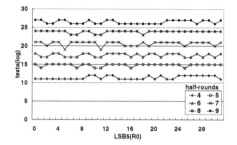

Fig. 1. #texts required for χ^2-value > 44.99 in each $lsb_5(R_0)$ (on the average of 100 keys)

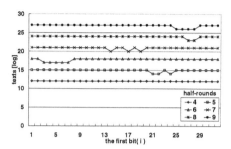

Fig. 2. #texts required for χ^2-value > 44.99 in each consecutive 5 bits of L_{h+1} (on the average of 100 keys)

by using the least square method. On the other hand, the number of available plaintexts in Test 1(Test 2) is $2^{54}(2^{59})$. By substituting the number of available plaintexts, we conclude that the case of Test 1, or Test 2 is estimated to be distinguishable from a random sequence by 18 half-rounds, or 20 half-rounds, respectively. As a result, Test 2 is more advantageous than Test 1.

4.2   Test 4 and 5

We observe the experimental results of Test 4 in Fig. 1. As we have discussed the above, setting $lsb_5(R_0) = 0$ means to fix the rotation amount in the first round. Note that the rotation amount is not necessarily equal to 0. Therefore, the same effect as $lsb_5(R_0) = 0$ would be expected if only fixing $lsb_5(R_0)$. Test 4 examines the hypothesis. In Fig. 1, the horizontal line corresponds to the fixed value of $lsb_5(R_0)$ and the vertical line corresponds to the number of plaintexts required for χ^2-value exceeding 44.99. From Fig. 1, we see that any $lsb_5(R_0)$ can be distinguished from a random sequence in almost the same way as $lsb_5(R_0) = 0$. To sum up, we do not have to set $lsb_5(R_0) = 0$ in order to increase the χ^2-value. We can use any plaintext (L_0, R_0) with any R_0 by just classifying it into the same $lsb_5(R_0)$.

We observe the experimental results of Test 4 in Fig. 1. In Test 5, we compute the χ^2-value in each consecutive 5 bits of L_{h+1}, and how many plaintexts are required in order to exceed the threshold χ^2-value of 44.99. Fig. 2 shows the experimental results. The horizontal line corresponds to the first bit of consecutive 5 bits of L_{h+1}, and each plot presents the number of plaintexts required for χ^2-value exceeding 44.99 for each consecutive 5 bits. For example, the case

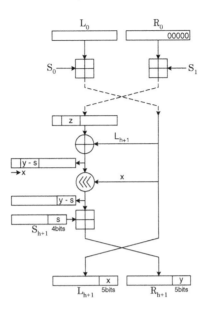

Fig. 3. Outline of Main algorithm

of $i = 1$, or $i = 32$ corresponds to $L_{h+1}^{[5,1]}$, or $\{L_{h+1}^{32}, L_{h+1}^{[4,1]}\}$. From Fig. 2, we see that any consecutive five bits can be distinguished from a random sequence in almost the same way as $L_{h+1}^{[5,1]}$. Correlations are observed on any consecutive five bits of L_{h+1}.

5   A Chosen Plaintext Correlation Algorithm

In this section, we present a key recovery algorithm, called Main algorithm. Main algorithm is designed by making use of the results of tests in Sect. 4 as follows:

1. Only $lsb_5(R_0)$ is fixed to 0 (**Test 1, 2**);
2. The parts measured by χ^2-statistic are not fixed to $lsb_5(L_{h+1})$ (**Test 5**);
3. The χ^2-value is computed on z to which consecutive 5 bits, y, is exactly decrypted by 1 half-round (see Fig. 3);
4. The decrypted, z, is classified into 32 cases according to $lsb_5(L_{h+1}) = x$, and the χ^2-value is computed on each distribution of z for each $lsb_5(L_{h+1}) = x$.

Algorithm 3 (Main algorithm)
This algorithm recovers $lsb_4(S_{h+1})$. Set $(lsb_5(L_{h+1}), lsb_5(R_{h+1})) = (x, y)$, and $lsb_4(S_{h+1}) = s$, where x is the rotation amount in the h-th half-round.
1. Choose a plaintext (L_0, R_0) with $lsb_5(R_0)$=0, and encrypt it.
2. For each $s(s = 0, 1, \cdots, 15)$, set $S_{h+1}^5 = 0$, and decrypt R_{h+1} by 1 half-round. Note that, from the rotation amount x in the h-th half-round, we exactly know where y is decrypted by 1 half-round, which is set to z.

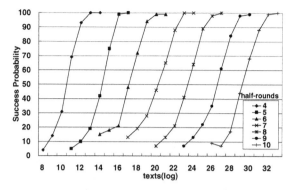

Fig. 4. Success probability of Main algorithm (in 100 trials)

3. For each value s, x, and z, we update each array by incrementing $count[s][x][z]$.
4. For each s and x, compute $\chi^2[s][x]$.
5. Compute the average $ave[s]$ of $\{\chi^2[s][x]\}$ for each s, and output s with the highest $ave[s]$ as $lsb_4(S_{h+1})$.

Main algorithm computes the χ^2-value on z, to which y is decrypted by the final round subkey. Therefore, the χ^2-value in $lsb_5(S_{h+1}) = 1s$ is coincident with that in $lsb_5(S_{h+1}) = 0s$ in the following reason. For $s = lsb_4(S_{h+1})$, we set two candidates of $lsb_5(S_{h+1})$, $t = 1s$ and $t' = 0s$. So $t = t' + 16 \pmod{32}$. We also set each decrypted value of $y = lsb_5(R_{h+1})$ by using each key, t or t', to z or z', respectively. Then $z^{[4,1]} = z'^{[4,1]}$, and $z^5 = \overline{z'^5}$, and thus $z = z' + 16 \pmod{32}$. Two distribution of $count[t][x][z]$ and $count[t'][x][z']$ in Algorithm 3 satisfy

$$count[t][x][z] = count[t'][x][z' + 16 \pmod{32}].$$

So, we get $\chi^2[t][x] = \chi^2[t'][x]$ from the definition of χ^2-value. This is why χ^2-value in $S_{h+1}^{[5,1]} = 1s$ is coincident with that in $S_{h+1}^{[5,1]} = 0s$ for $s = S_{h+1}^{[4,1]}$.

Figure 4 and Table 5 show the success probability among 100 trials for RC5 with $4 - 10$ half-rounds. More precise experimental results are shown in Table 6. All experiments are calculated to the second decimal place. In our policy, we design all experiments as precisely as possible in order to estimate the efficiency of algorithm strictly. From Table 6, the number of plaintexts required for recovering a key in r rounds(h half-rounds) with the success probability of 90%, $log_2(\#text)$, is estimated[1],

---

[1] Our estimation is computed by using the results of 5-10 half-rounds except for 4 half-rounds because key recovery with 4 half-rounds depends deeply on the choice of S_0, and plaintexts R_0. The key recovery with 4 half-rounds examines a bias for distribution of consecutive 5 bits in L_4, which is coincident with that in L_2. A bias for distribution of L_2 depends only on the 2nd half-round operation, S_0, and plaintexts R_0. In fact, the coefficient of determination ([12]) for approximation polynomial including the result of 4 half-rounds is worse than that for approximation polynomial without the result of 4 half-rounds.

Table 5. Success probability of Main algorithm (in 100 trials)

4 half-rounds		5 half-rounds		6 half-rounds	
#texts	#keys	#texts	#keys	#texts	#keys
2^{10}	31	2^{14}	42	2^{17}	48
2^{11}	69	2^{15}	75	2^{18}	72
2^{12}	93	2^{16}	99	2^{19}	94

7 half-rounds		8 half-rounds		9 half-rounds		10 half-rounds	
#texts	#keys	#texts	#keys	#texts	#keys	#texts	#keys
2^{20}	46	2^{23}	41	2^{26}	35	2^{29}	44
2^{22}	88	2^{25}	89	2^{28}	84	2^{31}	88
2^{23}	100	2^{26}	98	2^{29}	98	2^{32}	99

Table 6. # texts required for recovering a key with the success probability 90% and 45% in Main algorithm

#half-rounds	4	5	6	7	8	9	10
$\log_2(\#text)$ (90%)	11.89	15.43	18.78	22.07	25.15	28.28	30.92
$\log_2(\#text)$ (45%)	10.76	13.94	17.23	19.96	23.22	26.59	29.30

$$log_2(\#text) = 6.23r + 0.07, \ (log_2(\#text) = 3.12h + 0.07),$$

by using the least square method. In the case of success probability of 45%, the number of required plaintexts is estimated as follows,

$$log_2(\#text) = 6.18r - 1.47 \ (log_2(\#text) = 3.09h - 1.47).$$

By substituting $log_2(\#text) = 59$, Main algorithm can break RC5-32 with 18 rounds with $2^{56.23}$ plaintexts with a probability of 90%. With a success probability of 30%, Main algorithm can break RC5-32 with 19 half-rounds by using $2^{57.24}$ plaintexts. From these results, we see that it is indispensable to increase available plaintexts in order to break RC5 with higher round.

6   A Known Plaintext Correlation Algorithm

In this section, we present a key recovery algorithm, called Extended algorithm, which applies Main algorithm to a known plaintext attack.

6.1   Extended Algorithm

Extended algorithm classifies any plaintext (L_0, R_0) into the same $lsb_5(R_0)$, and applies Main algorithm as follows.

Algorithm 4 (Extended algorithm)
This algorithm recovers $lsb_4(S_{h+1})$. Set $s = lsb_4(S_{h+1})$, and $x = lsb_5(L_{h+1})$ in the same way as Algorithm 3.

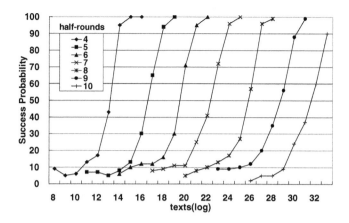

Fig. 5. Success probability of Extended algorithm (in 100 trials)

Table 7. Success probability of Extended algorithm (in 100 trials)

4 half-rounds		5 half-rounds		6 half-rounds	
#texts	#keys	#texts	#keys	#texts	#keys
2^{13}	43	2^{16}	30	2^{19}	30
2^{14}	95	2^{17}	65	2^{20}	71
2^{15}	100	2^{18}	94	2^{21}	95

7 half-rounds		8 half-rounds		9 half-rounds		10 half-rounds	
#texts	#keys	#texts	#keys	#texts	#keys	#texts	#keys
2^{22}	41	2^{25}	27	2^{28}	35	2^{31}	37
2^{23}	72	2^{26}	57	2^{30}	88	2^{32}	60
2^{24}	96	2^{27}	96	2^{31}	99	2^{33}	90

1. Given n known plaintexts (L_0, R_0), set $l = lsb_5(R_0)$, and encrypt it.
2. For each l, compute $\chi^2[l][s][x]$ according to Step 2-4 in Algorithm 3.
3. Compute the average $ave[l][s]$ of $\{\chi^2[l][s][x]\}_x$ for each s and each l.
4. Compute $sum[s] = \sum_{l=0}^{31} ave[l][s]$ for each s, and output s with the highest $sum[s]$ as $lsb_4(S_{h+1})$.

Figure 5 and Table 7 show that the success probability among 100 trials for RC5 with $4 - 10$ half-rounds. More precise experimental results are shown in Table 8. From Table 8, the number of plaintexts required for recovering a key on r rounds(h half-rounds) with the success probability of 90%, $log_2(\#text)$, is estimated,

$$log_2(\#text) = 6.14r + 2.27 \ (log_2(\#text) = 3.07h + 2.27),$$

by using the least square method. By substituting $log_2(\#text) = 64$, we conclude that our algorithm is estimated to recover a key on RC5 with 20 half-rounds

Table 8. # texts required for recovering a key with the success probability 90%, 45%, and 30% in Extended algorithm

#half-rounds	4	5	6	7	8	9	10
$log_2(\#text)$ (90%)	13.96	17.73	20.63	23.71	26.64	30.01	33.00
$log_2(\#text)$ (45%)	13.09	16.39	19.41	22.14	25.62	28.43	31.52
$log_2(\#text)$ (30%)	12.53	15.94	18.81	21.63	25.07	27.53	30.70

Table 9. The χ^2-value of correct keys and wrong keys in 4 half-rounds (in 100 trials)

	# texts	Key recovering probability	Correct keys Average	Variance	Wrong keys* Average	Variance
Main algorithm	2^{12}	93%	40.50	3.69	38.42	2.41
Extended algorithm	2^{14}	95%	32.31	0.10	32.08	0.09

* The highest χ^2-value among wrong keys is used for each trial.

with $2^{63.67}$ plaintexts in the success probability of 90%. In the case of success probability of 30%, the number of required plaintexts is estimated as follows,

$$log_2(\#text) = 5.90r + 1.12 \ (log_2(\#text) = 2.95h + 1.12).$$

Therefore, our algorithm can recover a key on RC5 until 21 half-rounds with $2^{63.07}$ plaintexts in the success probability of 30%.

6.2   Further Discussion

Here we discuss the difference between Extended algorithm and Main algorithm. Extended algorithm requires about 2^2 times as many plaintexts as Main algorithm in order to recover correct keys as we have seen in Table 7 and 5. Since all plaintexts in our experiments are randomly generated by m-sequences, $lsb_5(R_0)$ of plaintexts in Extended algorithm are roughly estimated to be uniformly distributed in $\{0, 1, \cdots, 31\}$. As a result, the χ^2-value in Extended algorithm is computed by using about 2^{-3} times as many plaintexts as Main algorithm because the χ^2-value in Extended algorithm is computed for each $lsb_5(R_0)$. We investigate experimentally the relation between χ^2-value of correct keys and that of wrong keys in both algorithms. Table 9 shows each average and variance in 100 keys. As for wrong keys, the highest χ^2-value among wrong keys is shown, which often causes to recover a wrong key. From Table 9, we see that the average among χ^2-value of correct keys in Extended algorithm is lower and the variance is much lower than that in Main algorithm. We expect that Extended algorithm reduces the variant of χ^2-value by using not specific $lsb_5(R_0)$ but all $lsb_5(R_0)$, and recovers a key with the lower χ^2-value. Thus, Extended algorithm can recover a key efficiently with the lower χ^2-value.

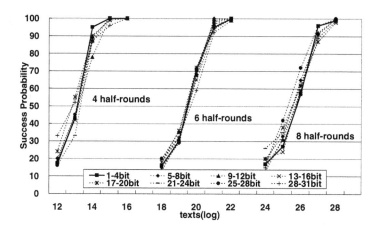

Fig. 6. Success probability of serial key recovery (in 100 trials)

7   31-Bit Key Recovery in the Final Round

Here discusses two algorithms that recover 31-bit-final-round subkey: the serial key recovery algorithm, and the parallel key recovery algorithm.

7.1   The Serial Key Recovery Algorithm

The serial key recovery algorithm recovers each 4 bits sequentially from $S_{h+1}^{[4,1]}$ to $S_{h+1}^{[28,25]}$ and $S_{h+1}^{[31,28]}$ by using Algorithm 4. For example, in the case of recovering $S_{h+1}^{[8,5]}$, we set $S_{h+1}^{[4,1]}$ to the value recovered before and apply Algorithm 4 by setting $s = S_{h+1}^{[8,5]}$ and $y = R_{h+1}^{[9,5]}$. As for the final 3 bits, $S_{h+1}^{[31,29]}$, we apply Algorithm 4 by using $s = S_{h+1}^{[31,28]}$ and recover $S_{h+1}^{[31,28]}$. After repeating the above procedures by eight times, $S_{h+1}^{[31,1]}$, that is, all bits of S_{h+1} except for MSB are recovered. The experimental results of serial key recovery are shown in Fig. 6. Compared with Fig. 6 and Table 7, we see that any key of any interval from $S_{h+1}^{[31,28]}$ to $S_{h+1}^{[8,5]}$ can be recovered with almost the same high probability as $S_{h+1}^{[4,1]}$. In the case of 4 half-rounds, 6 half-rounds, or 8 half-rounds, we can recover $S_5^{[31,1]}$, $S_7^{[31,1]}$, or $S_9^{[31,1]}$ with a success probability of about 99%, 97%, and 92% by using about 2^{15}, 2^{21}, or 2^{27} plaintexts on the average, respectively.

7.2   The Parallel Key Recovery Algorithm

We have seen that the serial key recovery algorithm can recover the final half-round key $S_{h+1}^{[31,1]}$ with the significantly high success probability. However, unfortunately, the serial key recovery algorithm can not work in parallel. This section investigates how to recover each subkey of $S_{h+1}^{[31,28]}$,..., $S_{h+1}^{[4,1]}$ in parallel. Before showing our parallel key recovery algorithm, we conduct the next experiment.

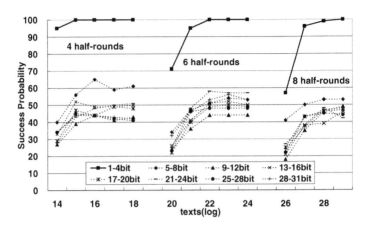

Fig. 7. Success probability of Test 6 (in 100 trials)

Test 6: Apply Algorithm 4 to $S_{h+1}^{[4+4i,1+4i]}$ ($i = 0, 1, \cdots, 6$) or $S_{h+1}^{[31,28]}$ by setting lower bits of S_{h+1} than $S_{h+1}^{[4+4i,1+4i]}$ or $S_{h+1}^{[31,28]}$ to 0. Compute the probability with which a correct key can be recovered.

Figure 7 shows the experimental results in Test 6. The result of $S_{h+1}^{[4,1]}$ in Test 6 is the same as that in Extended algorithm (Table 7). Figure 7 shows that Test 6 suffers completely from error bridging of lower bits. More importantly, the probability of Test 6 converges to about 50%: however many plaintexts are used, the probability has never become higher than an upper bound. From this, we put forward a hypothesis that some specific keys are not recovered. In fact, we see experimentally that, in the case of recovering keys of $S_{h+1}^{[8,5]}$, $S_{h+1}^{[8,5]}$ with the lower bits $S_{h+1}^{[4,1]} = 8, \cdots, 15$ can not be almost recovered. Especially, any $S_{h+1}^{[8,5]}$ with $S_{h+1}^{[4,1]} = 13, \cdots, 15$ can not be recovered at all however many plaintexts are used. The success probability of recovering keys in $S_{h+1}^{[8,5]}$ deeply depends on the lower bits $S_{h+1}^{[4,1]}$.

For simplicity, let us investigate the case of recovering $S_{h+1}^{[8,5]}$. The same discussion also holds in other cases of $S_{h+1}^{[31,28]}, \cdots, S_{h+1}^{[12,9]}$. In Test 6, we set lower 4 bits than $S_{h+1}^{[8,5]}$ to be 0. To make the discussion clear, we denote the real value by $S_{h+1}^{[4,1]}$, and the assumed value by β. In Tests 6, $\beta = 0$. For any R_{h+1}, $(R_{h+1} - S_{h+1})^{[9,5]}$ is determined by $S_{h+1}^{[9,5]}$, $R_{h+1}^{[9,5]}$, and the bridging on $(R_{h+1}^{[4,1]} - S_{h+1}^{[4,1]})$, which is estimated by $(R_{h+1}^{[4,1]} - \beta)$. This is why key recovering is failed if and only if the bridging on $(R_{h+1}^{[4,1]} - \beta)$ is not coincident with that on $(R_{h+1}^{[4,1]} - S_{h+1}^{[4,1]})$. The probability that bridging on $(R_{h+1}^{[4,1]} - \beta)$ is not coincident with that on $(R_{h+1}^{[4,1]} - S_{h+1}^{[4,1]})$ is different for each $S_{h+1}^{[4,1]}$. For example, in the case of $S_{h+1}^{[4,1]} = 0$, a bridging on $(R_{h+1}^{[4,1]} - 0)$ is apparently coincident with

Table 10. Error-bridging $R_{h+1}^{[4,1]}$ and the probability for $S_{h+1}^{[4,1]}$ ($\beta = 0$)

$S_{h+1}^{[4,1]}$	0	1	2	3	4	5	6	7
error-bridging $R_{h+1}^{[4,1]}$	-	0	0,1	0,1,2	$0,\cdots,3$	$0,\cdots,4$	$0,\cdots,5$	$0,\cdots,6$
Probability	0	1/16	2/16	3/16	4/16	5/16	6/16	7/16

8	9	10	11	12	13	14	15
$0,\cdots,7$	$0,\cdots,8$	$0,\cdots,9$	$0,\cdots,10$	$0,\cdots,11$	$0,\cdots,12$	$0,\cdots,13$	$0,\cdots,14$
1/2	9/16	10/16	11/16	12/16	13/16	14/16	15/16

Table 11. Success probability of each key recovering in $S_{h+1}^{[8,5]}$ with each error-bridging probability ($\beta = 0$, in 100 keys)

Error-bridging probability	$S_{h+1}^{[4,1]}$	4 half-rounds			6 half-rounds			8 half-rounds		
		2^{14}	2^{17}	2^{20}	2^{20}	2^{23}	2^{26}	2^{26}	2^{29}	2^{32}
$< 1/2$	$0,\cdots,7$	33/55	55/55	55/55	30/51	51/51	51/51	31/51	50/51	51/51
$\geq 1/2$	$8,\cdots,15$	7/45	4/45	5/45	4/49	3/49	2/49	10/49	3/49	2/49
$\geq 13/16$	$13,\cdots,15$	0/13	0/13	0/13	0/23	0/23	0/23	0/22	0/22	0/22

that on $(R_{h+1}^{[4,1]} - S_{h+1}^{[4,1]})$ for any $R_{h+1}^{[4,1]}$. Here, $R_{h+1}^{[4,1]}$ is called an error-bridging for $S_{h+1}^{[4,1]}$ if the bridging on $(R_{h+1}^{[4,1]} - \beta)$ is different with that on $(R_{h+1}^{[4,1]} - S_{h+1}^{[4,1]})$. For each $S_{h+1}^{[4,1]}$, we compute $R_{h+1}^{[4,1]}$ that is an error-bridging, and the error-bridging probability,

$$\frac{\#\{R_{h+1}^{[4,1]} \ni t \mid t \text{ is an error-bridging.}\}}{\#\{R_{h+1}^{[4,1]}\}}.$$

Table 10 shows the results. From Table 10, in the case of $S_{h+1}^{[4,1]} = 8, \cdots, 15$, a bridging on $(R_{h+1}^{[4,1]} - 0)$ is not coincident with that on $(R_{h+1}^{[4,1]} - S_{h+1}^{[4,1]})$ with the probability of $1/2$ and over. In these keys, the χ^2-value is computed by using invalid value with the probability of $1/2$ and over. Therefore, it is expected that recovering such keys is difficult even if many plaintexts are used. To observe this, we conduct an experiment on the relation between the success probability of a key recovering and the error-bridging probability. Table 11 shows the success probability of keys with the error-bridging probability of less than $1/2$, that of $1/2$ or above, and that of $13/16$ or above. In Table 11, we see that: 1. keys with the error-bridging probability of less than $1/2$ can be recovered correctly by using enough many plaintexts; 2. keys with the error-bridging probability of $1/2$ and over cannot almost be recovered correctly even if many plaintexts are used; 3. any key with the error-bridging probability of $13/16$ and over cannot be recovered correctly. From these observation, we estimate the lower bound of probability to recover a correct key, $\Pr(\beta)$, as the probability of keys with the error-bridging-probability of less than $1/2$,

Table 12. Error-bridging $R^{[4,1]}_{h+1}$ and the probability for $S^{[4,1]}_{h+1} (\beta = 8)$

$S^{[4,1]}_{h+1}$	0	1	2	3	4	5	6	7
error-bridging $R^{[4,1]}_{h+1}$	$0,\cdots,7$	$1,\cdots,7$	$2,\cdots,7$	$3,\cdots,7$	$4,\cdots,7$	5, 6, 7	6, 7	7
Probability	1/2	7/16	6/16	5/16	4/16	3/16	2/16	1/16

8	9	10	11	12	13	14	15
–	8	8, 9	8, 9, 10	$8,\cdots,11$	$8,\cdots,12$	$8,\cdots,13$	$8,\cdots,14$
0	1/16	2/16	3/16	4/16	5/16	6/16	7/16

Table 13. Success probability of each key recovering in $S^{[8,5]}_{h+1}$ with each error-bridging probability ($\beta = 8$, in 100 keys)

Error-bridging probability	$S^{[4,1]}_{h+1}$	4 half-rounds			6 half-rounds			8 half-rounds		
		2^{14}	2^{17}	2^{20}	2^{20}	2^{23}	2^{26}	2^{26}	2^{29}	2^{32}
$< 1/2$	$1,\cdots,15$	55/92	92/92	92/92	37/92	90/92	91/92	40/95	93/95	95/95
$1/2$	0	4/8	3/8	4/8	3/8	4/8	3/8	1/5	4/5	5/5

$$Pr(\beta) = \frac{\#\{S^{[4,1]}_{h+1} \ni t \mid \text{the error-bridging probability of } t < 1/2\}}{\#\{S^{[4,1]}_{h+1}\}}.$$

We get $Pr(0) = 1/2 = 0.50$, which reflects the experimental results in Fig. 7 and Table 11.

In order to improve the parallel attack, we have searched all available $\beta(0 \le \beta \le 15)$ to find β with the maximum $Pr(\beta)$. The maximum $Pr(\beta)$ is $15/16 = 0.9375$, which is given by $\beta = 7, 8$. We have also investigated a type of two-valued β such as

$$\beta = \begin{cases} 4 & \text{if } R^{[4,1]}_{h+1} < 8, \\ 11 & \text{otherwise.} \end{cases}$$

However, even in this type, the maximum $Pr(\beta)$ is $15/16$. There are total 256 kinds of β including two-valued, out of which 87 kinds give the maximum $Pr(\beta)$.

We discuss the improved parallel attack by using $\beta = 8$, in which $Pr(8)$ is just $15/16$. Table 12 shows error-bridging $R^{[4,1]}_{h+1}$ and the error-bridging probability for each $S^{[4,1]}_{h+1}$. Table 13 shows the success probability of keys with the error-bridging probability of less than $1/2$, and that of keys with the error-bridging probability of $1/2$. From Table 13, in the same way as $\beta = 0$, keys with the error-bridging probability of less than $1/2$ can be recovered correctly by using enough many plaintexts. More precise experimental results are shown in Table 14 and Fig. 8. From Table 14, we see that the parallel algorithm can recover a 31-bit key with the success probability of about 90% by using roughly the same plaintexts as that of about 50% in Test 6, and about twice as many plaintexts as that of the serial algorithm (Fig. 6).

Table 14. Success probability of improved parallel key recovery (the average of 7*100 trials of $S_{h+1}^{[31,28]}, S_{h+1}^{[28,25]}, \cdots, S_{h+1}^{[8,5]}$ (100 trials an interval))

4 half-rounds		6 half-rounds		8 half-rounds	
#texts	#keys	#texts	#keys	#texts	#keys
2^{13}	30.1	2^{19}	18.1	2^{25}	20.7
2^{14}	59.1	2^{20}	40.7	2^{26}	42.4
2^{15}	88.6	2^{21}	77.3	2^{27}	71.6
2^{16}	95.7	2^{22}	93.1	2^{28}	90.9
2^{17}	97.1	2^{23}	97.1	2^{29}	95.3

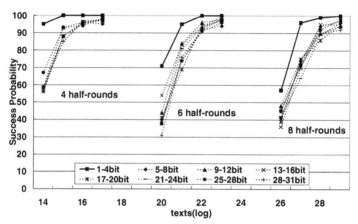

Fig. 8. Success probability of improved parallel key recovery (in 100 trials)

8   Conclusions

In this paper, we have proposed a known plaintext correlation attack on RC5. Our attack can break RC5-32/r with a success probability of 90% by using $2^{6.14r+2.27}$ plaintexts. Therefore, our attack can break RC5-32 with 20 half-rounds(10 rounds) by $2^{63.67}$ plaintexts. In a success probability of 30%, our attack can break RC5-32 with 21 half-rounds by using $2^{63.07}$ plaintexts.

We have also shown a parallel key recovery algorithm which recover 31-bit subkey of the final half-round on RC5. A parallel key recovery algorithm can work in parallel, and recover 31-bit keys with the high success probability.

References

1. A. Biryukov, and E. Kushilevitz, "Improved Cryptanalysis of RC5", *Advances in Cryptology-Proceedings of EUROCRYPT'98*, Lecture Notes in Computer Science, **1403**(1998), Springer-Verlag, 85-99.
2. J. Borst, B. Preneel, and J. Vandewalle, "Linear Cryptanalysis of RC5 and RC6", *Proceedings of Fast Software Encryption*, Lecture Notes in Computer Science, **1636**(1999), Springer-Verlag, 16-30.

3. J. Hayakawa, T. Shimoyama, and K. Takeuchi, "Correlation Attack to the Block Cipher RC5 and the Simplified Variants of RC6", submitted paper in Third AES Candidate Conference, April 2000.

4. B. Kaliski, and Y. Lin, "On Differential and Linear Cryptanalysis of the RC5 Encryption Algorithm", *Advances in Cryptology-Proceedings of CRYPTO'95*, Lecture Notes in Computer Science, **963**(1995), Springer-Verlag, 171-184.

5. J. Kelsey, B. Schneier, and D. Wagner, "Mod n Cryptanalysis, with applications against RC5P and M6", *Proceedings of Fast Software Encryption*, Lecture Notes in Computer Science, **1636**(1999), Springer-Verlag, 139-155.

6. L. Knudsen, and W. Meier, "Improved Differential Attacks on RC5", *Advances in Cryptology-Proceedings of CRYPTO'96*, Lecture Notes in Computer Science, **1109**(1996), Springer-Verlag, 216-228.

7. L. Knudsen, and W. Meier, "Correlations in RC6 with a reduced number of rounds", *Proceedings of Fast Software Encryption*, Lecture Notes in Computer Science, Springer-Verlag, to appear.

8. D. Knuth, *The art of computer programming, vol. 2, Seminumerical Algorithms*, 2nd ed., Addison-Wesley, Reading, Mass. 1981.

9. A. Menezes, P. C. Oorschot and S. Vanstone, *Handbook of applied cryptography*, CRC Press, Inc., 1996.

10. R. Rivest, M. Robshaw, R. Sidney and Y. Yin, "The RC6 Block Cipher. v1.1", 1998.

11. R. Rivest, "The RC5 Encryption Algorithm", *Proceedings of Fast Software Encryption*, Lecture Notes in Computer Science, **1008**(1995), Springer-Verlag, 86-96.

12. S. Shirohata, *An introduction of statistical analysis*, Kyouritu Syuppan, 1992, (in Japanese).

Micropayments Revisited

Silvio Micali and Ronald L. Rivest

Laboratory for Computer Science, Massachusetts Institute of Technology,
Cambridge, MA 02139
rivest@mit.edu, silvio@tiac.net

Abstract. We present new micropayment schemes that are more efficient and user friendly than previous ones.
These schemes reduce bank processing costs by several orders of magnitude, while preserving a simple interface for both users and merchants. The schemes utilize a probabilistic deposit protocol that, in some of the schemes, may be entirely hidden from the users.

1 Introduction

A payment scheme consists of a set of protocols involving (at least) three basic parties: the buyer or user, the merchant, and the bank. These could be individual entities – such people, devices, computer programs – or collections of entities[1].

ELECTRONIC CHECKS. The simplest form of payment scheme is an electronic check. Very informally, this consists of a check that is digitally signed rather hand-signed. In essence, a user pays a merchant for a transaction by digitally signing a piece of data that identifies the transaction (the user, the merchant, the "merchandise," the amount to be paid, the time, etc.) and possibly other data (such as account and credit information). The merchant deposits an electronic check by sending it to the bank. Having verified that the check is genuine and that has been deposited for the first time, the bank credits the merchant with the proper amount and charges the buyer with the same amount.

IT GOES WITHOUT SAYING. A few side details are supposed to be included in all payment schemes discussed herein. For instance, the parties' digital signature capabilities may supported by digital certificates (proving that they own their public keys, or that they are authorized to conduct electronic payments). These certificates can be sent alongside other messages, and their current validity may be verified before accepting an electronic payment. (See [13] for very efficient ways to verify certificates' current validity.)

When we say that the bank credits the merchant with x and charges the user with the same amount, we do not mean that these amounts are exactly equal: the bank may deduct fees from the merchant's credit or add fees to the

---

[1] Of course, in a payment system there may be a plurality of users and merchants, and there may be too a plurality of banks. Indeed, in each transaction the user may have his own bank and the merchant his own, separate bank.

B. Preneel (Ed.): CT-RSA 2002, LNCS 2271, pp. 149–163, 2002.
© Springer-Verlag Berlin Heidelberg 2002

user's charge. The bank may also impose entry or subscription fees to users and merchants for their participation in the payment system.

A payment scheme per se does not provide the assurance that the merchant will deliver the relevant "merchandise" (goods, services, information, etc.). Other schemes such as that of [12] may guarantee that a "fair exchange" takes place. Also, a payment scheme per se does not guarantee that the user has enough money to pay. A separate mechanism may insure the merchant against this risk (e.g., the bank may guarantee payment if the user had a valid certificate).

Anonymity is not explicitly added to our payment schemes, though it can be added to them[2].

MICROPAYMENTS. Payment schemes that emphasize the ability to make payments of small amounts are called *micropayment* schemes. Several micropayment schemes have been suggested, including *Millicent*[10] by Manasse et al., the *Payword* and *MicroMint*[16] schemes of Rivest and Shamir, Anderson's *NetCard*[2] scheme, Jutla and Yung's *PayTree*[7], Hauser et al.'s *Micro iKP*[5], the "Micropayment Transfer Protocol (MPTP)" of the W3C [4,18], the probabilistic polling scheme[6] of Jarecki and Odlyzko, the "Electronic Lottery Ticket" proposal[15] of Rivest, Wheeler's similar "Transactions Usings Bets"[19], Pedersen's similar scheme[14], and the related proposal for micropayments by efficient coin-flipping by Lipton and Ostrovsky[9] (this last paper includes an excellent survey of the field).

Applications of micropayments include paying for each web page visited, and for each minute of music or video as it is streamed to the user.

In principle, micropayments could be implemented by electronic checks. Merchant and user are otherwise engaged in the transaction, and the computation time they may devote to digital signing is not a real problem. The real problem lies with the *bank's processing cost* of any form of payment. Indeed, it seems likely that the cost of doing any transaction with the bank will be many times larger than the value of the micropayment itself. For example, processing a credit card transaction costs about 25 cents today, while a typical micropayment may be worth 1 cent. This processing cost seems unlikely to diminish dramatically in the near future.

Fortunately, while users and merchants are expected to be involved in every transaction, there is no essential reason why the bank must do work proportional to the number of transactions made. Micropayment schemes thus try to aggregate many small payments into fewer, larger payments, whose processing costs (by the bank) are relatively small.

1.1   Two Old Micropayment Schemes

Let us quickly recall two well-known micropayment schemes: (1) Rivest and Shamir's "PayWord" [16], and (2) Rivest's "electronic lottery tickets as micro-

---

[2] Payment schemes that emphasize user anonymity are often called *electronic coin* schemes. See Law et al. [8] for an excellent overview of electronic coin schemes.

payments" [15]. Indeed, our proposed micropayment schemes retain some of their ideas, and fix some of their drawbacks.

PAYWORD. Let H be a one-way function; that is, a function easy to evaluate but hard to invert. A user computes a "H-chain" consisting of values

$$x_0, x_1, x_2, \ldots, x_n$$

where

$$x_i = H(x_{i+1}) \quad \text{for } i = 0, 1, \ldots, n-1,$$

and commits to the entire chain by sending his signature of the root x_0 to the merchant. After that, each successive payment of the user is made by releasing the next consecutive value in the chain, which can be verified by checking that it hashes to the previous element. This allows the merchant to conveniently aggregate the buyer's payments as follows. Assume that the user has made i micropayments, and the merchants feels that, taken together, they constitute a sizable enough macropayment. Then, the merchant can make a single deposit for i cents by giving the bank only two values: x_i and the user's signature of x_0. The bank can verifies the user's signature of x_0 and iterates H on x_i i times, to verify that this operation yields x_0.

PAYWORD'S PROBLEM. PayWord suffers from a main problem:

> *A merchant cannot aggregate micropayments of different users.* Each user has established his own H-chain with the merchant, and there is no way of "merging" different chains. Thus, if a user spends only 1 cent with a given merchant, to deposit that cent the merchant or the bank would lose money, because the deposit's processing cost will exceed the payment value.

RIVEST'S LOTTERY SCHEME. In this scheme, there is a known *selection rate, s,* between 0 and 1. For each micropayment user and merchant interact according to a pre-determined protocol so as to *select* it with probability s: a non-selected micropayment is worthless and can be discarded, while a selected one can be deposited for an amount $1/s$ times bigger than the originally specified amount. For instance, if $s = 1/1000$ and each micropayment is for 1 cent, then on the average, out of 1000 micropayments, 999 will be discarded and 1 will be deposited for 1000 cents (i.e., for \$10), thus incurring only a single processing cost. This approach minimizes processing costs, while on the average every one pays and receives what he should.

A fast implementation of the lottery scheme also uses H-chains as in Pay-Word. In a first step, the merchant gives the user the root $w = w_0$ of a H-chain

$$w_0, w_1, w_2, \ldots, w_n \, .$$

where

$$w_i = H(w_{i+1}) \quad \text{for } i = 0, 1, \ldots, n-1,$$

In a second step, the user includes w into his commitment to x_0, and thus into his commitment to his own H-chain. Assuming that the selection rate is

1/1000, then the user's i-th is payment is selected if $(x_i \bmod 1000)$ is the same as $(w_i \bmod 1000)$. (A mathematically rigorous version of this approach can be found in [9].)

Note that, at the end of the selection protocol, the merchant learns whether a given micropayment has been selected. In principle therefore, he might deny giving the user the expected merchandise if the payment was not selected. But this cheating possibility is not worrysome: since we are dealing with a micropayment, the merchandise involved may be worth 1 cent; while the merchant's reputation and his ability to continue to conduct business through electronic micropayments is worth much more.

THE LOTTERY SCHEME'S PROBLEMS. The scheme suffers from two main problems:

1. *Interaction.* The user and the merchant must interact to select micropayments. This interaction slows down the whole process and makes it impractical for some applications.
2. *User risk.* The scheme burdens the user with the risk that he may have to pay more than he should. For instance, though the selection rate is 1/1000, 10 (rather than 1) of his first 1000 micropayments may be selected for a macropayment! This may be a rare problem, because the probability of its happening is small, and its relative impact decreases dramatically with the number of micropayments made. Nonetheless, it may constitute a strong psychological obstacle to the wide acceptance of the scheme, because ordinary users are not accustomed to managing risk.

The lottery scheme can also be implemented using an external entity. For instance, in the example above, the w_i may not be committed in advance by the merchant, but may consist of the winning numbers of the state lottery. This too, however, poses problems. Namely, the merchant may have to store a large number of payments, because he must wait for the next lottery outcome to determine which of them will be selected. In addition, the merchant may conspire with the external entity against the user. For instance, he may arrange for certain values w_i to "pop-up" so as to guarantee that a given user's payments are always selected. Alternatively, the user may conspire with the external entity against the merchant.

1.2   Three New Micropayment Schemes

We put forward three micropayment schemes that solve the above mentioned problems.

MR1 Scheme "MR1", presented in Section 2, improves on Rivest's lottery scheme by making it *non-interactive* while *allowing the merchant to learn immediately whether or not a check is selected for payment.* We achieve non interaction by (among other ways) properly utilizing the merchant's public key in the payment protocol.

MR2 Scheme "MR2", presented in Section 3, solves the problem that probabilistic schemes such as Rivest's lottery scheme and MR1 may – by bad luck
– charge the user more than the total value of the checks he has written. We
solve this problem by modifying the charging protocol to depend properly
on the serial numbers of the user's checks.

MR3 Scheme "MR3", presented in Section 4, is a variant which trades off the
immediacy (for the merchant) of finding out which checks are payable in favor
of giving the bank greater control and flexibility over the deposit process.

In all our three schemes, each micropayment requires the computation of a
digital signature, just as for an electronic check. A few years ago, this computational requirement was significant. But digital signature technology has continued to improve, and signature schemes that are both more secure and significantly faster are currently available. Moreover, the computational cost of a
public-key signature has continued to decrease due to the deployment of more
powerful processors. Furthermore, we note that the use of "on-line/off-line" digital signatures (as proposed in [3] and recently improved in [17]) may be a good
choice for micropayment schemes[3]. In sum, we now feel free to utilize public-key
computations even in most micropayment schemes.

2   The MR1 Scheme

In this section we improve Rivest's lottery scheme. As before, payments will be
selected to be deposited according to a selection rate s, but the MR1 scheme
does not require interaction between user and merchant, and yet *the merchant
learns right away which payments are selected.*

These features make the MR1 scheme eligible for new uses. For example,
it could be used by a packet to pay for its bandwidth as it travels along a
precomputed route, where the intermediate routers have known public keys.
(Note that having the router interact with the user sending the packet is not a
realistic option!)

The idea is that when a user sends a (micro) check C to a merchant, C
will be selected for (a macro) deposit if a given property holds between C and
a unique quantity dependent on C that is easily computable by the merchant,
but unpredictable to the user. Therefore, when the user sends a check C to the
merchant, he has no idea whether C will be selected for deposit (and thus the
user cannot avoid a possible deposit).

Notice that our proposal differs significantly from Rivest's lottery scheme not
only at an implementation level, but also at a high level. In Rivest's case, a check

---

[3] Such digital signature are in fact computed in two steps: first an "off-line" step,
computationally more demanding but performed before the message to be signed
is chosen or available, and then a light-weight "on-line" step, performed once the
message to be signed becomes available. Thus, for example, we imagine that the
off-line step could be performed while a human user is deciding what to do next;
once she decides, the signature for the micropayment can be completed quickly.

sent by the user to the merchant is not determined to be payable or unpayable. Rather, its "payability" will be determined by the execution of the selection protocol that ensues, and thus by the random choices that merchant and user will make. In our case, instead, any check sent by the user to the merchant is already "pre-selected" for payability[4]. In some sense, each check carries its own special mark, "payable" or "unpayable", so as to guarantee the following three properties (1) the fraction of checks marked payable is approximately s, (2) the user cannot read the mark of the checks he sends, (3) the merchant can read the mark and can make it visible to others.

Let us first describe our preferred, practical implementation of the MR1 scheme, and then some of its variants, including a theoretical one for which a rigorous proof of correctness could be provided.

2.1   The Preferred Embodiment of MR1

PRELIMINARIES. Let T denote (the encoding of) a transaction. We assume that T specifies all quantities deemed useful – such as the user, the merchant, the bank, the "merchandise", the transaction time, the transaction's (monetary) value, etc.

For simplicity, we assume that each transaction has a fixed value (equal to 1 cent) and that there is a fixed *selection rate*, denoted by s. (I.e., s is the fraction of payments that is expected to be selected for deposit.)

We let $F(\cdot)$ denote a fixed public function that takes arbitrary bit strings as input and returns as output a number between 0 and 1, inclusive. For example, F might operate by taking the input string e and pre-pending a zero and a point, and interpreting the result as a binary number, so that for example "011" becomes "0.011" which is interpreted as the number $3/8$. The function F might also apply a standard hash function to the input as an additional first step.

If X is a party and Y a message, we denote by $SIG_X(Y)$ X's digital signature of Y. For simplicity, we assume that each message is one-way hashed prior to being signed and is explicitly included in its own signature[5].

THE PREFERRED MR1 SCHEME.

Set up: Each user and each merchant establishes his own public key (with the corresponding secret key) for a secure digital signature scheme. The merchant's digital signature scheme must be *deterministic*.

Payment: A user U pays a merchant M for a transaction T (with selection rate s) by sending M the check $C = SIG_U(T)$.

Check C is actually *payable* if $F(SIG_M(C)) < s$. If C is payable, then M sends bank B both C and $SIG_M(C)$ for deposit. (Out of courtesy, M may inform U whether C was payable.)

---

[4] If you want, a check's probability of being selected depends on choices made by the merchant during a set-up phase – e.g., on which public signature key the merchant actually chose.

[5] For example, $SIG_X(Y)$ could be implemented as $(Y, SIG_X(H(Y)))$, where H is a fixed public one-way hash function.

Selective Deposit: If U's and M's signatures are correct, and C is a previously undeposited payable check, then B credits M's account with $1/s$ cents and debits U's account with the same amount (and may justify its action by providing U with $SIG_M(C)$).

BASIC PROPERTIES.

- *The set-up phase is simple and general.* There is no need of a separate set-up for each user-merchant pair as in PayWord.
- *The payment phase is non-interactive.* The user simply sends a signed message to the merchant, to which the merchant needs not to respond.
- *The selection rate is s.* In fact, $SIG_M(C)$ is a quantity unpredictable to U, because U does not know M's secret signing key. Thus, practically speaking, even if U may control C in any way he wants (e.g. by choosing the transaction T), $SIG_M(C)$ will essentially be a random number. Therefore, $F(SIG(C))$ is a random and long enough number between 0 and 1, and thus will be less than the selection rate s essentially for a fraction s of the checks C.
  (Note that for a reasonable selection rate, such as $1/1024$, it would be sufficient for $F(SIG_M(C))$ to be 10-bit long. A typical signature is instead hundreds of bits long, which is an "overkill.")
- *Bank B is called into action only for a fraction s of the micropayments, and once it acts it does so only on macropayments.* The merchant can immediately verify whether a check C is payable, because he can easily evaluate $F(SIG_M(C))$ and compare it to the selection rate. Thus, the every check forwarded by the merchant to the bank is payable, and each such check results in a "macropayment" because it has a selected value of $1/s$ cents. (e.g., if $s = 1000$, it has a value of \$10). Thus, the system generates only relatively negligible transaction costs.
- *No two parties can successfully cheat the third one.* Even with B's help, U cannot write a check that has "less than s chance" of being payable. Indeed, the value $SIG_M(C)$ is unpredictable to both B and U, even if they share information about $SIG_M(C')$ for any prior check C' and even if they jointly choose C. Similarly, M and B together cannot defraud U. Informally, once the public signature key of M is chosen in the set-up stage (by M alone or by M and B together), because the signature scheme is deterministic, there is only one possible value $SIG_M(C)$ for every check C, and thus no amount of conspiracy may change that value. Moreover, when M's public key is chosen, M and B do not know what U's checks look like. Even if B and M were capable of guessing or controlling which transactions U will execute, they cannot choose M's public key so as to guarantee that U's checks will be payable with probability greater than s. In fact, U's check for a transaction T consists of $SIG_U(T)$, and is thus unpredictable to both M and B. Finally, notice that the bank cannot be defrauded in the sense that, each time that B pays M, B withdraws the same amount from U's account. (The problem that U may be unable to pay his checks arises also for ordinary checks, and is therefore independent of our cryptographic scheme and should be handled as usual.)

2.2   Variants of the Preferred Embodiment of MR1

THEORETICAL VARIANTS. The basic scheme can be modified a bit so as to formally achieve security.

In our analysis of the MR1 scheme, it is crucial that $F(SIG_M(C))$ be a ("random") number of sufficient precision unpredictable to the user. Whether this security condition holds may depend on the signature scheme and on the definition of F.

For example, consider an F that returns the binary fraction whose representation is the low-order 20 bits of its input. The security condition is immediately true if one models digital signature schemes as random oracles; it is however more traditional to model only one-way hash functions as random oracles. The condition nonetheless holds if the merchant digitally signs using a suitable signature scheme such as RSA. In fact RSA is a deterministic signature scheme and it has been proven in [1] that, relative to a randomly chosen RSA public key of L bits, the last $c \cdot \log L$ bits of the signature of a random message are computationally indistinguishable from a random $c \cdot \log L$-bit string, where c is any constant greater than 1. As a consequence of this result, if merchant M randomly chose his RSA keys as 1024-bit long strings, then, letting $c = 2$, the last 20 bits of $SIG_M(C)$ provide a 20-bit string Y essentially indistinguishable from a random 20-bit string, no matter how C was chosen. (Recall in fact that rather than signing C directly, merchant M actually signs $H(C)$. Modeling the one-way hash function H as a random oracle, $H(C)$ would be random even if C were specially selected.) Notice that using even as few as 20 bits for Y enables one to easily implement a selection rate as low as 2^{-20} and provides sufficient resolution for most purposes; with $s = 2^{-20}$ selected 1-cent micropayments are transformed into macropayments worth \$10,000 each, whose processing costs would be quite negligible.

The same crucial point can also be formally solved without recourse to any random oracle model. Namely, it would suffice for the merchant to use a verifiable random function (VRF) rather than an ordinary digital signature scheme. As introduced and exemplified by Micali, Rabin and Vadhan [11], a VRF comprises a pair of keys and a pair of algorithms: a public key PK, a matching secret key SK, an evaluation algorithm E, and a verification algorithm V. Key PK totally specifies a function F ($= F_{PK}$), from arbitrary bit strings to k-bit strings, such that it is hard to compute $F(x)$ on input x and PK. Key SK enables one to evaluate F easily; that is, on inputs x, PK and SK, E returns $F(x)$ together with a proof P_x that indeed $F(x)$ is the correct value of F at point x. Proof P_x is accepted by V on additional inputs PK and x. The crucial property of a VRF is that $F(x)$ is polynomial-time indistinguishable from a random k-bit string for any input x for which a proof P_x has not been seen. (This remains true even if one is allowed to request and obtain $F(x')$ and $P_{x'}$ for any input $x' \neq x$ of his choice.) Thus, in the MR1 scheme, the merchant can select his own PK and SK and establish PK as his public VRF key, so that a check C becomes payable if $F_{PK}(C) < s$. Note that the merchant can immediately determine whether C is payable, because he knows SK and thus easily evaluates F_{PK}. Moreover he can enable the bank to verify that C is payable by releasing the proof P_C.

As for a different technical point, the user and merchant may choose their public signature keys by means of a mutually independent commitment scheme.

PRACTICAL VARIANTS. Different variants are possible that maintain the same (non-interactive) spirit of the MR1 scheme. In particular,

- *Time.* The basic scheme allows a merchant to deposit a payable check at any time. However, the bank may refuse to credit the merchant's account during the deposit phase unless he presents a payable check which has a sufficiently correct time. (E.g., if the transaction T to which a check C refers happened in day i, then the merchant should deposit C within the end of day i, or by day $i + 1$.) This gives an extra incentive to the merchant to verify the time accuracy of the checks he receives (which he should do anyway). Indeed, if the time is wrong, he could refuse to provide "the merchandise" requested. Timely deposit ensures that the user is not charged "too late," when he has no longer budgeted for that possible expenditure.
- *Functions F and G.* The functions F and G may not be fixed, but vary. For instance, a check or a transaction may specify which F or G should be used with it.
- The check-payability condition, $F(SIG_M(C)) < s$, could be replaced by $F(SIG_M(G(C))) < s$, where G is a given function/algorithm. So, rather than signing C itself, the merchant may sign a quantity dependent on C, denoted by $G(C)$. In particular, because C is U's signature of a transaction T, and because we assume that such a signature also specifies T, $G(C)$ may be a function of T alone, for instance a substring of T, such as T's date/time information. As for another example, T may also specify a user-selected string W, preferably unique to the transaction and selected at random, and $G(C)$ may just consist of W (so that the merchant will sign W, or W together with time information).
- The check-payability condition may be chosen in a rather different way. For instance, a check C may be payable if a given property holds between C and a quantity dependent on C that is computable only by the merchant, such as the property that the last 10 bits of C (or some specific 10 bits of T) equal the last 10 bits of $SIG_M(G(C))$.
- To minimize the merchant's number of signatures, rather than using $F(SIG_M(G(C)))$ to determine check payability, one may use $F(SIG_M(G(V_i)))$, where $\{V_i\}$ is a sequence of values associated to a sequence of times. For instance, V_i is a daily value and specifies the day in question (e.g., $V_i = 02.01.01$, $V_{i+1} = 02.02.01$, etc.) and a check C relative to a transaction T on day i may be payable if $F(SIG_M(V_i)) < s$ (or if some other property holds between C and a quantity computable from V_i only by M, such as whether the last 10 bits of C – or some specific 10 bits in T – equal the last 10 bits of $SIG_M(V_i)$).
  Note that the merchant may evaluate $F(SIG_M(G(V_i)))$ at the beginning of day/time interval i, so that, upon receiving a check C on that day/time interval, M may immediately discard C if it is not payable, and set C aside for proper credit otherwise.

Note too that it is better in this variant for the merchant to hide all information about which checks he has discarded and which checks he has set aside for credit during a given day/time interval. Else malicious users may predict or infer somewhat $F(SIG_M(G(V_i)))$, and give M checks that are not payable or have less probability of being payable. For this reason, if the merchant uses a V_i-approach, we recommend that he stores all his payable checks of a given day/time interval, and then send all of them to the bank at the end the day/time interval. This way even a malicious bank cannot collude with a user so as to enable him to defraud the merchant.

– A special way to implement the above approach consists of utilizing a hash/ one-way function chain. That is, the merchant computes a sequence of values

$$x_0, x_1, x_2, \ldots, x_n$$

where

$$x_i = H(x_{i+1}) \quad \text{for } i = 0, 1, \ldots, n-1,$$

where H is a one-way function/hash, and puts x_0 in his public file, or otherwise publicizes x_0 (e.g., by steps that include digitally signing it). Then one can use x_i, rather than $F(SIG_M(G(V_i)))$ on day/time interval i.

– It is easy to extend the basic MR1 scheme to handle checks of different values; everything is simply scaled appropriately for each check.

3   The MR2 Scheme

Recall that Rivest's lottery scheme suffered from two problems: (1) interaction in the payment process, and (2) the possibility of user's excessive payments. The MR1 scheme solved the first problem, but did not address the second one. Of the two problems, we regard the first one to be a real one, and the second to be mostly a "psychological" one. Indeed, the possibility that the user may be debited substantially more than the micropayments he makes is very small, and will decrease with the number of micropayments made. Nonetheless, user acceptance is key to making micropayments widely used.

Accordingly, in this section we present a selective-deposit micropayment scheme that solves both problems. In particular, it guarantees that a honest user is never charged more than he actually spends. The small risk of excessive payment is shifted from the user to the bank. Note that this is much preferable for two reasons. First, as we said, excessive payment occurred only rarely (i.e., for few users) and in moderate amounts. Now if this may have bothered users, it will not bother banks who are actually accustomed to managing substantial risks, never mind the rare risk of a small excessive payment! Second, the relative risk becomes less and less probable in the long run, and thus is less probable for the bank, given that it will experience much higher volumes than a single user.

Another main attraction of the scheme is its extreme simplicity. Accordingly, rather than trying hard to prevent cheating, it simply punishes cheating parties, or purges them from the system before they can create any substantial damage.

PRELIMINARIES. We adopt the same simplifying assumptions and notations (about transactions, fixed monetary value, fixed selection rate, and digital signatures, etc.) as in the MR1 scheme.

THE BASIC SCHEME.

Set up: Each user and each merchant establishes his own public key (with the corresponding secret key) for a secure digital signature scheme; the merchant's signature scheme must be deterministic.

Payment: A user U pays a merchant M for a transaction T (with selection rate s) by sending M the check $C = SIG_U(T)$. The user includes the time and a serial number SN in every check/transaction. (The serial numbers should start at 1 and be assigned sequentially.)

Check C is actually *payable* if $F(SIG_M(C)) < s$. If C is payable, then M sends bank B both C and $SIG_M(C)$ for deposit.

Selective Deposit: Let $maxSN_U$ denote the maximum serial number of a payable check of U processed by B so far (initially $maxSN_U = 0$). Assume that C is a new, payable check, and that U's and M's signatures are correct. Then the bank B credits M's account with $1/s$ cents. Furthermore, if the serial number SN of the check is greater than $maxSN_U$, the bank B debits U's account by $SN - maxSN$ cents, and sets $MaxSN_U \leftarrow SN$ – and may justify its action by providing U with $SIG_M(C)$.

(An exception to the above rules is made if the bank notices that the new check has the same serial number as a previously processed check, or if the new check's serial number and time are "out of order" somehow with respect to previously processed checks, or if the amount of the check is excessive, or if other bank-defined conditions occur. In such exceptional cases the bank may fine the user and/or take other actions as it deems appropriate.)

Selective Discharge: The bank may keep statistics and throw out of the system (e.g., by revoking their certificates) users whose payable checks cause exceptions (as noted above) because they are inconsistently numbered and/or dated, or whose checks are "more frequently payable than expected." It may similarly throw out merchants with whom such problematic or "more frequently payable" checks are spent.

BASIC PROPERTIES. As for the MR1 scheme, the set-up phase is simple and general, the payment phase is non-interactive, and the selection rate is s.

Let us now argue that the scheme is fair for the honest user. At any time t, an honest user U has been charged $maxSN$ cents if $maxSN$ is the highest serial number of U's successfully deposited checks. Assume now that, by time t, U has made n transactions. Then, because an honest user numbers his checks sequentially starting with 1, n will also be the highest serial number of any check that U has written, and thus $maxSN \leq n$. That is, U will have been charged at most (rather than exactly) 1 cent per transaction.

Out of courtesy, M may inform U when a check was payable, but in this scheme it is preferable that M does not so inform the user, and certainly unnecessary for him to do so. It is better to keep the user ignorant of which serial numbers have turned out not to be payable. Note that the user's cumulative charges do *not* depend much on which checks turned out to be payable, but only on the number of checks he has written.

To incur lesser charges, a malicious user U' may try to lower artificially $maxSN$ by using twice at least one serial number SN. Thus U' can be caught by B in at least two ways: (1) two checks of U' are deposited whose serial numbers and times are inconsistent, or (2) two checks of U' with the same serial number are deposited[6]. Thus, if a suitably high fine or punishment is imposed on users caught cheating (something that is preferably agreed-upon beforehand), then cheating would be counterproductive.

A malicious user U', however, may collude with a malicious merchant M', so as to ensure that a check of U' spent with M' is always payable. Indeed, for each potential check C, M' can tell U' the value of $SIG_{M'}(C)$, so it is no longer unpredictable to U' whether the check would be payable. With a little trial and error, U' only writes payable checks. This way, U' will always pay just 1 cent to B, while B will always pay $1/s$ cents (i.e., \$10 if $s = 1/1000$) to M'. U' and M' may then share their illegal proceeds: indeed, U' may coincide with M' if he sets himself up as a merchant! Nonetheless, U' and M' may only make a modest illegal gain: if they try to boost it by repeating it several times, they are likely to be thrown out of the system. (This is a high price to pay, particularly if M' also has legitimate gains in the system.) If it is not easy for thrown-out users and merchants to come back in the system (e.g., under a new identity), or if the price to get into the system (e.g., that of getting an initial certificate) is sufficiently high, this illegal game pays little or even has negative returns to the user, and its cost may be easily absorbed by the bank.

A small probability exist that a honest user may look malicious because he makes n checks and significantly more than n/s of them become payable. In this case, he may be thrown out. With appropriate parameter settings, there will be very few such users. In addition, they can be convinced that they unintentionally caused losses to the bank (e.g., because the bank presents them with the relevant $SIG_M(U)$ values for their checks). Therefore, they may accept being kept on the system under different conditions – for instance, as users of an MR1 system; that is, they may agree to be debited 1000 cents, from then on, for each payable check. (Such a transition to an MR1 system might even be an automatic feature of the original agreement between the user and the bank.)

VARIANTS. In general, variants of the MR1 scheme apply here too.

We note that to handle checks of different values needs a bit of care. In general, a check worth v cents should be treated as a bundle of v one-cent checks (with consecutive serial numbers). A bit more efficiently, the user may write a single check that, rather than having a traditional serial number has a serial-number interval, $[SN, SN + v]$. We leave as an exercise how to modify the MR2 scheme (and its "penalty system") so t handle properly such checks.

---

[6] Note that way 1 can occur even if honest merchants check for the time accuracy of the checks they receive: there cannot be perfect accuracy. Note that way 2 may occur because U' does not control which of his checks spent with honest merchants become payable.

4   The MR3 Scheme

This scheme differs from both MR1 and MR2 in that the bank determines, probabilistically and fairly, which checks are payable. Again, the small risk of excessive payment is shifted from the user to the bank, which is accustomed to risk management. And again simplicity is a main attraction: rather than trying hard to prevent cheating, the bank simply punishes/eliminates cheating parties before they can create any substantial damage.

THE BASIC SCHEME.

Set up: Each user and each merchant establishes his own public key for a secure digital signature scheme.

Payment: A user U pays a merchant M for a transaction T by sending M the check $C = SIG_U(T)$. The user includes in every check/transaction a progressive serial number SN.

Selective Deposit: Let t' and t denote, respectively, the time of M's last and current deposit. M groups all checks dated between t' and t into n lists, $L_1,...,L_n$. Denote by V_i the total value of the checks in L_i, and by V the sum of the V_i's. M computes a commitment C_i to list L_i, preferably together with V_i (e.g., practically speaking by one-way hashing them so that $C_i = H(L_i, V_i)$), and then sends $C_1, \ldots, C_n$ to B, preferably signed and with an indication of deposit time. For instance, M sends $SIG_M(t, n, V, H(L_1, V_1), \ldots, H(L_n, V_n))$ to B.

B verifies M's latest deposit time, and selects k indices, $i_1, i_2, ..., i_k$, and sends them to the merchant.

M responds by de-committing $C_{i_1}, \ldots, C_{i_k}$.

B credits M's account with V cents, and debits the users whose checks belong to $L_{i_1}, \ldots, L_{i_k}$ according to the serial numbers used – e.g., as in the MR2 scheme.

(An exception to the above rules is made if the bank notices that something is wrong. For instance, if the sum of the checks in L_{i_j} is not V_{i_j}, if one check in L_{i_j} has the wrong time, if a newly processed check has the same serial number as a previously processed check, or if the new check's serial number and time are "out of order" somehow with respect to previously processed checks, or if the amount of the check is excessive, or if other bank-defined conditions occur. In such exceptional cases the bank may fine and/or throw out of the system the merchant and/or the user, or take other actions as it deems appropriate.)

Selective Discharge: B may keep statistics and throw out of the system (e.g., by revoking their certificates) users or merchants who misbehave, those users U whose checks cause B to pay merchants more than it is entitled to receive from U, and the merchants with whom those users spend their checks.

BASIC PROPERTIES. As for the MR1 and MR2 schemes, the set-up phase is simple and general and the payment phase is non-interactive. Moreover, the present scheme is very understandable and looks very fair to the merchants.

Notice that the value of k is arbitrary and up to the bank. When there is more attempted fraud, or there is suspicion of a particular merchant, a larger value of k may be used. Indeed, B may ask the merchant to de-commit *all* of his commitments. (Failure to de-commit, in particular, may trigger a fine or a discharge of the merchant.) Choosing $k > 1$ is recommendable in order to having a chance to catch two checks from the same user with the same serial number (rather than throwing out such a user later on "statistical evidence").

Notice that the merchant may deposit at prescribed times, $t_1, t_2, \ldots$, or at times of his choice. For instance, at a time t in which he has new checks totaling a given value (so that he does not want to delay payment any further), or when he has sufficiently many new checks (and does not want to store them any more).

As in MR1 and MR2, users and merchant may collude, but again with little or no benefit, since the bank may adopt the same defense mechanisms. Honest users who look suspicious may be treated similarly too.

VARIANTS. Variants of the MR1/MR2 schemes may also be applied here. In addition, the merchant may make use of Merkle trees to commit to $L_1, \ldots, L_n$. Check value information may be authenticated within the tree or alongside with it. In particular, the Merkle tree may authenticate at each node the total value of the checks "stored below it," as well as the total value of the check stored below each child. The same holds for check time information. The root of the Merkle tree may be digitally signed by the merchant – possibly together with other data, such as (partial or total) check value and time information.

The value of n may be variable or fixed. The bank may choose k out the n lists to have the merchant decommit, and pay the merchant an amount that depends on k, n and the value in the checks contained in the k de-committed lists. For instance, if the total value of these checks is TV, the bank may pay nTV/k.

Banks and merchant may agree not to process deposits whose checks total more than a given value, or deposits containing a list totaling more than a given value. (This discourages a single cheating attempt with the goal of getting either a high payoff or going bust.)

5   Conclusions

We believe that the schemes presented here provide effective solutions to the micropayments problem. Malicious behavior of a single player is generally prevented by design, while malicious behavior of a coalition of players is dealt with by a penalty system backed up by hard evidence. From a user's point of view, the interface is beautifully simple: it is just like writing (small) checks. From the bank's point of view, it is just like processing (large) checks. And the merchant is happy, because he can efficiently aggregate small payments from many users.

References

1. W. Alexi, B. Chor, O. Goldreich, and C. P. Schnorr. Rsa/rabin functions: Certain parts are as hard as the whole. *SIAM Journal on Computing*, 17(2):194–209, June 13 1988.

2. Ross Anderson, Harry Manifavas, and Chris Sutherland. NetCard – A practical electronic cash system. In *Proceedings Fourth Cambridge Workshop on Security Protocols*, volume 1189 of *Lecture Notes in Computer Science*. Springer, 1996.

3. Shimon Even, Oded Goldreich, and Silvio Micali. On-line/off-line digital signatures. In Gilles Brassard, editor, *Advances in Cryptology - Crypto '89*, pages 263–277, Berlin, 1989. Springer-Verlag. Lecture Notes in Computer Science Volume 435.

4. Phillip Hallam-Baker. W3C payments resources, 1995.
   http://www.w3.org/hypertext/WWW/Payments/overview.html.

5. Ralf Hauser, Michael Steiner, and Michael Waidner. Micro-payments based on iKP. Technical Report 2791 (# 89269), June 1996.

6. Stanislaw Jarecki and Andrew Odlyzko. An efficient micropayment scheme based on probabilistic polling. In *Proceedings 1997 Financial Cryptography Conference*, volume 1318 of *Lecture Notes in Computer Science*, pages 173–191, Springer, 1997.

7. Charanjit Jutla and Moti Yung. PayTree: "amortized-signature" for flexible MicroPayments. In *Proceedings of the Second USENIX Workshop on Electronic Commerce*, pages 213–221. USENIX, 1996.

8. Laurie Law, Susan Sabett, and Jerry Solinas. How to make a mint: the cryptography of anonymous electronic cash. National Security Agency, Office of Information Security Research and Technology, Cryptology Division, June 1996.

9. Richard J. Lipton and Rafail Ostrovsky. Micro-payments via efficient coin-flipping. In *Proceedings of Second Financial Cryptography Conference, '98*, volume 1465 of *Lecture Notes in Computer Science LNCS*, pages 1–15, February 1998.

10. Mark S. Manasse. Millicent (electronic microcommerce), 1995.
    http:// www.research.digital.com/SRC/personal/Mark_Manasse/uncommon/ucom.html.

11. S. Micali, M. Rabin, and S. Vadhan. Verifiable random functions. In *Proc. 40th Symp. on Foundations of Computer Science*, pages 120–130, October 1999.

12. Silvio Micali. Certified e-mail with invisible post offices. In Proceedings RSA97, San Francisco, CA, January 1997. Also, U.S. Patent No. 5,666,420.

13. Silvio Micali. Efficient certificate revocation. In Proceedings RSA97, San Francisco, CA, January 1997. Also U.S. Patent No. 5,666,416.

14. Torben P. Pedersen. Electronic payments of small amounts. Technical Report DAIMI PB-495, Aarhus University, Computer Science Department, Århus, Denmark, August 1995.

15. Ronald L. Rivest. Electronic lottery tickets as micropayments. In *Proceedings of Financial Cryptography '97*, volume 1318 of *Lecture Notes in Computer Science*, pages 307–314. Springer, 1997.
    (Available as http://theory.lcs.mit.edu/~rivest/lottery.pdf).

16. Ronald L. Rivest and Adi Shamir. PayWord and MicroMint–two simple micropayment schemes. In Mark Lomas, editor, *Proceedings of 1996 International Workshop on Security Protocols*, volume 1189 of *Lecture Notes in Computer Science*, pages 69–87. Springer, 1997. (Also available in Crypto-Bytes, volume 2, number 1 (RSA Laboratories, Spring 1996), 7–11, and at http://theory.lcs.mit.edu/~rivest/RivestShamir-mpay.pdf).

17. Adi Shamir, 2001. Personal communication.

18. W3C. Micropayments overview. http://www.w3.org/ECommerce/Micropayments/.

19. David Wheeler. Transactions using bets. In Mark Lomas, editor, *Security Protocols*, volume 1189 of *Lecture Notes in Computer Science*, pages 89–92. Springer, 1996. (Also available by ftp from the server ftp.cl.cam.ac.uk as /users/djw3/tub.ps.).

Proprietary Certificates

(Extended Abstract)

Markus Jakobsson[1], Ari Juels[1], and Phong Q. Nguyen[2]

[1] RSA Laboratories
{mjakobsson,ajuels}@rsasecurity.com
[2] CNRS/École normale supérieure
pnguyen@ens.fr

Abstract. Certificates play an essential role in public-key cryptography, and are likely to become a cornerstone of commerce-related applications. Traditional certificates, however, are not secure against *certificate lending*, i.e., a situation in which a certificate holder voluntarily shares with others the rights bestowed upon him through a certificate. This type of abuse is a concern in several types of applications, such as those related to digital rights management.

In this paper, we introduce the notion of *proprietary* and *collateral* certificates. We present a scheme whereby one certificate, known as a *proprietary* certificate, may be linked to another, known as a *collateral* certificate. If the owner of the proprietary certificate shares the associated private key, then the private key of the collateral certificate is simultaneously divulged.

Certificates in our scheme can be integrated easily into standard PKI models and work with both RSA and discrete-log-based keys (such as those for DSS). Our scheme leaks no significant information about private keys, and leaks only a small amount of information about certificate ownership. Thus, use of proprietary certificates still allows users to maintain multiple, unlinkable pseudonyms, and adds functionality without posing any threats to user privacy.

Keywords: certificates, collateral key, digital rights, fair encryption, proprietary key

1 Introduction

A digital certificate assigns an identity or a right to its holder, that is, to the possessor of the associated private key. This assignment is made by way of a digital signature that a certificate authority (CA) applies to the corresponding public key and to a description of the certificate's scope of use. Certificates may be employed for such broad purposes as binding a user identity to a public key for purposes of encryption, or to assign an entity the authority to sign legal documents on its own behalf. More specific situations in which certificates may be useful include the granting of access rights to a building or to a subscription-based service.

It is implicitly assumed that certificates, and the rights that come with them, belong solely to the person or entity to which they were issued. The issue of *non-transferability* of certificates and their associated rights, however, has been only

B. Preneel (Ed.): CT-RSA 2002, LNCS 2271, pp. 164–181, 2002.
© Springer-Verlag Berlin Heidelberg 2002

superficially addressed in the literature. In this paper we propose the notion of *proprietary certificates*, which are certificates with the property that their rights cannot be transferred (corresponding to giving somebody the private keys associated with the certified public keys) without punitive leakage of collateral information.

We believe that proprietary certificates may be important for three reasons. One is the likely future dependence on certificates for applications relating to commerce, and the benefit of incorporating new functionality into certificates. A second is the need for user privacy, which our solution preserves to a high degree. The third and perhaps most critical reason is the likely possibility of a proliferation of fair charging and access control mechanisms for information-based Web services in the near future. With many forms of Web advertising in decline (see, e.g., [19]), content providers have expressed a growing need to turn to subscription fees for revenue at some point.

We therefore see the main contribution of our paper as the concept of proprietary certificates, with its possible impact on the development of new services. Our primary focus is to exhibit secure and reasonably efficient structures for proprietary certificates. In doing so, we rely to a large extent on a combination of cryptographic components introduced for other purposes. Our current proposal causes an increase of the certificate size of between 384 and 768 Bytes, depending on the cryptosystems used. Alternatively, this certificate augmentation may be kept externally, indexed by the proprietary certificate it relates to. Such an approach would allow the certificates to retain their exact format while extending their functionality by means of this external record. We note that a further study of appropriate mechanisms – with a focus on their use in proprietary certificates – may result in more compact certificates.

We develop a mechanism to ensure that users can produce multiple unlinkable certified pseudonyms, with the certificates issued by one or many certificate authorities, such that it is impossible for a user to give away the right (to another user) to issue one or more types of signatures or other secret functions related to the certificates. This holds unless the user gives away the right to issue all kinds of signatures for all kinds of certificates and pseudonyms he holds, which means a total impersonation of the "lender". Also, it is possible to produce a system in which some keys (but not all) are released, were some keys to be given out. Thus, if a user is not willing to give away the right to sell his home to a second user, or to sign other legal documents for the first user, he can also not give away or share the right to access a subscription, or to enter a building, etc. (This assumes off-line collaboration, which is a good model for many scenarii.) In other words, we show how we can construct certificates on unlinkable pseudonyms such that the disclosure of one private key (which we will call the *proprietary* key) automatically implies the disclosure of a second private key (what we call the *collateral* key). If two keys are each others' collateral (directly or indirectly), we call the relationship *symmetric*; otherwise *asymmetric*.

A trivial approach would be to have the user employ the same private and public keys for every certificate (or for many). This approach has several drawbacks. First, it immediately and publicly links the identity of the holder to all of the associated certificates, thereby undermining the privacy guarantees afforded by unlinkability. Second, this approach is crude in that it does not permit the establishment of asymmetric relationships. In the trivial approach, disclosure of private keys is all or nothing, whereas our approach allows for a great deal more refinement, as we show below. Finally, in the trivial approach, there is no clear way to link certificates employing different cryptosystems. In contrast, we demonstrate in this paper how to offer this kind of flexibility.

Technically, this can be achieved by incorporating a ciphertext corresponding to the collateral private key (or a representation thereof), either in the proprietary certificate, or in an external database. (Onwards, we assume the ciphertext to be part of the certificate, for simplicity, but note that the options are technically equivalent.) Given that the encryption of the collateral key would be performed using the proprietary public key, we have that a party with knowledge of the proprietary private key will be able to derive the collateral private key by decryption of this ciphertext. Assuming the use of semantically secure encryption, the ciphertext will not reveal any information about the link between the two public keys or their certificates. It is important to note that it is not sufficient simply to encrypt one private key using another public key and incorporate the result in the certificate. Namely, it is important to guarantee robustness (i.e., to allow the CA to be certain about the contents of the ciphertext) without having to give the encrypted private key to any party or set of parties. Another technical difficulty is to provide the above functionality for schemes supported by standards, as opposed to schemes and structures designed solely for the purpose of the paper. While the "interior" of our solution contains schemes that are not currently supported by standards (such as Paillier's encryption scheme), it is important to note that the "exterior" of our solution relates to standard schemes, namely RSA and DSS. This means that any RSA or DSS key can be used as a proprietary or collateral key.

Outline. We begin in section 2 by describing the related work, followed in section 3.1 by an informal description of our goals and a statement of the contributions of the paper. We then describe our technical approach in section 3.2, but keep the discussion on a detail-free and intuitive level. In section 4, we then introduce denotation and outline the structure of our modified certificates, and review the building blocks we will use. In section 5 we then describe our solution in technical detail, using the previously introduced building blocks to describe our protocols. We state the properties of our solution in section 6; Appendix A contains proofs of these claims.

2   Related Work

The notion of non-linkability and independence among signatures arises frequently in the literature on digital payments, while the issue of non-transferability of access rights has been investigated from one perspective by Dwork, Lotspiech, and Naor [9], and from another by Goldreich *et al.* [14]. The combination of the two properties, however, has to our knowledge not been considered yet, and poses interesting technical questions as well as the possibility of new applications.

Our work is conceptually related to the work on *signets* by Dwork, Lotspiech and Naor [9], in which a secret, such as a credit card number, is incorporated in a private key to prevent the latter from being given away. Similarly, our aim is to some extent related to that of digital watermarking, as surveyed in [17]. In a digital watermarking system, identifying information of some kind is embedded in an indelible way in an image so as to discourage illicit copying. In neither of these proposals, though, is the embedded private (the collateral key in our terminology) hidden from the party who wishes to verify that it is there. In contrast, this is precisely what occurs in our solution.

The problem we study is also spiritually related to a problem previously studied by Goldreich *et al.* [14]. In their paper, a user owns a certificate associated with some rights, and wishes to delegate a certain portion of these to himself. This allows him to delegate rights for use on a laptop, with the benefit that if this gets stolen, then the damage is limited to the delegated portion of the rights (and the lost machine.) Thus, their scenario is the following: A user has a primary (long-term) key associated with some personalized access rights, some of which he wishes to delegate to some secondary (and short-term) keys. If sufficiently many secondary keys are disclosed, then the primary key can be recovered from these, thereby preventing the user from giving away his secondary keys. On the other hand, if few secondary keys are disclosed (fewer than a certain threshold), then the primary key remains secure. If the threshold is set to one (as it is in our scheme), however, then their primary and secondary keys are identical. (In other words, the issue of user privacy, or unlinkability of certificates, is not addressed in [14].) On the other hand, if the threshold is set higher, a corrupt user can give out some keys without any risk. Therefore, their solution – which was not intended for securing intellectual property – is also not very well suited for this task.

The problem we address in this paper is related to that studied by Camenisch and Lysyanskaya [4], who produce a credential scheme in which signatures (and other authentication elements) are generated from one and the same private key without being linkable to each other. Additionally, and similar to what is achieved in our scheme, they allow different private keys associated with a user to be tied to each other in a way that prevents users from sharing some private keys without sharing others as a result of this. However, while conceptually related from a birds-eye view, the two results are different on several counts. First, we do not provide unlinkability on a signature-by-signature basis. In our scheme, all signatures associated with one public key can be linked to this public key, as is normal for standard signatures and a standard PKI infrastructure. While this linkability is highly undesirable for, e.g., group signatures (whose very goal is for the opposite to hold), it is desirable in an infrastructure with standard signatures where each public key and all its signatures get associated with its owner. (However, it remains desirable that signatures from *different* public keys remain unlinkable, which we provide.) Another difference between the schemes is that the leakage of one key in their scheme immediately results in the leakage of *all* other keys, while our approach allows a tighter control of the inferrable relations between keys – namely, we can employ any graph of symmetric or asymmetric relationships between nodes / keys. This results in many practical advantages. Furthermore, and more importantly, we allow the linking of private keys for *standard* signature schemes (such as DSS and RSA), while the methods in [4] relate only to a new signature scheme introduced in their paper. It is worthwhile to notice that the employment of our methods to existing signature schemes is not only of potential technical value, but also of practical value in any legacy system (of which digital signatures may be one of the best examples).

Technically speaking, our solution depends most importantly on methods for key escrow, for which similar cryptographic building blocks are employed. Young and Yung [23] recently showed how to obtain a software key escrow system if users provide ciphertexts to certification authorities that permit the recovery of users' private keys. These ciphertexts are encrypted under the public key of an escrow authority. The structure that is used to assure the CA that the ciphertext is of the correct format has been called *fair encryption*. Thus, the encryption key in an escrow application is the public key of the escrow authority, while in our system, it is the public proprietary key. Similarly, the *encrypted* key in a escrow scheme is the user's private key, while it corresponds to the collateral key in our scheme for proprietary certificates.

The fair encryption scheme of [23] could be used for purposes of proprietary certificates. In fact, proprietary certificates constitute a new application for fair encryption. In order to allow for compatibility with more common crypto systems – namely RSA and standard discrete-log based schemes – we do not employ their methods, which are based on a "double decker" structure.

We do, however, make direct use of the rather efficient fair encryption scheme of Poupard and Stern [21] for some of the protocols of our scheme, namely those where the proprietary key is an RSA key. We develop and propose new schemes for the case where the proprietary key is a discrete-log key. These new schemes constitute an extension of previous results for fair encryption and are thus of independent interest.

Finally, we employ methods from [12] for proving equality of discrete logs over composite integers. These, in turn, are related to proof methods of Chaum [7] for proving equality of discrete logs over prime-order fields.

3   Goals

3.1   Overview

Just as a person may carry several identifying tokens for access to various resources and rights (such as a driver's license, a passport, and various credit and debit cards), he or she may need several public keys, each one of which may be associated with a different sets of rights. Different public keys (and their certificates) may also be associated with different policy requirements, such as requirements on the methods used to verify the identity of the certificate owner at the time the certificate is issued; the possible escrowing of keys; and the acceptable uses of the certified public key.

Our aim is to construct a proprietary certificate system that respects these requirements for heterogeneity and flexibility in a public key infrastructure. We assume that certificate authorities publish directories containing public information on the certificates they have issued. To make our goals precise, let us consider a case in which a certificate authority CA_1 wishes to issue a proprietary certificate C_1 to a certain user. The user is to provide a second certificate C_2, issued by a (possibly) distinct entity CA_2, as collateral. Informally stated, our goals in creating the proprietary/collateral relationship between C_1 and C_2 are as follows:

1. **Non-transferability:** With high probability, any player who learns the private key for C_1 will learn the private key for C_2, and be able to locate public information for C_2 in the directory maintained by CA_2. Thus, given that the user does not wish to relinquish control over C_2, the private keys associated with C_1 are non-transferable.

2. **Unlinkability:** CA_1 learns that the user knows the private key associated with C_2, and that CA_2 issued C_2. CA_1 learns no additional information about certificates held by the user, and no other player learns any information about the certificates of the user.

3. **Cryptosystem agility:** C_1 and C_2 can be based on different cryptosystems: either can make use of an RSA key or a discrete-log key.

4. **Locality:** CA_1 needs to interact only with the user, and not CA_2.

5. **Security:** CA_1 learns only a negligible amount of information about the private keys associated with C_1 and C_2. No other party learns any information relating to the certificates.

6. **Efficiency:** The certificate C_1 is not substantially larger than a conventional certificate of its type. Moreover, the computational and communication requirements on CA_1 and the user in establishing the proprietary/collateral relationship are reasonable.

7. **PKI compatibility:** We require that the modified certificates allow for easy integration into standard PKI models. (While we require use of so-called "safe" RSA keys, these are fully compatible with most existing mechanisms.)

We may view the collection of certificates belonging to a particular user as a collection of nodes $C = \{C_1, C_2, \ldots, C_t\}$ in a directed graph $G = (C, E)$. An arc $(C_i, C_j) \in E$ in this graph represents a binding of a proprietary certificate C_i to a collateral one C_j. If there is an additional reverse arc (C_j, C_i), then the relationship between C_i and C_j is called *symmetric*; otherwise, it is called *asymmetric*. Nodes may have degrees of arbitrary size. In our system, the size of a certificate C_i is linear in its out-degree, as are the computational and communication requirements to establish outgoing arcs. As an illustration of exactly what purpose an arc serves, we present the following brief example.

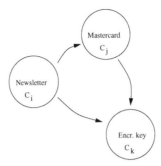

Fig. 1. Example certificate relationships.

Example. Say that a user wants to obtain a certificate for an on-line newsletter. This newsletter requires that the certificate C_i associated with one particular user of their service serve as a proprietary one, with the user's Mastercard certificate C_j as collateral. Moreover, to guard against a situation in which the user closes her Mastercard account, the newsletter may require that the user's public encryption key and corresponding certificate C_k also serve as collateral. (Alternatively, the newsletter may blacklist the certificate C_i once it detects that Mastercard blacklists C_j.) Additionally, Mastercard may employ the user's decryption key as collateral for its own certificate. Thus, the newsletter creates arcs (C_i, C_j) and (C_i, C_k) in the graph G. See figure 1 for a graphical depiction of this scenario.

Privacy note. It is important to note that the CA, while not learning either the collateral or proprietary secret key, learns the association between the two public keys. Namely, he learns that the public key of the proprietary certificate is associated with the same person as the public key of the collateral certificate. For many purposes, this may not be so outlandish, as long as the public cannot infer the same relation. To hide the association of keys from the CA, it is possible to use more heavy-weight protocols, in which the user proves correct encryption with respect to an *unspecified* public key belonging to some set of potential collateral public keys. The likely drawback of such solutions, though, would be the resulting reduction of the efficiency. Alternatively, if we *do* allow the CA to learn the association, standard techniques [16] can be employed to prevent him from convincing others.

3.2   What Does an Arc Look Like?

Our approach is to include in a proprietary certificate a ciphertext on the collateral keys. This ciphertext (while not necessarily of a standard format) may be decrypted using the public keys of the proprietary private keys, thereby yielding the collateral keys. In order to reach this goal, we need secure protocols for a certificate holder to prove to the CA that the generated ciphertexts are of the right format (namely that they contain valid representations of the collateral private keys). This must be done in a manner that is both efficient and which limits leaks of information to the CA. This will be achieved by basic fair encryption techniques, including use of zero-knowledge proofs and semantically secure encryption. Apart from including an encryption of the private collateral key in the certificate, the certificate authority may additionally include a pointer to the directory entry for the collateral key. For reasons of privacy, this would also be encrypted, using a semantically secure encryption scheme and the public key of the proprietary certificate (making it possible to decrypt given the private key associated with the proprietary certificate). Technically, the encryption of the pointer is straightforward, as the plaintext information does not need to be hidden from the CA issuing the proprietary certificate. Thus, we will focus on the encryption of the private key instead of that of the public key. Practically, it is worth mentioning, though, as it allows the retrieval of private keys as well as an *understanding of what was retrieved*, should the proprietary key be given away.

In a fair encryption system, a user holds a private/public key pair (PK, SK), and a public key PK_T is published for some trusted third party. The user constructs a ciphertext $\Gamma_{1\to2}$ and a non-interactive proof that $\Gamma_{1\to2}$ is an encryption under PK_T of a representation of SK or data that enable efficient reconstruction of SK. In our system, $\Gamma_{1\to2}$ is a ciphertext on the private key for C_2 (or something equivalent) under the public key associated with C_1. A critical difference in our system from conventional use of fair encryption is our assumption that CA_1 is responsible for ensuring that Γ is correctly constructed. (This is the case in all of the applications we envisage, as it is CA_1 that wishes to prevent abuse of C_1.) Hence, the owner of C_2 must prove correct construction of Γ only once to CA_1. In consequence, the proof may be interactive, and the size of the proof is of less importance than in a typical fair encryption system, as it has no impact on the size of C_1, which only carries $\Gamma_{1\to2}$.

In section 5, we detail the various protocols for creating an arc (C_1, C_2) between two certificates C_1 and C_2. As explained above, we use the Poupard-Stern fair encryption system as the basis for protocols in which C_1 is an RSA-key-based certificate and C_2 is either RSA or DL-based. An important contribution of our paper is a pair of novel fair encryption protocols for the case where C_1 is instead a discrete-log-based certificate and C_2 is either RSA or discrete-log based.

4   Notation and Building Blocks

4.1   Notation

We define a cryptosystem CR in the broadest sense to include a suite of five algorithms $\mathsf{keygen}_{CR}, \mathsf{sign}_{CR}, \mathsf{verify}_{CR}, \mathsf{encrypt}_{CR}$ and $\mathsf{decrypt}_{CR}$ for the respective operations of key generation, signing, verification, encryption, and decryption. (Thus, where we consider a signing algorithm such as, e.g., DSS, we assume a corresponding encryption/decryption algorithm over the same algebraic structures, e.g., ElGamal.) We assume implicitly that a secure suite of algorithms of this kind are available for all cryptosystems under consideration. To produce a certificate C_1 on a public/private key pair (PK_1, SK_1), a given certificate authority applies an algorithm sign_{CR} to PK_1 and possibly to some additional policy information aux_1.

Our contribution in this paper is a set of protocols for arc creation, and a corresponding set of protocols for key extraction, i.e., for computation of a collateral key given the corresponding proprietary one.

Arc creation: Let (PK_1, SK_1) and (PK_2, SK_2) be the public/private key pairs respectively for proprietary certificate C_1 and collateral certificate C_2. Additionally, let C_1 be a certificate on a public key for cryptosystem type CR_1 and C_2 a certificate for cryptosystem type CR_2. We let $\text{Arc}_{CR_1 \rightarrow CR_2}$ denote an arc creation protocol on the proprietary/collateral certificate pair (C_1, C_2).

The protocol $\text{Arc}_{CR_1 \rightarrow CR_2}$ takes as input from the prover the two public/private key pairs (PK_1, SK_1) and (PK_2, SK_2), and relevant security cryptosystem security parameters. Input from the verifier is a security parameter specifying protocol soundness. The output of the protocol is the pair of public keys (PK_1, PK_2), the collection of security parameters, and a ciphertext $\Gamma_{1 \rightarrow 2}$.

We require two security properties on a given algorithm $\text{Arc}_{CR_1 \rightarrow CR_2}$:

- **Soundness:** With overwhelming probability over the coin flips of the prover and verifier, the ciphertext $\Gamma_{1 \rightarrow 2}$ is well formed. In other words, given input SK_1 and $\Gamma_{1 \rightarrow 2}$, the protocol $\text{Extract}_{CR_1 \rightarrow CR_2}$ described below yields output SK_2.
- **Privacy:** The full transcript of the protocol is simulatable in a computationally indistinguishable manner by a player with knowledge of PK_1 and PK_2 only. (Thus, e.g., the verifier learns no non-negligible information about SK_1 or SK_2.)

We consider two cryptosystem types in this paper, namely RSA and discrete log (DL). In other words, either CR_1 or CR_2 can be a cryptosystem in which the public key is an RSA modulus, e.g., RSA encryption and signing or Paillier encryption with RSA signing, or, alternatively, a cryptosystem based on discrete log, such as (ElGamal / DSS), (ElGamal / Schnorr,) etc. Thus, in our paper, we specify four generic arc creation protocols: $\text{Arc}_{RSA \rightarrow DL}$, $\text{Arc}_{RSA \rightarrow RSA}$, $\text{Arc}_{DL \rightarrow RSA}$, and $\text{Arc}_{DL \rightarrow DL}$.

Key extraction: Corresponding to each arc creation algorithm $\text{Arc}_{CR_1 \rightarrow CR_2}$ is a key extraction algorithm $\text{Extract}_{CR_1 \rightarrow CR_2}$. This algorithm takes as input a ciphertext $\Gamma_{1 \rightarrow 2}$ and the keys PK_2 and SK_1. If successful, it outputs SK_2. Involved here are some standard decryption operations in combination with additional cryptographic operations such as lattice reduction and gcd algorithms.

4.2   Building Blocks

Discrete-log-based signature schemes. If the signature scheme associated with key pair (PK, SK) is discrete-log based (such as DSS [18] or Schnorr [22]), then

$PK = (p, q, g, y)$, for primes p, q, such that $p = kq + 1$, an element g of order q, and a value $y = g^x \bmod p$. Here, the private key $SK = x$ is chosen uniformly at random from $\mathbb{Z}_q$. Example sizes of the parameters are $(|p|, |q|) = (1024, 160)$. We refer to [18,22] for more details on the schemes.

ElGamal encryption. Let g be a generator of a large subgroup $\mathcal{G}$ of $\mathbb{Z}_n$. Often, the integer n is chosen to be prime, but we will also consider the use of strong RSA moduli, and we assume that all computation is performed modulo n, where applicable. Let s denote the size of the subgroup generated by g, and let $y = g^x$ be the public key used for encryption, where $x \in \mathbb{Z}_s$.

To encrypt a message m, one picks a random element $\alpha \in_u \mathbb{Z}_s$ and computes the pair $(a, b) = (y^\alpha m, g^\alpha)$. (We note that s can be determined by a party who knows the factorization of n, which will not be a restriction in our setting.) To decrypt a ciphertext (a, b), one computes $m = a/b^x$.

It is well-known that the ElGamal scheme is semantically secure under the Decision Diffie-Hellman (DDH) assumption on the subgroup $\mathcal{G}$, and given that messages are chosen from $\mathcal{G}$ (see [1]). If messages are chosen from another set, the ciphertext may leak some information, such as the Jacobi symbol of the message.

Proof of equality of discrete logs. Let $y_i = g_i^{x_i}$, for $i \in \{1, 2\}$, where $y_1, y_2, g_1, g_2 \in \mathcal{G}$ for some group $\mathcal{G}$. We let $\mathsf{EQDL}_1(y_1, y_2, g_1, g_2)$ denote a zero-knowledge proof protocol demonstrating that $\log_{g_1} y_1 = \log_{g_2} y_2$. There are many methods proposed in the literature for implementing EQDL_1, [7,22]. In the appendix, we exhibit a version of [22] modified for use with RSA moduli, and discuss its security.

A useful variant employed in our protocols involves elements spanning two groups $\mathcal{G}_1$ and $\mathcal{G}_2$. In particular, we let $\mathsf{EQDL}_2(y_1, y_2, g_1, g_2, n_1, n_2)$ denote a proof that $y_1 = g_1^{x_1} \bmod n_1$ and $y_2 = g_2^{x_2} \bmod n_2$ for $x_1 = x_2$. (In general, $\mathcal{G}_1$ and $\mathcal{G}_2$ need not be modular multiplicative groups, but these are the only type used here.) We use the very efficient proof technique for EQDL_2 introduced in [5].

Both EQDL_1 and EQDL_2 are zero-knowledge. While the soundness of both protocols relies on the discrete log assumption, we note that the soundness of the efficient, one-round version of the protocol for EQDL_2 depends additionally on the *strong RSA* assumption. See [5] for more detailed discussion.

Let $(a, b) = (my^k, g^k)$ represent an ElGamal ciphertext under public key y. Observe that a prover with knowledge of the private key $x = \log_g y$ can prove in zero-knowledge that (a, b) represents a valid ciphertext on plaintext m simply by proving $\mathsf{EQDL}_1(y, a/m, g, b)$.

Paillier encryption. The Paillier cryptosystem was introduced in [20]. It uses the Carmichael lambda function $\lambda(N)$ defined as the largest order of the elements of $\mathbb{Z}_N^*$. Let $N = PQ$ be an RSA modulus such that $\varphi(N)$ is coprime to N. Recall that $\lambda(N) = \mathsf{lcm}(p - 1, q - 1)$. The general Paillier's cryptosystem, as defined in [20], uses an integer G of order multiple of N modulo N^2. It was noticed in [8,6] that the simplest choice is probably $G = 1 + N$, because $(1 + N)^M \equiv 1 + MN \bmod N^2$. Thus, in this paper, we only use $G = 1 + N$, which slightly simplifies the description of the scheme and has no impact on the semantic security: we refer to [20] for a general description. The public key is N and the secret key is $\lambda(N)$.

To encrypt a message $M \in \mathbb{Z}_N$, randomly choose $u \in \mathbb{Z}_N^*$ and compute the ciphertext $c = (1 + MN)u^N \bmod N^2$. To decrypt c, compute:

$$M = \frac{L(c^{\lambda(N)} \bmod N^2)}{\lambda(N)} \bmod N,$$

where the L-function takes as input an element congruent to 1 modulo N, and outputs $L(u) = \frac{u-1}{N}$.

The Paillier public-key cryptosystem is semantically secure under the hardness of distinguishing N-th residues modulo N^2 (see [20]).

Fair encryption methods. Assume that users have pairs of public and private keys and give an encryption E of their private key (or something allowing the recovery of the private key) using the public key PK_T of a trusted third party. A *fair encryption* is a publicly verifiable proof that the third party is able to recover the private key using his own private key and the ciphertext. Poupard and Stern proposed practical fair encryption [21] using the Paillier cryptosystem (meaning that the third party's public key is a Paillier public key). Poupard-Stern proposed two protocols: One to encrypt ElGamal-type keys, and one for RSA keys. To the best of our knowledge, no fair encryption protocol in which the third party uses a discrete log system exists in the literature. There exist other fair encryption protocols (see [10,2] for instance), but they do not seem to be as efficient as the Poupard-Stern protocols for our application, so we do not use them.

We will use fair encryption as a proof that any person knowing the private key corresponding to PK_T can recover the private key encrypted in E. In other words, in the setting of proprietary certificates, the "third party" is any possessor of the proprietary private key, and the fair encryption is the proof of collateral private key recovery.

5   Arc Creation Protocols

We will now consider how one can perform the various proofs of ciphertext correctness, with the various types of encryption needed. We will denote the various protocols by the types of proprietary and collateral keys they relate to. Thus, a $DL \rightarrow RSA$ protocol is a protocol for proving that given a ciphertext and the correct discrete log private key (the proprietary key), one can decrypt and obtain the correct RSA private key (the collateral key). We note that we will use Paillier's encryption scheme in lieu of RSA – however, since one can perform Paillier encryption and decryption using an RSA public versus private key, this is not a restriction.

In the following, we let (y, x) be a public key / private key pair for a discrete-log-based scheme, as described previously, and (e, d) be the public versus private keys of an RSA scheme with public modulus N. We use the same moduli and generators as previously shown. For $\mathsf{Arc}_{DL \to RSA}$ and $\mathsf{Arc}_{DL \to DL}$, it is necessary to include in the certificates a generator G as described below. We note that this does not impact the unlinkability properties, since G relates to the public key in whose certificate it is included.

5.1   RSA → DL

Let (e_1, d_1) denote the proprietary public and private keys corresponding to a public modulus N_1, and (y_2, x_2) the collateral public and private keys, with associated modulus p. His Paillier public key is N_1, and his private key is $\lambda(N_1)$.

In the protocol $\mathsf{Arc}_{RSA \to DL}$, the user randomly chooses $u \in \mathbb{Z}_{N_1}^*$ and computes the ciphertext $\Gamma_{1 \to 2} = (1 + x_2 N_1) u^{N_1} \bmod N_1{}^2$, and a non-interactive proof (to the CA) of the "third party"'s ability to compute x_2 from y_2 and $\Gamma_{1 \to 2}$, using the Poupard-Stern fair encryption of ElGamal keys [21, Sect. 3.1].

Extraction of keys. The algorithm $\mathsf{Extract}_{RSA \to RSA}$ involves application of the key recovery process of the Poupard-Stern fair encryption [21, Proof of Theorem 1], based on Gauss lattice reduction algorithm (note: a simple Paillier decryption presumably does not always enable to recover the private key, due to some cheating strategy, as explained in [21]; the proof refers to this key recovery process and not Paillier decryption). This yields x_2.

5.2   RSA → RSA

Let (e_1, d_1) denote the proprietary public and private keys associated with a public modulus N_1, and (e_2, d_2) the collateral public and private keys associated with a public modulus N_2. His Paillier public key is N_1, and his private key is $\lambda(N_1)$.

In the protocol $\mathsf{Arc}_{RSA \to RSA}$, the user computes $x = N_2 - \varphi(N_2)$, randomly chooses $u \in \mathbb{Z}_{N_1}^*$ and the ciphertext $\Gamma = (1 + x N_1) u^{N_1} \bmod N_1^2$. He proves to the CA that a party with knowledge of the decryption key (i.e., our proprietary key) is able to factor N_2 using $\Gamma_{1 \to 2}$ and his Paillier private key, using the Poupard-Stern fair encryption of RSA keys [21, Sect 3.2].

Extraction of keys. To recover the collateral private key using $\mathsf{Extract}_{RSA \to RSA}$, one must apply the key recovery process of the Poupard-Stern fair encryption [21, Proof of Theorem 2] to obtain the factorization of N_2 from Γ and the Paillier private key.

5.3   DL → RSA

Let (y_1, x_1) be the public/private key pair for the DL (i.e., proprietary) certificate, and let N_2 be the modulus for the RSA (i.e., collateral) certificate. For the user to ensure privacy of her private keys, we require that N be the product of two safe primes. Namely, we should have $N_2 = PQ$ where P, Q, $(P-1)/2$ and $(Q-1)/2$ are all large primes. (thus, in particular, P and Q are congruent to 3 modulo 4). The use of safe primes can be proved using [5]. To ensure soundness of the protocol, however, the user need only prove about N_2 that it is the product of (at most) two primes. This can be accomplished with practical computational and communication requirements by combining protocols from [11] or [15] with those in [3], as shown in, e.g., [13].

Apart from a proof that N_2 is a well-formed RSA modulus, there are two key components to the protocol $\mathsf{Arc}_{DL \to RSA}$. The first is that of *key translation*. This is a procedure whereby the user constructs a generator G with large order in $\mathbb{Z}_{N_2}$ and a public ElGamal key $Y = G^{x_1} \bmod N_2$. Since x_1 is the private key for the DL certificate of the user, a player with access to this private key will be able to decrypt any ElGamal ciphertext under public key[1] Y. The user proves correct translation through straightforward use of EQDL_2.

The second key component in the protocol is encryption of a non-trivial root r of unity in $\mathbb{Z}_{N_2}^*$. In particular, the user constructs an ElGamal ciphertext (a, b) on such a root r under the public key Y. Given r, it is easy to compute a factor of N_2, and thus compute any private key for the RSA certificate (provided that N_2 is a well-formed RSA modulus). To prove that the plaintext r corresponding to (a, b) is indeed a root of unity, the user must prove that (a^2, b^2) has plaintext 1. To see that r is a non-trivial root, i.e., not equal to 1 or -1, the CA must verify the following three Jacobi quantities:

- The integer -1 has Jacobi symbol 1. (This is always the case if N is a product of two large safe primes.)
- The value b has Jacobi symbol 1.
- The value a has Jacobi symbol -1.

Together, these three checks ensure that (a, b) has a plaintext r with Jacobi symbol -1, and thus that $r \notin \{-1, 1\}$ and is thus non-trivial. With all of the other proofs given above, this ensures that a player with knowledge of x can use the ciphertext (a, b) to factor N_2 and obtain any private keys associated with the RSA certificate.

Here is our protocol in detail. If any of the verification performed by the CA fails, or any of the sub-protocols fails, then the protocol is aborted.

Protocol $\mathsf{Arc}_{DL \to RSA}$

1. The user selects an element $G \in \mathbb{Z}_{N_2}^*$ of Jacobi symbol 1 such that $G^2 \neq 1$ and $G^2 - 1$ is coprime to N_2. Thus, G has multiplicative order of either $(P-1)(Q-1)/4$ or $(P-1)(Q-1)/2$ (see for instance [12]). Note that the DDH problem in the subgroup spanned by G is believed to be hard (see [1]). The user sends G to the CA.

---

[1] Note that this public key will have order at least (P-1)(Q-1)/4 with overwhelming probability. Thus, with overwhelming probability, the choice of public key will itself leak no information about the plaintext root.

2. The user performs the key translation. She computes $Y = G^{x_1} \bmod N_2$ and proves $\mathsf{EQDL}_2[g, y, G, Y, n, N]$.

3. The user computes a non-trivial root r of unity with Jacobi symbol -1. This is easy to accomplish given knowledge of P and Q and use of the Chinese Remainder Theorem.

4. The user selects an encryption factor $\alpha \in \mathbb{Z}_{(P-1)(Q-1)/2}$ uniformly at random. She constructs an ElGamal ciphertext on r of the form $(a, b) = (Y^\alpha r, G^\alpha)$. She sends this to the CA.

5. The user proves that (a, b) is a ciphertext under Y of a root of unity. In particular, she proves $\mathsf{EQDL}_1[G, Y, b^2, a^2]$.

6. CA verifies that -1 and b have Jacobi symbol 1, and that a has Jacobi symbol -1.

Key extraction. The algorithm $\mathsf{Extract}_{DL \to RSA}$ interprets the proprietary key x_1 as a key $X = x_1$ for the composite ElGamal ciphertext $E = (a, b)$, one can compute $r = a/b^X \bmod N_2$. One derives the factorization of N_2 by simple gcd: Indeed, $r^2 = 1 \bmod N_2$ implies $(r-1)(r+1) = 0 \bmod N_2$ where $r \neq \pm 1 \bmod N_2$, so that $\gcd(r-1, N_2)$ is a non-trivial factor of N. This yields the private collateral key.

5.4   DL → DL

In order to use a discrete-log proprietary key and a discrete-log collateral key – although possibly over different group structures, we introduce the use of *intermediary keys*. This is a key whose only use is to act as a connector between existing protocols for putting up collateral and performing extraction. Namely, when performing extraction, the proprietary key is used to obtain the intermediary key (serving as a collateral), and then the intermediary key is used as proprietary key to obtain the real collateral key.

Thus, in the protocol $\mathsf{Arc}_{DL \to DL}$, the user selects a strong RSA modulus N' as an intermediary public key (whose corresponding private key is $\varphi(N')$). He then uses the $DL \to RSA$ protocol above to establish N' as the collateral key of his proprietary key. Then, he uses N' as the proprietary key in a $RSA \to DL$ protocol (Poupard/Stern). The result is two sets of ciphertexts, one containing the intermediary key, and using the proprietary key for encryption/decryption; the second containing the collateral key, and using the intermediary key for encryption/decryption.

Key extraction. The protocol $\mathsf{Extract}_{DL \to DL}$ is an obvious composition of the previous key extraction protocols $\mathsf{Extract}_{DL \to RSA}$ and $\mathsf{Extract}_{RSA \to DL}$. This is a two-step process in which one first obtains the intermediary key and then the collateral key.

6   Claims

We prove in appendix A that our solution satisfies *non-transferability*, *unlinkability* and *security*. It is clear from our protocol description that it satisfies *cryptosystem agility; locality;* and *PKI compatibility.* We address the efficiency of our scheme in appendix A as well.

References

1. D. Boneh. The decision Diffie-Hellman problem. In *Proc. of ANTS-III*, volume 1423 of *LNCS*, pages 48–63. Springer-Verlag, 1998.
2. F. Boudot and J. Traoré. Efficient publicly verifiable secret sharing schemes with fast or delayed recovery. In *ICICS '99*, volume 1726 of *LNCS*, pages 87–102. Springer-Verlag, 1999.
3. J. Boyar, K. Friedl, and C. Lund. Practical zero-knowledge proofs: Giving hints and using deficiencies. *Journal of Cryptology*, 4(3):185–206, 1991.
4. J. Camenisch and A. Lysyanskaya. An efficient system for non-transferable anonymous credentials with optional anonymity revocation. In B. Pfitzmann, editor, *Eurocrypt '01*, volume 2045 of *LNCS*, pages 93–117. Springer-Verlag, 2001.
5. J. Camenisch and M. Michels. Separability and efficiency for generic group signature schemes. In M. Wiener, editor, *Crypto '99*, volume 1666 of *LNCS*, pages 413–430. Springer-Verlag, 1999.
6. D. Catalano, R. Gennaro, N. Howgrave-Graham, and P. Q. Nguyen. Paillier's cryptosystem revisited. In P. Samarati, editor, *8th ACM Conference on Computer and Communications Security*. ACM Press, 2001. To appear.
7. D. Chaum and H. Van Antwerpen. Undeniable signatures. In G. Brassard, editor, *Crypto '89*, volume 435 of *LNCS*, pages 212–216. Springer-Verlag, 1989.
8. I. Damgård and M. Jurik. A generalisation, a simplification and some applications of Paillier's probabilistic public-key system. In *PKC '01*, volume 1992 of *LNCS*, pages 119–136. Springer-Verlag, 2001.
9. C. Dwork, J. Lotspiech, and M. Naor. Digital signets: Self-enforcing protection of digital information. In *Proc. of 28th STOC*, pages 489–498. ACM, 1996.
10. E. Fujisaki and T. Okamoto. A practical and provably secure scheme for publicly verifiable secret sharing and its applications. In K. Nyberg, editor, *Eurocrypt '98*, volume 1403 of *LNCS*, pages 32–46. Springer-Verlag, 1998.
11. Z. Galil, S. Haber, and M. Yung. Minimum-knowledge interactive proofs for decision problems. *Siam J. of Computing*, 18(4):711–739, 1989.
12. R. Gennaro, H. Krawczyk, and T. Rabin. RSA-based undeniable signatures. In B. Kaliski, editor, *Crypto '97*, volume 1294 of *LNCS*, pages 132–149. Springer-Verlag, 1997.
13. R. Gennaro, D. Micciancio, and T. Rabin. An efficient non-interactive statistical zero-knowledge proof system for quasi-safe prime products. In *5th ACM Conference on Computer and Communications Security*, pages 67–72. ACM Press, 1998.
14. O. Goldreich, B. Pfitzmann, and R. L. Rivest. Self-delegation with controlled propagation - or what if you lose your laptop. In H. Krawczyk, editor, *Crypto '98*, volume 1462 of *LNCS*, pages 153–168. Springer-Verlag, 1998.
15. J. van de Graaf and R. Peralta. A simple and secure way to show the validity of your public key. In B. Kaliski, editor, *Crypto '87*, volume 293 of *LNCS*, pages 128–134. Springer-Verlag, 1987.
16. M. Jakobsson, K. Sako, and R. Impagliazzo. Designated verifier proofs and their applications. In U. Maurer, editor, *Eurocrypt '96*, LNCS, pages 143–154. Springer-Verlag, 1996.

17. S. Katzenbeisser and F.A.P. Petitcolas, editors. *Information Hiding Techniques for Steganography and Digital Watermarking.* Artech House, 1999.
18. National Institute of Standards and Technology (NIST). *FIPS Publication 186: Digital Signature Standard,* May 1994.
19. Bloomberg News. Ad-revenue worries weigh down Yahoo. 1 September 2000. URL: http://yahoo.cnet.com/news/0-1005-200-2670551.html.
20. P. Paillier. Public-key cryptosystems based on composite degree residuosity classes. In J. Stern, editor, *Eurocrypt '99,* volume 1592 of *LNCS,* pages 223–238. Springer-Verlag, 1999.
21. G. Poupard and J. Stern. Fair encryption of RSA keys. In B. Preneel, editor, *Eurocrypt '00,* volume 1807 of *LNCS,* pages 173–190. Springer-Verlag, 2000.
22. C. P. Schnorr. Efficient signature generation by smart cards. *Journal of Cryptology,* 4:161–174, 1991.
23. A. Young and M. Yung. Auto-recoverable auto-certifiable cryptosystems. In K. Nyberg, editor, *Eurocrypt '98,* volume 1403 of *LNCS,* pages 17–31. Springer-Verlag, 1998.

A   Analysis

Non-transferability. The scheme satisfies non-transferability if the CA can be guaranteed that for any certificate he has issued, knowledge of its proprietary private key allows the corresponding collateral private key to be computed with an overwhelming probability, and in polynomial time. Thus, this directly corresponds to the soundness of the protocol for proving that the ciphertext in an encryption of the appropriate plaintext (the collateral private key, or a representation thereof), and under the appropriate public key (the proprietary public key.)

The soundness of the fair encryption schemes used for arc $\mathsf{Arc}_{RSA \to DL}$ and $\mathsf{Arc}_{RSA \to RSA}$ has already been proven (see [21]). Since $\mathsf{Arc}_{DL \to DL}$ is composed of $\mathsf{Arc}_{DL \to RSA}$ and $\mathsf{Arc}_{RSA \to DL}$ (the latter which we know is sound), we see that only the soundness of $\mathsf{Arc}_{DL \to RSA}$ remains to be proven.

The soundness of EQDL_2 was proven in [5]. Thus, step two is sound, and establishes that $X = x$ for $Y = G^X \bmod N$, $y = g^x \bmod p$. Furthermore, step five is sound, given the soundness of Schnorr signatures (see [22]) and their extension to composite moduli (see appendix B.) Thus, this step establishes that (a^2, b^2) is a valid encryption of 1, using public key Y and modulus N. Thus, $(a^2, b^2) = (Y^\beta, G^\beta)$. This implies that the plaintext must be a root r of unity, and that $(a, b) = (Y^\alpha r, G^\alpha)$, where $\beta \equiv_{\varphi(N)} 2\alpha$. The CA verifies (in step six) that b have Jacobi symbol 1. Therefore, since Y is a power of G (as established in step 2), a/r is a power of b. Since a has Jacobi symbol -1, so must r. In step 6, it is established that -1 has Jacobi symbol 1, and (obviously), the same holds for the value 1. Therefore, the plaintext r must be a non-trivial root of 1. As was outlined in the key extraction protocol for $\mathsf{Arc}_{DL \to RSA}$, knowledge of such a value allows straightforward factoring of N. Since knowledge of the proprietary discrete log private key x implies knowledge of the decryption key X (as established in step 2), we see that anybody with knowledge of the proprietary key can compute the private collateral key, which concludes the proof.

Security. The security of the RSA $\rightarrow$ DL and the RSA $\rightarrow$ RSA arc creation protocols is the same as in the Poupard-Stern fair encryption protocols [21]. Namely the proofs are zero-knowledge, and the ciphertext is with respect to the Paillier cryptosystem which is semantically secure under the hardness of distinguishing N-th residues modulo N^2 (see [20]). For the DL $\rightarrow$ RSA protocol, the proofs are zero-knowledge.

One needs to assume, however, that the key-translation protocol does not weaken the hardness of the discrete log problem. For this, we rely on a variant of the DDH assumption. Normally, this assumption is applied over a single group $\mathcal{G}$ of order q. It states that for generators μ_1 and μ_2 drawn uniformly at random, and exponents a, b drawn uniformly at random from $\mathbb{Z}_q$, it is computationally infeasible for a polynomial-time entity to distinguish between the two distributions $D_1 = \{\mu_1, \mu_2, \mu_1^a, \mu_2^a\}$ and $D_2 = \{\mu_1, \mu_2, \mu_1^a, \mu_2^b\}$.

We introduce a variant assumption that we call the *cross-group DDH assumption*. We consider two groups $\mathcal{G}_1$ and $\mathcal{G}_2$, where the order of $\mathcal{G}_1$ is q, and that of $\mathcal{G}_2$ is at least q. The distributions D_1 and D_2 are constructed exactly as above, except that μ_1 is a generator of $\mathcal{G}_1$ and μ_2 is a generator of $\mathcal{G}_2$. In other words, the cross-group DDH assumption as applied to $\mathcal{G}_1$ and $\mathcal{G}_2$ states that it is infeasible to test equality of discrete logs across groups. We apply this assumption in our paper to two groups for which the conventional DDH assumption is believed to be hard. The cross-group DDH assumption may be seen to arise in implicit form in earlier literature such as, e.g., [5], and seems a potentially important assumption for a wide range of protocols.

The ciphertext E is an ElGamal encryption of a non-trivial root r of unity, and such an r does not belong to the subgroup $\mathcal{G}$ spanned by G because it has Jacobi symbol -1. But the semantic security of ElGamal under the DDH assumption over $\mathcal{G}$ relates to plaintext in $\mathcal{G}$. However, one can easily notice that if ElGamal with plaintexts chosen in the kernel of the Jacobi symbol is semantically secure (which is equivalent to the DDH assumption in that kernel, which itself is believed to be true), then ElGamal with plaintexts having Jacobi symbol -1 is also semantically secure. Indeed, if an attacker is able to build two particular plaintexts m_0 and m_1 having Jacobi symbol -1, and to determine with non-negligible advantage if a challenge ciphertext (of either m_0 and m_1) is an encryption of m_0 or m_1, then he could also determine with non-negligible advantage if a challenge ciphertext c (of either m_0^2 and $m_0 m_1$) is an encryption of m_0^2 or $m_0 m_1$ (by division). Thus, since both m_0^2 and $m_0 m_1$ have Jacobi symbol $+1$, this would break the semantic security of ElGamal for plaintexts in the kernel of the Jacobi symbol. Note that the DDH problem for the kernel of the Jacobi symbol is believed to be hard when the modulus is a product of two safe primes (see [1]).

Unlinkability. We see that unlinkability follows from the fact that we use semantically secure encryption of the collateral private keys and the pointers to the collateral public keys and their associated CA; and that no information about *other* keys associated with a user is used or included in a signature using one particular public key.

Efficiency. All of the protocols require the inclusion of a ciphertext describing or pointing to the collateral public key. Additionally, $\mathsf{Arc}_{DL \to RSA}$ and $\mathsf{Arc}_{DL \to DL}$ require the inclusion of a generator G as described above. Assuming (probabilistically padded) RSA encryption is used when the proprietary key is an RSA key, and ElGamal encryption used when it is a discrete log key, the encryption of the "pointer" has size $|N|$ resp. $2|p|$.

The two arc establishment protocols that are directly based on Paillier encryption result in ciphertexts of size $2|N|$. The protocol using composite ElGamal alone results in ciphertexts of that same size, while the protocol using both ElGamal encryption and Paillier encryption naturally results in ciphertexts of size $4|N|$.

Thus, for $|N| = |p| = 1024$ bits, the total certificate expansion is between $3|N| = 384$ and $4|N| + 2|p| = 768$ Bytes.

Stateless-Recipient Certified E-Mail System Based on Verifiable Encryption

Giuseppe Ateniese and Cristina Nita-Rotaru

Department of Computer Science
The Johns Hopkins University
{ateniese,crisn}@cs.jhu.edu

Abstract. In this paper we present a certified e-mail system which provides fairness while making use of a TTP only in exceptional circumstances. Our system guarantees that the recipient gets the content of the e-mail if and only if the sender receives an incontestable proof-of-receipt. Our protocol involves two communicating parties, a sender and a recipient, but only the recipient is allowed to misbehave. Therefore, in case of dispute, the sender solicits TTP's arbitration without involving the recipient. This feature makes our protocols very attractive in real-world environments in which recipients would prefer to assume a passive role rather than being actively involved in dispute resolutions caused by malicious senders. In addition, in our protocol, the recipient can be *stateless*, i.e., it does not need to keep state to ensure fairness.

1 Introduction

The Internet has revolutionized the computer and communications world like nothing before and it is today an important global resource for millions of people. With its continue growing, it generated tremendous benefits for the economy and our society. However, the Internet does not provide all the services required by the business communication model such as secure and fair electronic exchange or certified electronic delivery.

A fair electronic exchange protocol ensures that, at the end of the exchange, either each player receives the item it expects or neither part receives any information about the other's item. The classical solution to the fair exchange problem is based on the idea of *gradually* exchanging small parts of the items. However, practical solutions to the problem require a trusted third party (TTP) as arbitrator. More specifically, in *on-line* protocols the trusted party is employed as a delivery channel whereas in *off-line* protocols the trusted party is involved only in case of dispute. On-line protocols require the presence of the TTP in every transaction and, usually, do not provide confidentiality of the items exchanged. In addition, in some cases, the sender receives a receipt signed by the TTP rather than by the original recipient of the message. In off-line protocols, the TTP is invoked only under exceptional circumstances, for example in case of disputes or emergencies.

B. Preneel (Ed.): CT-RSA 2002, LNCS 2271, pp. 182–199, 2002.
© Springer-Verlag Berlin Heidelberg 2002

In a certified e-mail scheme the intended recipient gets the mail content if and only if the mail originator receives an irrefutable proof-of-delivery from the recipient.

In this paper we present a certified e-mail system which implements an off-line protocol that makes use of *verifiable encryption* of digital signatures as building block. A *verifiable encryption* of a digital signature represents a way to encrypt a signature under a designated public key and subsequently prove that the resulting ciphertext indeed contains such a signature.

The rest of the paper is organized as follows. The next section discusses related work done in the areas of fair exchange and certified e-mail protocols. Section 3 outlines the certified e-mail protocol used by our system. We analyze the protocol in Section 4. Finally, we discuss the implementation of the system in Section 5.

2   Related Work

One approach in solving the *fair exchange* problem consists of gradually exchanging information about the items between the two parties. Works in this direction generally rely on the unrealistic assumption that the two parties have equal computational power ([12]) or require many rounds to execute properly ([5]).

Another approach focuses on increasing the overall efficiency by using TTPs. Notable works in this direction are the three-message off-line protocol for certified e-mail presented in [18] and the efficient off-line fair exchange protocols in [1,2,7]. The protocol in [1] makes use of verifiable escrow schemes implemented via a *cut and choose* interactive proof. Although expensive, the protocol provides timely termination assuming only resilient channels.

A on-line certified e-mail protocol is presented in[24]. The protocol uses as TTP a number of replicated servers. This has the drawback that each server must be trusted in order to have the protocol working properly. Having only one single compromised server would invalidate the entire scheme. Bahreman and Tygar [6] present an on-line scheme using six messages. The scheme does not address confidentiality from the TTP. An optimal on-line certified e-mail protocol using only four messages is described in [11].

Schneier and Riordan [21] present a protocol where the TTP acts as a public publishing location (which might be implemented as a secure database server). The authors describe both an on-line and an off-line version of the protocol. We note that the off-line solution requires a *visible TTP*, since the form of the receipt changes depending on whether the trusted entity was invoked or not. Moreover, the TTP must be directly involved in any secondary adjudication as it must provide, in the case involving dispute resolution, an additional signed proof-of-mailing with each query or deposit.

Recently, Ateniese et al. [4] have shown how to realize hybrid schemes that combine the strengths of both the on-line and off-line approaches to achieve efficiency while involving parties that are semi-trusted rather than fully trusted.

3 An Efficient Off-Line Protocol

In this section, we describe the setting in which we operate and the certified e-mail protocol built via verifiable encryption of RSA-based digital signatures.

In the rest of the paper, we will assume that the communication is carried over private and authenticated channels.

A certified e-mail protocol should minimally provide:

- **Fairness:** No party should be able to interrupt or corrupt the protocol to force an outcome to his/her advantage. The protocol should terminate with either party having obtained the desired information, or with neither one acquiring anything useful.
- **Monotonicity:** Each exchange of information during the protocol should add validity to the final outcome. That is, the protocol should not require any messages, certificates, or signatures to be revoked to guarantee a proper termination of the protocol. This is important, because if revocation in needed to ensure fairness, then the verification of the validity of the protocol outcome becomes a bottleneck as it requires TTP's active participation.
- TTP **invisibility:** A TTP is *visible* if the end result of an exchange makes it obvious that the TTP participated during the protocol.
- **Timeliness:** It guarantees that both parties will achieve their desired items in the exchange within finite time.

Occasionally, it is desirable to keep the content of confidential e-mails secret from trusted parties acting as intermediaries. Thus, an optional feature is:

- **Confidentiality:** In case the exchange is deemed confidential, the protocol should not need to disclose the message contents to any other party excepting the sender and the recipient.

3.1 Our Setting and Verifiable Encryption of RSA Signatures

Let n be the product of two large primes p and q, such that factoring n is computationally infeasible without knowing p or q. It is generally convenient to work inside some cyclic subgroup of large order. For that reason, we generate p and q as *safe* primes, i.e., $p = 2p' + 1$ and $q = 2q' + 1$ where p' and q' are primes. We then consider the subgroup $QR(n)$ of quadratic residues, i.e., $QR(n)$ contains all the elements y such that there exists x with $y = x^2$. It is easy to see that $QR(n)$ is a cyclic group of order $p'q'$. Finding a generator is also straightforward: randomly select $\bar{g}$ and compute the generator $g = \bar{g}^2$. Since elements in $QR(n)$ have orders p', q', or $p'q'$, the order of g will be $p'q'$ with overwhelming probability.

We can also describe a proof of knowledge that allows a prover to convince a verifier of the equality of discrete logarithms. Let $g, h \in QR(n)$ be publicly known generators. The prover selects a secret x and computes $y_1 = g^x$ and $y_2 = h^x$. The prover must convince the verifier that:

$$\text{Dlog}_g y_1 = \text{Dlog}_h y_2.$$

The protocol, drawn from [9], is run as follows:

1. The prover randomly chooses t and sends $(a, b) = (g^t, h^t)$ to the verifier.
2. The verifier chooses a random challenge $c \in \{0, 1\}^{160}$ and sends it to the prover.
3. The prover sends $s = t - cx \bmod p'q'$ to the verifier.
4. The verifier accepts the proof if:

$$a = g^s y_1^c \quad \text{and} \quad b = h^s y_2^c.$$

To turn the protocol above into a signature on an arbitrary message m, the signer can compute the pair (c, s) as:

$$c = \mathcal{H}(m\|y_1\|y_2\|g\|h\|g^t\|h^t), \ \ s = t - cx \bmod p'q'.$$

where $\mathcal{H}(\cdot)$ is a suitable hash function. To verify the signature (c, s) on m, it is sufficient to check whether $c' = c$, where

$$c' = \mathcal{H}(m\|y_1\|y_2\|g\|h\|g^s y_1^c\|h^s y_2^c).$$

Following the notation in [3], we will denote an instance of this signature technique by $EQ\_DLOG(m; g_1^x, g_2^x; g_1, g_2)$. Substantially, $EQ\_DLOG(\cdot)$ is a Schnorr-like signature [23] based on a proof of knowledge performed non-interactively making use of an ideal hash function $\mathcal{H}(\cdot)$ (à la Fiat-Shamir [14]).

In [3] it is shown that it is possible to define very efficient protocols for verifiable encryption of several digital signature schemes. Given an instance S of a digital signature on an arbitrary message, we say that $VE(S)$ is a TTP-verifiable encryption of S, if such an encryption can be verified to contain S in a way that no useful information is revealed about S itself. Only TTP is able to recover the signature from the encryption $VE(S)$.

We focus our attention on RSA signatures [22], that is, if (e, n) is a public key with e prime then the secret key is d such that $ed \equiv 1 \bmod 2p'q'$. To sign a message m, it is sufficient to compute $C = R(m)^d$, where $R(\cdot)$ is a publicly known redundancy function as defined in PKCS#1, ISO/IEC 9796, etc. (see [17] p. 442). For the sake of simplicity, we employ the hash-and-sign paradigm and assume that $R(\cdot)$ is a suitable hash function such as SHA-1. The signature is accepted only if $C^e = R(m)$. The encryption algorithm used to encrypt the RSA signature is the ElGamal algorithm: given a secret key x and a corresponding public key g^x, a message s is encrypted by generating a random r and computing $K_1 = sg^{xr}$, $K_2 = g^r$. The value s can be recovered by computing $s = K_1/(K_2)^x$.

Each user first runs a one-time *initialization phase* by which the user and the trusted third party $\mathcal{T}$ agree on common parameters. More specifically (see [3] for details):

1. The user, U say, sends (e, n) to $\mathcal{T}$;
2. $\mathcal{T}$ authenticates the user then selects a random base $\bar{g}$ and a random exponent x. It computes $g = \bar{g}^2$ and sends $CERT_{\mathcal{T}:U} = Sign_{\mathcal{T}}(g, y = g^x, U, (e, n))$ to the user, where $Sign_{\mathcal{T}}(\cdot)$ denotes a signature computed by $\mathcal{T}$.

It is assumed that the user provides a proof of n being a product of safe primes (see [8]). The signature $CERT_{\mathcal{T}:U}$ is a publicly known certificate and, in a real-world implementation, will contain relevant information including protocol headers, timestamp, transaction ID, and certificate lifetime. Notice that, the trusted party $\mathcal{T}$ does not need to store the secret exponent x for each user. In fact, such a secret can be inserted into $CERT_{\mathcal{T}:U}$ encrypted via a symmetric encryption algorithm. Thus, $\mathcal{T}$ needs to store only one value, the symmetric key, for all the users.

The computation of a verifiable encryption of a RSA signature on a message m is performed as shown in [3]. In particular, the user U encrypts via ElGamal the signature $R(m)^d$ by computing a random r and:

$$K_1 = R(m)^{2d}y^r \text{ and } K_2 = g^r.$$

Notice that, the signature $R(m)^d$ is squared to make sure that the value encrypted belongs to the set of quadratic residues $QR(n)$. Then, the user provides evidences that the encryption has been correctly computed by showing that:

$$\text{Dlog}_{y^e}(K_1^e/R(m)^2) = \text{Dlog}_g(K_2),$$

and this is done via $EQ\_DLOG(\cdot)$ w.r.t. the message m. (Observe that the verifier should recover the bases $y = g^x$ and g from $CERT_{\mathcal{T}:A}$.) We will denote the verifiable encryption of a RSA signature, $R(m)^d$, with $VE_{\mathcal{T}}(R(m)^{2d})$. Let U denote a generic user, the value $VE_{\mathcal{T}}(R(m)^{2d})$ contains: $CERT_{\mathcal{T}:U}$, the ElGamal encryption of $R(m)^{2d}$, and the signature of knowledge of the equality of discrete logarithms $(EQ\_DLOG(\cdot))$.

3.2   The Protocol

A certified e-mail protocol using verifiable encryption is shown in Figure 1. Consider a scenario in which the sender A sends a message to B and wants a receipt signed by B in exchange. The recipient B has to generate and sign the receipt before being able to read the content of the message. The protocol has to provide fairness, specifically, it must ensure that the sender receives the receipt if and only if the recipient can read the message. The protocol is designed so that the TTP is invoked only in case of dispute. As long as both A and B are honest, there is no need to involve the trusted entity in the protocol. This is a big advantage compared to on-line protocols in which a trusted entity is needed for each transaction.

Moreover, the protocol is designed to make sure that A cannot misbehave. Only B is allowed to cheat by not sending the message in the last step. This feature is highly desirable in the setting of certified e-mail, as the recipient would prefer to assume a *passive role* rather than being actively involved in dispute resolutions. Notice that, in a certified e-mail protocol, the sender initiates the exchange process, thus it is natural to desire that the recipient of the message be relieved by any burden caused by malicious senders.

User B receives the certificate $CERT_{T:B}$ by engaging in an initialization phase with the trusted party T as explained in the previous section. Similarly, B's public key is (e, n) with e prime and n product of safe primes and $QR(n)$ is the subgroup of squares in which we operate. The protocol consists of the following steps (operations are taken modulo n):

- **Step 1** The sender A selects a random r, computes $y = r^e R(m)$, and signs it including a protocol header PH. Such a signature, denoted by S_A, is sent to B.
- **Step 2** The recipient B squares y and computes $(y^2)^d = r^2 R(m)^{2d}$. It then computes the verifiable encryption of y^{2d}, $VE_T(y^{2d})$, and sends the result to A. However, B has to sign the result in order to include a protocol header $\overline{PH}$ and the sender's signature S_A. More importantly, B's signature (S_B) makes it possible to neutralize *malleability* attacks against the ElGamal encryption and also preserves B's *protocol view* at that specific point in time.
- **Step 3** After receiving the message from B, A verifies the signature and that the encryption contains the correct receipt. If that is the case, A sends m to B.
- **Step 4** The recipient B reads the message m and sends the receipt $Rec = R(m)^d$ to A.

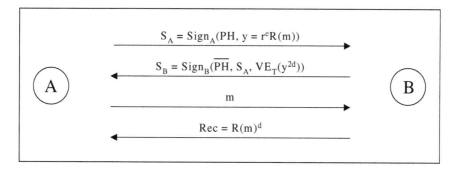

Fig. 1. Off-line certified e-mail protocol via verifiable encryption of RSA signatures

Notice that, a certified e-mail protocol is not a simultaneous exchange of items but rather a *asymmetric* exchange since the message has to be sent first to allow the recipient to compute a corresponding receipt based on the message received. This fact has some positive side effects on our scheme. For instance, the recipient does not need to include any *time limit* into the signature S_B since the decisions of sending a particular message and when this has to happen are taken exclusively by the sender.

If B does not send the receipt in Step 4, then A contacts the trusted entity and both run the following protocol:

- **Step 1** A sends B's signature, S_B, to the TTP along with r and m.
- **Step 2** The TTP verifies first the signatures S_A and S_B (S_A is contained in S_B). Then, it recovers y^{2d} from the verifiable encryption and computes y^d (see Remark 2 below). Finally, the TTP checks whether the value $s = y^d/r$ is indeed a valid signature of the message m under B's public key (i.e, it checks whether $s^e = R(m)$). If so, it sends s to A and forwards m to B.

The TTP has to forward the message m to B to nullify any attempts of the sender A to successfully retrieve a receipt without revealing the message m to B. Specifically, A may not have sent the message m in Step 3 above.

The protocol fairness is built around the assumption that the sender A can verify that the verifiable encryption indeed contains a valid receipt. Only the TTP can recover the receipt from the verifiable encryption.

Remark 1. The protocol headers, PH and $\overline{PH}$, contain relevant information such as the identities of the parties involved (A,B, and TTP), the cryptographic algorithms employed, timestamps and transaction IDs to prevent replay attacks, and other pertinent information about the protocol.

Remark 2. Notice that, the recipient B squares the value $y = r^e R(m)$ sent by A to make sure that it is signing an element of the set $QR(n)$ (d is odd since e is prime). Only the recipient B knows the factorization of n and it is usually infeasible to compute square roots modulo n without knowing the factors of n. However, given $z = y^{2d}$, the TTP can efficiently compute y^d by employing the following well-known method, based on the Euclidean algorithm, which we report here for convenience:

1. observe that $z^e = y^2$ and that $gcd(e, 2) = 1$;
2. we can use the extended Euclidean algorithm to compute two integers u, v such that $u2 = 1 + ve$ in $\mathbb{Z}$;
3. raising both terms of the equation $z^e = y^2$ by u, we obtain: $z^{ue} = y^{u2} = y^{1+ve} = yy^{ve}$;
4. thus we have: $z^{ue}y^{-ve} = y$, or $(z^u y^{-v})^e = y$;
5. it is now clear that the term $z^u y^{-v}$ is congruent to y^d (modulo n).

Remark 3. In our protocol, the sender A has to reveal the message m to the TTP in case of dispute. If message privacy has to be preserved, it is sufficient to substitute m with $\overline{\overline{PH}}\|P_B(m)$ in the protocol above, where $P_B(\cdot)$ represents the public-key encryption under B's public key and $\overline{\overline{PH}}$ is a protocol header. Notice that the receipt assumes a new format:

$$Rec = R(\overline{\overline{PH}}\|P_B(m))^d,$$

which has to be interpreted in a special way: it is considered a valid receipt of the message m only when accompanied by m and $\overline{\overline{PH}}$ such that:

$$(Rec)^e = R(\overline{\overline{PH}}\|P_B(m)).$$

The public-key encryption $P_B(\cdot)$ should be deterministic or, if randomized, the sender A must reveal the random parameters used to encrypt the message. The approach we have taken for the implementation of the protocol is to encrypt the message m as $E_k(MAC_l(m)||m)$, $P_B(k||l)$, where: k, l are random secret values; $MAC_l(\cdot)$ is a MAC function, such as HMAC-SHA-1; $P_B(\cdot)$ is a deterministic public-key encryption algorithm, such as plain RSA; $E_k(\cdot)$ is a symmetric-key encryption algorithm, such as AES in CBC mode.

The new protocol header $\overline{\overline{PH}}$ has to be checked, either by B or the TTP, to contain the correct information relevant to the protocol. Moreover, it has to clearly state that the receipt Rec has to be interpreted in the special way described above.

Remark 4. The certified e-mail protocol presented above works for RSA signatures but can easily be extended to work for other schemes based on a similar setting such as Rabin and Guillou-Quisquater [16] signature schemes, or provably unforgeable signature schemes such as Cramer-Shoup [10] and Gennaro-Halevi-Rabin [15]. In [3], it is shown how to extend the verifiable encryption protocol to work with such popular signature schemes.

4   Analysis and Comparisons

In this section we present an analysis of our protocol and we compare it with the state-of-the-art in the field. Our claim is the following:

Claim: *The protocol above is a certified e-mail protocol which provides fairness, monotonicity, timeliness, and TTP invisibility. Moreover, the protocol optionally provides confidentiality of the message, i.e., the arbitration can be performed without revealing the e-mail content to the trusted intermediary.*

Sketch: Clearly our protocol provides TTP invisibility since the structure of the receipt does not indicate whether the TTP was involved or not in dispute resolutions. The protocol provides also monotonicity since any signature (including the receipt) will not be revoked in order to guarantee a proper termination of the protocol. Confidentiality is achieved by encrypting the actual message content in such a way that only the recipient can *open* it and this is achieved through standard encryption technology.

We assume only *resilient* channels. A resilient channel will eventually deliver a message sent through it within a time lapse which may be arbitrarily long, yet finite. Moreover, the recipient does not need to include any *time limit* into the signature S_B and the sender A has the ability to decide to abort the protocol and adopt a scheme for protocol resolution that can be executed in a finite period of time. Therefore, our protocol provides timeliness.

Regarding fairness, it is sufficient to prove that: a message is read by the recipient B if and only if the sender A gets the corresponding receipt. Observe that the first two messages of our protocol are used just to collect evidences by which the sender A can solve disputes by interacting with the TTP. If the

TTP is not invoked then the relevant protocol messages are only those in Step 3 and Step 4 where a message m is sent in exchange of the corresponding receipt. Therefore, fairness is preserved in this case. If the TTP is invoked (by A) then B's signature (S_B) will be sent to the TTP along with the message m and the blinding factor r. The TTP will compute y^d from the verifiable encryption and will check whether:

$$(y^d/r)^e = R(m).$$

If that is the case, then A and B receive y^d/r and m, respectively. Hence, even in this case fairness is preserved since the sender will receive a signature on a message which is forwarded to the recipient. □

It is interesting to notice that the recipient does not need to contact the TTP in case of dispute. This feature makes our protocols very attractive in real-world environments in which recipients would prefer to assume a passive role rather than being actively involved in dispute resolutions caused by malicious senders. More importantly, the recipient is *stateless* in the sense that it does not need to store state information regarding the transactions in which he is involved[1]. Indeed, the recipient may not store anything about the first two steps of the protocol and, in principle, the message embedded by the sender in the value y in Step 1 of the protocol could be different from the message sent in Step 3. Obviously, this does not violate the fairness property since the sender A cannot use the message in Step 2, since it is encrypted, unless he contacts the TTP. However, the TTP will always forward the corresponding message to B, thus neutralizing de facto any attempt of the sender to force an outcome of the protocol to his advantage.

We compare now our protocol with previously proposed protocols. Some of the off-line protocols are not monotonic, for instance, the protocol in [2] requires signatures to be revoked in order to guarantee fairness. Among monotonic and off-line protocols, we believe those in [18,1] represent the state-of-the-art in the field. The work of Micali [18] shows that it is possible to achieve a simple certified e-mail protocol with only three messages (one less than our protocol). However, it should be noticed that:

1. the recipient of the message has to be actively involved in the dispute resolution and is forced to keep state;
2. a time limit has to be incorporated into the message by the sender to force the recipient to send the receipt within a specified period of time. This has to be done in order to guarantee fairness;
3. a *reliable* channel (as opposed to a resilient channel) is required between the recipient and the trusted third party.

A channel is reliable when it is always operational and operates without delays. It is very difficult to build a reliable channel in some network environments, such as wireless networks. This fact may limit the applicability of the protocol in

---

[1] Notice however that implementations of the certified email scheme may require the recipient to store certain state information.

[18]. Furthermore, for each message received, the recipient is forced to communicate with the trusted intermediary in case of dispute and such a communication has to happen before the time limit expires.

The work in [1] presents a fair-exchange protocol which is provably secure in the random oracle model. The authors specialize their protocol to work as an off-line certified e-mail scheme that, similarly to our protocol, requires only resilient channels. Their protocol is based on *verifiable escrow schemes*, which essentially are verifiable encryptions where each encryption comes with an attached *condition* that specifies a decryption policy. As our scheme, their protocol works for a broad range of signature schemes. However, the scheme in [1] has some drawbacks, in particular:

1. it is expensive in terms of communication complexity, performance, and amount of data transmitted. This is mainly due to the *cut-and-choose* interactive proof technique employed to achieve a verifiable escrow.
2. the recipient has to keep state and both the sender and the recipient have to be actively involved in dispute resolutions;
3. the trusted third party needs to keep state.

Notice that both protocols [18,1] and the version of our protocol that provides confidentiality are *invasive*, that is, the receipt generated by the receiver is not a regular signature but has to be interpreted in a special way. Our original protocol, however, is *non-invasive* as the receipt is precisely the signature of the recipient on the message received.

We believe it is important to have a *stateless* recipient who is not involved in dispute resolution protocols. Imagine a scenario in which a user receives hundreds of messages and is forced to keep track of all of them, store state information, and engage in protocol resolutions with the TTP in case of dispute. This may be very unappealing for users, in particular for those operating in environments where servers may frequently *crash*, losing state information. In some case (such as in [18]), operations have to be made before a time limit expires which may make even impossible to guarantee fairness in some environments.

Stateless-recipient protocols may be very useful when users are equipped with mobile devices such as cellular phones or wireless PDAs. Indeed, mobile devices are often switched off, which may cause some time limits to expire without giving the possibility to run any dispute resolution protocol[2].

5   System Architecture

In this section we present a description of the system we implemented. First, we give a brief overview of the technology behind electronic mail systems, then we show how to establish forms of interaction between our system and other

---

[2] For instance, it is required to leave mobile devices off in proximity of hospitals or inside airplanes. Devices could also turn themselves off when, for instance, batteries are flat.

standard modules involved in the process of delivering electronic mail contents and, finally, we describe in detail our system. We called our prototype implementation TURMS, an Etruscan god, messenger of the gods and guide of the deceased to the underworld.

5.1   Overview of Electronic Mail

Electronic mail is today probably one of the most used services on the Internet. It provides support to send a message to a destination. The message is passed from one computer to another, often through computer networks and/or via modems over telephone lines. The process of sending, delivering and receiving e-mail is specified in some standards and makes use of three types of programs, each of them with a specific task.

A *Mail User Agent* (MUA) is a program that allows the user to compose and read electronic mail messages. It provides the interface between the user and the Mail Transfer Agent (MTA). Outgoing mail is passed to an MTA for delivery while the incoming messages are picked up from a MTA. A MUA can also pick up mail remotely from, for instance, a POP3 server, via POP3 protocol.

A *Mail Transfer Agent* (MTA) is a system program which accepts messages from the MUA and routes them to their destinations. It sometimes delivers mail into each user's system mailbox. A MTA can also communicate with other MTA programs via the SMTP protocol, in order to deliver mail remotely. MTAs are responsible for properly routing messages to their destination, using so-called Mail eXchanger (MX) records. Mail eXchanger records are maintained by domain name servers (DNS) and tell MTAs where to send mail messages. More precisely, they tell an MTA which intermediate hosts should be used to deliver a message to the target host. The MX records vary depending on the domain.

A *Mail Delivery Agent* (MDA) is used to place a message into a user's mailbox. When the message arrives at its destination, the MTA will give the message to the appropriate MDA, who will add the message to the user's mail-box.

Our system works at the transport level, in connection with Exim [13], a mail transport agent. This approach has the advantage that allows a TURMS user to handle mail using theoretically any MUA.

5.2   TURMS and Exim

Exim [13] is a mail transfer agent developed at the University of Cambridge designed to work efficiently on systems that are permanently connected to the Internet and are handling a general mix of mail. In addition, with special configuration, Exim can act as a mail delivery agent.

Exim was built having a decentralized architecture, so there is no central process performing overall management of mail delivery, but some DBM files are maintained to make the delivery more efficient in some cases. The system implements flexible retry algorithms, used for directing and routing addresses and for delivery.

The system can handle a number of independent local domains on the same machine and provides support for multiple user mailboxes controlled by prefixes or suffixes on the user name.

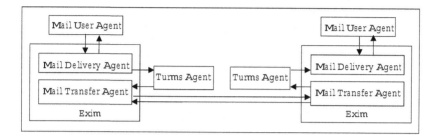

Fig. 2. TURMS and Exim communication

The main delivery processing elements (drivers) of Exim are called directors, routers, and transports.

A *director* is a driver that operates on a local address, either determining how to deliver the message, or converting the address into one or more new addresses (for example, via an alias file). A local address is one whose domain matches an entry in the list given in the 'local_domains' option, or has been determined to be local by a router. The fact that an address is local does not imply that the message has to be delivered locally; it can be directed either to a local or to a remote transport.

A *transport* is a driver that transmits a copy of the message from Exim to some destination. There are two kinds of transport: for a local transport, the destination is a file or a pipe on the local host, while for a remote transport the destination is some other host. A message is passed to a specific transport as a result of successful directing or routing. If a message has several recipients, it may be passed to a number of different transports.

A *router* is a driver that operates on an apparently remote address, that is an address whose domain does not match anything in the list given in 'local_domains'. When a router succeeds it can route an address either to a local or to a remote transport, or it can change the domain, and pass the address on to subsequent routers.

Our system takes advantage of the following features of Exim: the ability to allow a message to be piped to another program, the possibility to have multiple user mailboxes controlled by prefixes or suffixes on the user name and the ability to send messages (to remote or local e-mail addresses) using Exim.

These features allow a process, the Turms_agent, to intercept messages before they are delivered to a user and process them. Also, when needed, auxiliary messages can be generated by the TURMS agent and sent using Exim.

The message exchange mechanism between a local TURMS agent and Exim is presented in Figure 2. Special entries in Exim's configuration file indicate to

```
# directors configuration section
turms_handler:
     driver = pipe
  1
  ──▶ command =''/bin/turms_agent$ {local_part}''
     user = crisn
     group = users

# transport configuration section
turms:
     driver = smartuser
  2
  ──▶ transport = turms_handler
  3
  ──▶ prefix = turms-
```

Fig. 3. Exim configuration file - TURMS entries

Exim that mail delivered to an user containing a certain prefix, *turms-* in our case (see Figure 3, Mark3), to be delivered using a specific transport, *turms_handler* in our case (see Figure 3, Mark 2). The transport specified for that prefix is a pipe to a program, Turms_agent (see Figure 3, Mark 1). This way, the message send to a local or remote user is delivered to a local Turms_agent program.

When Turms_agent receives a message, it processes it according to the protocol specifications, it generates a new message and it sends it to the real user using Exim. If the message is sent to a domain for which Exim does not do local delivery, the message will be sent via SMTP to another Exim server on another machine. There, because of the prefix mechanism, the message will be delivered to a local Turms_agent process which in turn will send a message to the local user, by using Exim. If the message is sent to a domain for which Exim does local delivery, the message will be delivered either to the end user or the Turms_agent, depending on the protocol. The delivery address specifies to Exim if the destination is a Turms_agent process or a local user inbox. Note that all the Exim servers running the certified e-mail protocol need to be configured to deliver special messages to a local TURMS agent, responsible of implementing the certified e-mail protocol.

5.3   System Implementation

We have actually implemented a slightly modified version of the protocol in Section 3.2. In particular, the value y in Step 1 is now $MAC_k(m)$, for a suitable MAC function such as HMAC-SHA-1. In Step 3, the sender reveals m and also k and the recipient checks whether the message m is the same received in Step 1. The receipt is a signature on the MAC function.

The protocol itself requires an initialization phase which has to be done only once, and in case of dispute, a recovery phase. We provide support for all of these operations. The system consists of a web-based interface, Turms_CA, providing support for the initialization and recovery off-line phases, and of an MTA, Turms_agent, implementing the certified e-mail protocol.

One of the design features we considered was encapsulating all the cryptographic operations in a library, used by both Turms_CA and Turms_agent. The library makes use of the openssl [19] library and provides facilities such as: strong RSA keys generation and managing, TURMS certificate definition and management (generating, signing, verifying), operations on Schnorr-like objects used in our protocol, RSA verifiable encryption definition and management (generation, verifying, ability to extract the RSA signature out of the verifiable encryption). In addition, the library also provides conversion from the computation data format (openssl specific data structure) to communication format for both verifiable encryption and TURMS certificate entities. For each of these data structures, the library provides a simple and easy to use API.

Turms_CA provides an interface allowing users to register and to obtain a certificate. The user submits his public RSA key along with some additional information and will receive the corresponding TURMS certificate In addition, for each user, we have decided to save the corresponding secret of the CA rather than incorporating it into the certificate This secret is used to extract RSA signatures out of a verifiable encryptions.

Turms_CA can also solve disputes. A user can submit a claim including the verifiable encryption file and the message. The CA computes the signature out of the verifiable encryption information, then sends the signature to the user that submitted the claim and the message to the user whose verifiable encryption was submitted.

The core of the system is the *Turms_agent* program which implements the certified e-mail protocol. *Turms_agent* is a stateless program, a different instance of the program is invoked with every new message. The state of the protocol for different messages is saved on the disk. Every state of the protocol for a message has a correspondent file saved on disk. A message is uniquely identified by a concatenation of the process id, host id and current time. Every state also has a unique identifier associated with it. The security of the channel between two Turms_agents is achieved via Blowfish encryption with a size key of 16 bytes. We used an HMAC with a key size of 10 bytes to obtain a randomized one-way function of the message. Turms_agent also logs for each transaction information about the main important steps.

The protocol consists of a sequence of actions taken by a Turms_agent program upon receiving a message. Every message is associated with a specific transaction. The type of the message along with the transaction identifier are specified in the destination address. A transaction is opened when a user attempts to send a certified e-mail message and it ends in one of the following cases: the exchange was performed correctly, the exchange was canceled, or the exchange started but the recipient did not send the receipt. The protocol uses the following types of messages:

- Original_Mess is the message that has to be sent. It is generated by a MUA, no transaction identifier or type is associated with it.
- Hmac_Mess contains the value of an HMAC function applied on the body of an Original_Mess. It is generated by a *Turms_agent*.

- Invitation_Mess is the message that notifies a user about a certified e-mail message. It is generated by a *Turms_agent*.
- Cancel_Mess is an Invitation_Mess which has the Subject field consisting only of the word 'Cancel' indicating that the transaction was refused. The body of the message is ignored. It is generated by a MUA, in reply to an Invitation_Mess coming from a *Turms_agent*.
- Accept_Mess message is similar as structure with the Cancel_Mess, but the Subject field consists of the word 'Accept' indicating that the transaction was accepted. The body of the message is ignored. It is generated by a MUA, in reply to an Invitation_Mess coming from a *Turms_agent*.
- Transaction_Canceled_Mess indicates that a transaction was canceled. It is generated by a *Turms_agent*.
- Verifiable_Encryption_Mess contains a verifiable encryption message. It is generated by a *Turms_agent*.
- Original_Mess_and_Key contains the body of an Original_Mess and the key used to compute the HMAC value applied on the Original_Mess that was sent in the corresponding (has the same transaction identifier) Hmac_Mess. It is generated by a *Turms_agent*.
- Signature_Mess contains a RSA signature of an HMAC value of an Original_Mess message. It is generated by a *Turms_agent*.

In response to an event, a *Turms_agent* can take actions that will result in sending messages and/or saving additional data on the disk. Consider a scenario in which the sender A wants to send a certified e-mail to B. We make use of the following notation: A's domain is denoted by $domain_A$ and B's domain is denoted by $domain_B$, the *Turms_agent* programs corresponding to A's mail server and B's mail server are denoted by $turms\_agent_A$ and $turms\_agent_B$, respectively . Finally, $user_A$ and $user_B$ represent the MUAs at A's site and B's site, respectively.

The protocol consists of the following steps:

Step 1. $user_A$ sends the Original_Mess message. $user_A$ sends the Original_Mess message to an alias defined for B, $certified\_B$ say, which includes the address to which the mail is sent. The address should be prefixed with the prefix specified in the Exim configuration file (see Figure 3), should be sent to the local mail server and should contain enough information to allow to recover the actual remote address. Exim will deliver the message via the pipe transport to the $turms\_agent_A$ program. The result of this step is that $turms\_agent_A$ will receive the Original_Mess.

Step 2. $turms\_agent_A$ receives the Original_Mess message. When the message is received by $turms\_agent_A$, the agent creates a unique identifier for this new transaction: $timestamp - process\_id - host\_id$. Then it processes the message, saves the body of the message along with some header information. It generates a key that will be used to compute a HMAC of the body of the message, saves the key on the disk, computes HMAC of the message, saves the content of the message, recovers the real address, generates a Hmac_Mess with the HMAC value just computed and sends it to $turms\_agent_B$. The type of the message along with the identifier is specified in the address. Also the reply address is updated such that the mail appears as coming from a *turms_agent* (by adding the prefix). The result of this step is that $turms\_agent_B$ will receive a Hmac_Mess.

Step 3. $turms\_agent_B$ receives the Hmac_Mess message. When $turms\_agent_B$ receives the Hmac_Mess, it saves this information on the disk and generates an Invitation_Mess, and sends it to $user_B$, notifying about a certified e-mail message for him, and asking him to reply to this message with 'Accept' written in the subject, if he accepts the message, or 'Cancel' if he refuses it. The result of this step is that $user_B$ will receive an Invitation_Mess.

Step 4. $user_B$ receives the Invitation_Mess message. When $user_B$ receives the Invitation_Mess, he will reply either with 'Accept' or 'Cancel' in the subject. The result of this step is that $turms\_agent_B$ will receive either an Accept_Mess or a Cancel_Mess.

Step 5. $turms\_agent_B$ receives the Cancel_Mess message. When $turms\_agent_B$ receives the Cancel_Mess, it generates and sends two Transaction_Canceled-_Mess messages, one to $user_B$ and the other to $turms\_agent_A$. The transaction is closed. The result of this step is that both $user_B$ and $turms\_user_A$ will receive a Transaction_Canceled_Mess.

Step 6. $turms\_agent_B$ receives the Accept_Mess message. When $turms\_agent_B$ receives the Accept_Mess, it recovers the HMAC value of the message from the disk, then computes the verifiable encryption of B's signature, and generates a Verifiable_Encryption_Mess message and sends it to $turms\_agent_A$. The result of this step is that $turms\_agent_A$ will receive a Verifiable_Encryption_Mess.

Step 7. $turms\_agent_B$ receives the Transaction_Canceled_Mess message. When $turms\_agent_B$ receives the Transaction_Canceled_Mess, he sends it to $user_A$ and deletes the HMAC and the OriginalL_Mess information saved on the disk in step 2. The result of this step is that $user_A$ will receive a Transaction_Canceled_Mess.

Step 8. $turms\_agent_A$ receives the Verifiable_Encryption_Mess message. When $turms\_agent_A$ receives the Verifiable_Encryption_Mess, it verifies that the message indeed contains a RSA signature. If yes, it recovers from the disk both the body of the Original_Mess and the key used to compute HMAC value, generates the Original_Mess_and_Key message by concatenating the key to the message and then sends it to $turms\_agent_B$. Also the Verifiable_Encryption_Mess is sent to $user_A$. The result of this step is that $turms\_agent_B$ will receive an Original_Mess_and_Key message and $user_A$ will receive a Verifiable_Encryption_Mess.

Step 9. $turms\_agent_B$ receives an Original_Mess_and_Key message. Upon receiving an Original_Mess_and_Key message, $turms-\_agent_B$ computes a HMAC on the body the received message with the key he just received and it compares it with the HMAC data saved on the disk in Step 2. If they are the same, it computes B's RSA signature on the HMAC, generates a Signature_Mess message and sends it to $turms\_agent_A$. If the two HMAC values are not the same, then $turms\_agent_B$ generates and sends two Transaction_Canceled_Mess messages, one to $user_B$ and the other to $turms\_agent_A$. The result of this step consists of two messages: $user_B$ receives the Original_Mess and $turms\_agent_A$

receives a Signature_Mess, or both $user_B$ and $turms\_agent_A$ receive a Transaction_Canceled_Mess.

Step 10. $turms\_agent_A$ receives a Signature_Mess message. When $turms\_agent_A$ receives Signature_Mess, it sends it to $user_A$ and cleans all the auxiliary files used during the transaction.

In order to have the protocol described above working correctly in the case when the message contains one or more attached files, some additional processing needs to be done: the corresponding MIME information from the message header needs to be saved when the original message is processed. This information is used when the message is finally sent to the user (step 7), to make sure that MUA understands that the message carries some attached files.

We used the following set up for developing and testing our system. The web interface running on a Linux machine dual 450MHz Pentium II, 128MB RAM, running Apache Web Server. We tested our system in a configuration of two virtual domain names (securemail1.cs.jhu.edu and securemail2.cs.jhu.edu), each served by an Exim server version 3.14. The machines were 300MHz Pentium II, 256 Mb RAM, Linux boxes (2.16 kernel). The program PINE [20] was used as Mail User Agent.

6   Conclusions

We presented a very efficient off-line certified e-mail system. Both the recipient and the TTP can be set to be stateless and the recipient can assume a passive role without being involved in dispute resolutions so that the burden of solving a dispute is given only to the sender, the initiator of the protocol.

We implemented a prototype (TURMS) of the protocol and we reviewed some of the technology available today that could be employed to effectively build any certified e-mail system.

Acknowledgements

We would like to thank Theo Schlossnagle for his helpful suggestions on setting up the testing environment. Many thanks to the anonymous referees for their insightful comments.

References

1. N. Asokan, V. Shoup, and M. Waidner, "Optimistic fair exchange of digital signatures," *IEEE Journal on Selected Area in Communications*, 2000.
2. N. Asokan, V. Shoup, and M. Waidner, "Asynchronous protocols for optimistic fair exchange," in *Proceedings of the IEEE Symposium on Research in Security and Privacy* (I. C. S. Press, ed.), pp. 86–99, May 1998.
3. G. Ateniese, "Efficient verifiable encryption (and fair exchange) of digital signatures," in *Proceedings of the 6th ACM Conference on Computer and Communications Security*, ACM Press, 1999.

4. G. Ateniese, B. de Medeiros, M.T.Goodrich. TRICERT: Distributed Certified E-mail Schemes. In ISOC 2001 Network and Distributed System Security Symposium (NDSS'01), San Diego, CA, USA, 2001.

5. M. Ben-Or, O. Goldreich, S. Micali, and R. Rivest, "A fair protocol for signing contracts," *IEEE Transactions on Information Theory IT-36(1)*, pp. 40–46, 1990.

6. A. Bahreman and J. D. Tygar, "Certified electronic mail," in *Proceedings of Symposium on Network and Distributed Systems Security* (I. Society, ed.), pp. 3–19, February 1994.

7. F. Bao, R. H. Deng, and W. Mao. Efficient and Practical Fair Exchange Protocols with Off-line TTP. In *IEEE Symposium on Security and Privacy*, Oakland, California, 1998.

8. J. Camenisch and M. Michels. Proving in zero-knowledge that a number is the product of two safe primes. In *Advances in Cryptology – EUROCRYPT '99, Lecture Notes in Computer Science*, Springer-Verlag, 1999.

9. D. Chaum and T. Pedersen. Wallet databases with observers. In *Advances in Cryptology – Crypto '92*, pages 89-105, 1992.

10. R. Cramer and V. Shoup. Signature Schemes Based on the Strong RSA Assumption. In *6th ACM Conference on Computer and Communication Security*, ACM Press, 1999.

11. R. H. Deng, L. Gong, A. Lazar, and W. Wang, "Practical protocols for certified electronic e-mail," *Journal of Networks and Systems Management*, vol. 4, no. 3, pp. 279–297, 1996.

12. S. Even, O. Goldreich, and A. Lempel, "A randomized protocol for signing contracts," *Comm. ACM 28*, no. 6, pp. 637–647, 1985.

13. "http://www.exim.org."

14. A. Fiat and A. Shamir. How to prove yourself: practical solutions to identification and signature problems. In *Advances in Cryptology – CRYPTO '86*, volume 263 of *Lecture Notes in Computer Science*, pages 186–194, Springer-Verlag, 1987.

15. R. Gennaro, S. Halevi, and T. Rabin. Secure signatures, without trees or random oracles. In *Advances in Cryptology – EUROCRYPT '99*, volume 1592 of *Lecture Notes in Computer Science*, pages 123–139, Springer-Verlag, 1999.

16. L. C. Guillou and J. J. Quisquater. A paradoxical identity-based signature scheme resulting from zero-knowledge. In *Advances in Cryptology – CRYPTO '88*, volume 403 of *Lecture Notes in Computer Science*, pages 216–231, Springer-Verlag, 1988.

17. A. J. Menezes, P. C. van Oorschot, and S. A. Vanstone. *Handbook of applied cryptography*. CRC Press, 1996. ISBN 0-8493-8523-7.

18. S. Micali. Simultaneous electronic transactions. Technical Report 566420, http://www.delphion.com/cgi-bin/viewpat.cmd/US566420__, 1997.

19. OpenSSL Project team, "Openssl," May 1999. http://www.openssl.org/.

20. Pine Information Center, http://www.washington.edu/pine/.

21. J. Riordan and B. Schneier, "A certified e-mail protocol," in *13th Annual Computer Security Applications Conference*, pp. 100–106, December 1998.

22. R. L. Rivest, A. Shamir, and L. M. Adleman. A method for obtaining digital signatures and public-key cryptosystems. *Communications of the ACM*, 21(2):120–126, 1978.

23. C. P. Schnorr. Efficient signature generation by smart-cards. *Journal of Cryptology*, 4(3):161–174, 1991.

24. J. Zhou and D. Gollmann, "Certified electronic mail," in *Proceedings of Computer Security - ESORICS'96* (S. Verlag, ed.), pp. 55–61, 1996.

RSA-Based Undeniable Signatures
for General Moduli

Steven D. Galbraith[1], Wenbo Mao[2], and Kenneth G. Paterson[3]

[1] Pure Mathematics Department, Royal Holloway University of London,
Egham, Surrey TW20 0EX, UK
`Steven.Galbraith@rhul.ac.uk`
[2] Mathematics, Cryptography and Security Group
Hewlett-Packard Laboratories, Bristol
Filton Road, Stoke Gifford, Bristol BS34 8QZ, UK
`wm@hplb.hpl.hp.com`
[3] Information Security Group,
Mathematics Department, Royal Holloway University of London,
Egham, Surrey TW20 0EX, UK
`Kenny.Paterson@rhul.ac.uk`

Abstract. Gennaro, Krawczyk and Rabin gave the first undeniable signature scheme based on RSA signatures. However, their solution required the use of RSA moduli which are a product of safe primes. This paper gives techniques which allow RSA-based undeniable signatures for general moduli.

Keywords: Undeniable Signatures, RSA-based Undeniable Signatures.

1 Introduction

Undeniable signatures were introduced by Chaum and van Antwerpen [7,8]. They offer good privacy for the signer since signatures cannot be verified without interaction with the signer. Undeniable signatures (and generalisations of them, such as confirmer and convertible signatures) have various applications in cryptography [2,3,10].

The zero-knowledge undeniable signature scheme of Chaum [8] works in the multiplicative group of integers modulo a prime. Although Chaum, van Heijst and Pfitzmann [9] provided an undeniable signature scheme with security related to factoring, before 1997 there was not a scheme based on traditional RSA signatures. Gennaro, Krawczyk and Rabin [18] were the first to obtain an RSA-based undeniable signature scheme. Their scheme is closely related to the scheme of Chaum [8] and both schemes have similar security and efficiency.

One significant drawback with the scheme of [18] is that it requires the use of RSA moduli which are products of safe primes. It was explicitly stated as an open problem in [18] to provide an undeniable signature scheme based on RSA which does not require special moduli. The goal of the present paper is to solve this problem.

B. Preneel (Ed.): CT-RSA 2002, LNCS 2271, pp. 200–217, 2002.
© Springer-Verlag Berlin Heidelberg 2002

Of course, it is trivial to construct an undeniable signature scheme for general moduli where the confirmation protocol has soundness probability $1/2$, but we seek solutions where the confirmation protocol is more efficient (possibly at the expense of more demanding key certification). We must mention that general constructions due to Michels and Stadler [23] and Camenisch and Michels [5] also give solutions to this problem, however their systems require auxiliary tools ([23] utilises confirmer commitment schemes, while [5] requires a secure encryption scheme).

In the course of solving this problem we improve the efficiency and zero-knowledge property of the denial protocols for RSA-based undeniable signatures. The methods of this paper are therefore a useful addition to the protocol of [18], even when safe primes are being used.

1.1 Pros and Cons of Special Moduli

The undeniable signature scheme of Gennaro, Krawczyk and Rabin [18] is modelled on RSA [27]. Thus a signature on a message m is a number $s = m^d \pmod{N}$ where N is a product of two primes and where d is an integer coprime to $\varphi(N)$. The difference between usual RSA signatures is that the number e such that $de \equiv 1 \pmod{\varphi(N)}$ is not public, and so an interactive proof (preferably zero-knowledge) is required to confirm that s is a valid signature for m (i.e. that $s^e \equiv m \pmod{N}$). This protocol relies on a public key having previously been certified by an authority.

There are various reasons why Gennaro, Krawczyk and Rabin [18] restricted to the case where the RSA modulus N is a product of safe primes (i.e. primes p such that $(p-1)/2$ is also prime) but the most important one is that, for products of safe primes, the group $\mathbb{Z}_N^*$ does not have many elements of small order. If one runs the scheme of Gennaro, Krawczyk and Rabin [18] with a general modulus then there is a high probability that a dishonest signer can cheat (see Section 3 for details).

In general, restricting to moduli which are a product of safe primes makes many cryptographic issues easier to handle. However, there are several drawbacks of schemes which require special moduli. One major problem is that it is necessary for a certification authority to guarantee the properties of the public key. As we discuss next, none of the currently known protocols for allowing a user to prove to a certification authority that their modulus is a product of safe primes are fully satisfactory.

Gennaro, Micciancio and Rabin [17] have given a very nice protocol to prove the a number is a product of two quasi-safe primes (i.e. primes p such that $(p-1)/2$ is a power of a prime), but this is significantly less than the assurance we require. For instance, a prime of the form $2 \cdot 3^k + 1$ is a quasi-safe prime but a modulus constructed as a product of primes of this form would be vulnerable to attacks such as those outlined in Section 3.

Camenisch and Michels [4] have given a protocol to prove that a number is a product of safe primes. Their protocol requires performing the Miller-Rabin primality test in zero-knowledge on a hidden number. It therefore requires an enormous amount of communication between the prover and the certification authority. This protocol is unsuitable for practical applications.

Another problem is that choosing special moduli goes against the conventional wisdom in cryptography of avoiding special cases. Indeed, in Section 8.2.3 of [22] and in [29] it is explicitly stated that products of random primes are advisable for cryptography.

There are many other protocols which currently require moduli which are a product of safe primes [6,14,16,26,28] and it is of great interest in cryptography to provide solutions which do not require this assumption. Some recent papers in this direction include [13] and [15]. We hope that the new techniques introduced in this paper might be of wider applicability to solve other problems in the area.

1.2   Our Work

We provide an undeniable signature scheme for general RSA moduli. Our scheme is, in fact, a parameterised family of cryptosystems depending on three parameters B, K_1 and K_2.

The number 2^{-K_1} will be the probability that a dishonest signer Alice will be able to cheat the certification authority (CA) when certifying her public key. A value of K_1 should be agreed in advance between Alice and the CA and could form part of Alice's certified public key. Similarly 2^{-K_2} will be the probability that Alice can cheat a verifier Bob in either the confirmation or denial protocols. We allow different values of K_1 and K_2 for generality. Typical values that might be used in practice are $K_1 = K_2 = 100$.

The number $1/B$ will be the soundness probability for each iteration of the signature confirmation and denial protocols. The number of iterations required to obtain a cheating probability of 2^{-K_2} will be $\frac{K_2}{\log_2 B}$. A typical choice of B might be 2^{10} in which case 10 iterations of our protocols are needed for $K_2 = 100$.

The value of B also determines how 'special' the moduli must be, and accordingly, how expensive public key certification is. Essentially, with B chosen, the modulus N must have the property that $\varphi(N)$ is not divisible by any odd primes $p < B$. Alice will prove this to the CA during key certification.

Large values of B will give efficient signature and denial protocols, but the moduli N will be rather special (and there is necessarily a lot of work required in our process for public key certification). In some sense, moduli which are a product of safe primes as in [18] are a limiting case of our cryptosystem in which $B = N^{1/4}$. Our public key certification process has been designed with rather general RSA moduli in mind (i.e. for small values of B). If special RSA moduli (i.e. larger values of B) are to be used then certification protocols should be developed using techniques like those in [5]. Small values of B result in a scheme which does not require special moduli (and for which public key certification is relatively efficient), but the resulting confirmation and denial protocols require many rounds to achieve the desired soundness probability of 2^{-K_2}. We therefore have a tunable family of undeniable signature schemes. In particular, we do obtain an undeniable signature scheme which works for completely general RSA moduli (see Section 8.4). For a fuller discussion of the performance of our schemes, see Section 8.3. As we shall see, for the typical values $K_1 = K_2 = 100$ and $B = 2^{10}$, the protocols are all perfectly practical.

The next section sets up some notation for the rest of the paper. In Section 3 we review the scheme of [18] and indicate some of the pitfalls in adapting this scheme to general RSA moduli. Our process for public key certification is specified in detail in Section 4. We emphasise that the cost of certification is a one-time cost. Our signature confirmation and denial protocols are described in Section 5, with proofs of zero-knowledge and security against existential forgery appearing in the following two sections. One important innovation here is a new signature denial protocol which is more efficient and has a cleaner proof of zero-knowledge than the protocol used in [18]. In Section 8 we give variations of the scheme which provide confirmer signatures and convertible signatures. We also discuss the performance of our scheme there.

2   Preliminary Definitions and Notation

Let N be a positive integer. We write $\mathbb{Z}_N^*$ for the multiplicative group of integers modulo N. We write Q_N for the subgroup of quadratic residues (squares) in $\mathbb{Z}_N^*$. We write $\varphi(N)$ for the Euler phi function. A safe prime is an odd prime p such that $(p-1)/2$ is prime.

Given any $g \in \mathbb{Z}_N^*$ we define the order of g to be $\operatorname{ord}(g) = \min\{n \in \mathbb{Z} : n \geq 1 \text{ and } g^n \equiv 1 \pmod{N}\}$. When $N = p_1 p_2$ is a product of two distinct primes then every $g \in \mathbb{Z}_N^*$ has order dividing the least common multiple $\operatorname{lcm}(p_1 - 1, p_2 - 1)$.

3   The Scheme of Gennaro, Krawczyk and Rabin

In this section we briefly sketch the RSA-based undeniable signature scheme of Gennaro, Krawczyk and Rabin [18] for products of safe primes. We also indicate why it is nontrivial to adapt this to the case of a general RSA modulus.

Alice possesses a public RSA modulus N, which is assumed to be a product of two safe primes, and a pair of secret integers (e, d) such that $ed \equiv 1 \pmod{\varphi(N)}$. Alice's undeniable signature on a message $m \in \mathbb{Z}_N$ is $s = m^d \pmod{N}$, i.e. a standard RSA signature.

Since e is not public knowledge, it is not possible for Bob to verify the validity of the signature s without interacting with Alice. Instead, Alice the prover and Bob the verifier engage in a zero-knowledge protocol to show that $s^e \equiv m \pmod{N}$. For this signature confirmation protocol it is necessary to have some fixed commitment to the value e. This is achieved in [18] by taking a random element $g \in \mathbb{Z}_N^*$ (which can be shown to have large order in the case of special moduli) and publishing $h = g^d \pmod{N}$

The signature confirmation protocol of [18] (presented for simplicity in the case of honest verifiers) is the following:

1. Given the public key (N, g, h) and an alleged message-signature pair (m, s) the verifier chooses random integers $1 \leq i, j < N$, constructs a challenge $C = s^{2i} h^j \pmod{N}$, and sends C to the prover.

2. The prover sends the response $R = C^e \pmod{N}$ to the verifier.
3. The verifier checks whether $R \equiv m^{2i}g^j \pmod{N}$.

The signature denial protocol suggested in [18] is an adaptation of a protocol due to Chaum (originally developed for the case of finite fields $\mathbb{F}_q^*$). The denial protocol requires the prover to perform an exhaustive search over k values where k is a security parameter. The probability of successful cheating by a dishonest prover in this case is $1/k$. There is also a minor complication about how aborting the protocol affects the zero-knowledge properties, this is handled in [18] by using a commitment to zero.

3.1   Generalising to General Moduli, Problem I

In this and the next subsection we motivate the need for our more complex protocols by considering what happens if the protocol due to [18] is naively used with a general RSA modulus N. The problems we sketch should be seen as part of a general phenomenon, that protocols developed in the case of finite fields do not necessarily give rise to secure protocols when working with $\mathbb{Z}_N^*$.

Let Alice be a dishonest prover. Since Alice controls the factorisation of the modulus N she can choose N so that there is a small prime ℓ with $\ell|\varphi(N)$. She can also find an element $\alpha \in \mathbb{Z}_N^*$ such that α has order ℓ. Suppose Alice publishes a signature $s = \alpha m^d$ for a message m. What is the probability that Alice can fool a verifier Bob that this is a valid signature? In the confirmation protocol Alice receives a challenge $C = s^{2i}h^j \pmod{N}$. In general Alice does not know the value of i, but she can compute a response $R = \alpha^r C^e \pmod{N}$ where r is chosen at random. If $r + 2ie \equiv 0 \pmod{\ell}$ then the check performed by the verifier will be satisfied. Hence the probability of successful cheating is at least $1/\ell$. Since ℓ can be chosen to be 3 this probability is quite high. There is an analogous attack using elements of order 4 which has probability $1/2$ of success. Hence the confirmation protocol must be executed many times to give an assurance that the signature is valid. This is unsatisfactory.

Notice that when N is a quasi-safe prime product (see [17]), using a small ℓ as above will render N vulnerable to well-known factoring algorithms, such as Pollard's $P-1$ method or the elliptic-curve method. So if Alice's objective is to fool Bob with reasonable probability and she is not concerned about using a modulus that succumbs to these factoring algorithms, then she can choose to use a modulus that is a product of quasi-safe primes.

We will solve these issues by giving a method for Alice to certify that her public key N is such that there are no small (up to a bound B) odd primes dividing $\varphi(N)$.

3.2   Generalising to General Moduli, Problem II

There is a more subtle and devastating attack. Once again suppose Alice is a dishonest signer and suppose that (either by construction, or by accident) her public key element g does not have maximal order in $\mathbb{Z}_N^*$.

For simplicity of presentation we suppose that there is a prime q such that $q\|\varphi(N)$ (i.e. $q|\varphi(N)$ but $q^2 \nmid \varphi(N)$) and $q \nmid \text{ord}(g)$. We assume that q is not too large (less than 80 bits, say) so that the discrete logarithm problem in the subgroup of order q can be solved using standard methods. Let $\alpha \in \mathbb{Z}_N^*$ be an element of order q. Alice constructs her public key $h = g^d \pmod{N}$ as usual.

Let $m \in \mathbb{Z}_N^*$ be any message (it doesn't matter whether $q|\text{ord}(m)$ or not). Suppose Alice publishes $s = \alpha m^d \pmod{N}$ as her signature on m.

Consider the signature confirmation protocol. Alice receives the challenge $C = s^{2i} h^j$. By raising C to the power $\varphi(N)/q$ and solving a discrete logarithm problem to the base α Alice can determine the value of $i \pmod{q}$. Alice can therefore respond with $R = \alpha^r C^e \pmod{N}$ where $r = -2ie \pmod{q}$ is constructed so that the check by the verifier will always be satisfied. In other words, Alice can fool the verifier with probability one! Similarly, whenever Alice desires, she can successfully run a signature denial protocol on that signature.

This is an extremely severe attack on an undeniable signature scheme. We address this problem in our scheme by using a set of generators $g_1, \ldots, g_k$ where we take k to be large enough so that the group generated by all the g_i is overwhelmingly likely to contain Q_N.

4   Public Key Certification

Suppose Alice wants to be able to generate undeniable signatures. Let the parameters B and K_1 be fixed as in Section 2. The public key for Alice is a tuple $(N, g_1, \ldots, g_k, h_1, \ldots, h_k)$ where k is such that $g_1, \ldots, g_k$ generate a subgroup of $\mathbb{Z}_N^*$ which contains Q_N with probability at least $1 - 2^{-K_1}$. For example, for the typical values $B = 2^{10}$ and $K_1 = 100$ we can take $k = 11$. More generally, we should take k so that $\frac{2}{k-1}(B-1)^{1-k} < 2^{-K_1}$ (see below). The private key is a pair (e, d) (these values are also defined below). We emphasise that this is different from a standard RSA public key, which would include the signature verification exponent e. Alice must register her public key with a CA, who will issue a certificate which confirms that Alice's public key is suitable for the undeniable signature scheme we propose. The properties of the public key which must be guaranteed by this certificate are:

1. N is a product of two prime powers $p_1^{s_1} p_2^{s_2}$ such that each $p_i \equiv 3 \pmod{4}$. (See Section 8.4 for discussion of how to relax the assumption that $p_i \equiv 3 \pmod{4}$.)
2. $\gcd(\Delta, \varphi(N)) = 1$ where Δ is the product of all primes $2 < l < B$.
3. The g_i are chosen at random in a way which is not controlled by Alice.
4. The g_i and h_i are correctly related by $h_i = g_i^d \pmod{N}$ for some secret integer d which is coprime to $\text{lcm}(\text{ord}(g_1), \ldots, \text{ord}(g_k))$.

4.1   Construction of the Modulus

The first step of key generation for Alice is to construct an integer $N = p_1 p_2$ which is a product of two primes such that $p_i \equiv 3 \pmod{4}$. Let B be the integer specified in Section 1.2 and which determines the soundness probability of our confirmation and denial protocol. We demand that for all primes $2 < l < B$ one has $l \nmid (p_i - 1)$ for $i \in \{1, 2\}$. This means that $\varphi(N)$ is coprime to $\Delta = \prod_{\text{primes } 2 < l < B} l$.

Alice must prove to the CA that N is a product of primes p_i such that $p_i \equiv 3 \pmod 4$. This can be achieved using a protocol due to van de Graaf and Peralta [19]. The protocol of [19] proves that $N = p_1^{s_1} p_2^{s_2}$ where $p_i \equiv 3 \pmod 4$ and the s_i are odd integers. This is enough for our application (we do not need to assume that $s_1 = s_2 = 1$ for our protocols) and if a stronger assurance about the difficulty of factoring N is required then more advanced techniques may be used.

4.2 Proof That $\varphi(N)$ Does Not Have Small Prime Factors

Alice must prove to the CA that $\varphi(N)$ is not divisible by any odd primes $2 < l < B$. In Figure 1 we give a protocol to achieve this in the honest verifier case. Recall that Δ is the product of all primes $2 < l < B$. So Δ is approximately B bits in length (for $B = 2^{10}$, Δ has 1420 bits). The protocol involves exponentiations to the power Δ. For small B, say up to 2^{10}, this is a perfectly feasible computation. However, for large B (necessary when a modulus is being certified to be 'special'), this becomes infeasible. The largest B one might wish to use in practice would be, perhaps, $B = 2^{20}$.

Protocol Certification–1(N, Δ).

1. The CA chooses a random integer $x \in \mathbb{Z}_N^*$, computes the challenge $C = x^\Delta \pmod N$ and sends C to Alice.
2. Alice sends to the CA the response $R = C^{\Delta^{-1}} \pmod N$ which is unique when $\gcd(\varphi(N), \Delta) = 1$.
3. The CA accepts if $R = x$.

Fig. 1. Proof that $\gcd(\Delta, \varphi(N)) = 1$ in the honest verifier case.

This protocol can be made into a perfect zero-knowledge protocol (i.e. robust against dishonest verifiers) in a standard way: Replace the second move (i.e. where Alice sends her response to the CA) with the transmission of a perfectly-hiding commitment to R. The CA then must send x to Alice, so that she can check that the initial challenge was correctly formed. Alice can then open the commitment to R, allowing the CA to see that the response is correct.

Theorem 1. *Let (N, Δ) be as above. The protocol Certification–1 has the following properties:*

Completeness: *If $\gcd(\Delta, \varphi(N)) = 1$ then the CA will always accept Alice's proof.*

Soundness: *If* $\gcd(\Delta, \varphi(N)) \neq 1$ *then Alice, even computationally unbounded, cannot convince the CA to accept her proof with probability better than* $1/3$.

Zero-knowledge: *When Alice behaves correctly, the CA gains no information about Alice's private input apart from the validity of her proof.*

Proof. The proof of completeness is immediate from inspection of the protocol. We focus on the further two properties.

Soundness: Suppose that l is a prime such that $l | \gcd(\Delta, \varphi(N))$. Then there are at least l elements of $\mathbb{Z}_N^*$ of order l (there could be l^2 of them). Let $\alpha \in \mathbb{Z}_N^*$ be such an element. Let $x \in \mathbb{Z}_N^*$ be chosen at random and define $C = x^\Delta \pmod{N}$. Then for each $0 \leq i < l$ we have $C \equiv (x\alpha^i)^\Delta \pmod{N}$. Alice therefore has no information about which of the possibilities x is the one chosen by the CA. Hence Alice cannot respond with the correct value with probability better than $1/l$.

This discussion applies to all primes $2 < l < B$, but in the worst case (as far as the CA knows) we have $\gcd(\Delta, \varphi(N)) = 3$.

Zero-knowledge: Alice publishes a perfectly hiding commitment to the value R and only opens it if the CA already knows the value. Hence the CA learns nothing. □

The protocol must be repeated $\frac{K_1}{\log_2 3}$ times to achieve a probability of successful cheating as low as 2^{-K_1}. The flows of instances of the protocol may all be sent in parallel.

It would be very interesting to have an efficient protocol to prove that $\varphi(N)$ is coprime to the integer Δ which has a soundness probability smaller than $1/3$. It seems to be highly non-trivial to construct such a protocol.

4.3   Construction of Generators

The next step of key generation is to construct elements $g_1, \ldots, g_k \in \mathbb{Z}_N^*$. We cannot allow Alice to generate these elements as she might choose them to have small order, in which case attacks like those in Section 3 apply. Hence the values for the g_i should be generated using a protocol in which both Alice and the CA jointly contribute randomness (this is easy to achieve using commitment schemes). Another solution would be to let the CA choose the values g_i (indeed, they could even be fixed values for all users).

The reason for choosing many elements g_i is to ensure that the whole of the subgroup $Q_N \subset \mathbb{Z}_N^*$ is generated (see the attack in Section 3.2). A similar technique has been used in [25], [15]. This will be important in the soundness proof of our signature confirmation protocol.

With $N = p_1 p_2$ the techniques in [25] and [15] can be used to show that the number of k-tuples $g_1, \ldots, g_k$ generating all of $Q_N \subset \mathbb{Z}_N^*$ is equal to

$$\varphi_k((p_1 - 1)/2)\varphi((p_2 - 1)/2)$$

where

$$\varphi_k\left(\prod_{i=1}^{t} q_i^{e_i}\right) = \prod_{i=1}^{t} q_i^{e_i} \prod_{i=1}^{t}\left(1 - q_i^{-k}\right)$$

is a generalisation of the Euler phi function. From this, using estimates like those in [25], we can prove that the probability that $g_1, \ldots, g_k$ generate all of Q_N is at least

$$1 - \frac{2}{k-1}(B-1)^{1-k}$$

where B is the bound on the size of the prime factors of $(p_1-1)/2$ and $(p_2-1)/2$. We can make this probability arbitrarily close to one by taking k to be sufficiently large. For example, with $B = 2^{10}$ we can take $k = 11$ to guarantee that the g_i generate $Q_N \subset \mathbb{Z}_N^*$ with probability at least $1 - 2^{-100}$

In any case, we assume from now on that k has been chosen large enough that the g_i generate a subgroup of $\mathbb{Z}_N^*$ which contains Q_N.

The next step is for Alice to choose a secret key pair (e, d) such that $ed \equiv 1 \pmod{\varphi(N)}$. Alice should then construct $h_i = g_i^d \pmod{N}$ for $i = 1, \ldots, k$.

It is crucial that the h_i are correctly calculated and so Alice must provide a proof that this is the case. We give such a proof in Figure 2.

Protocol Certification–2$(N, g_1, \ldots, g_k, h_1, \ldots, h_k)$.

1. Alice chooses random integers $1 \leq k_1, k_2 \leq \varphi(N)$, constructs $u_i = h_i^{e+k_1} \pmod{N}, v_i = g_i^{d+k_2} \pmod{N}$ for $i = 1, 2, \ldots, k$, and sends the elements u_i and v_i to the CA.
2. The CA sends to Alice a random bit $b \in \{0, 1\}$.
3. If the bit is zero then Alice sends the two integers $n_1 = (e+k_1) \pmod{\varphi(N)}$ and $n_2 = (d+k_2) \pmod{\varphi(N)}$. The CA can then check whether $u_i \equiv g_i^{n_1} \pmod{N}$ and $v_i \equiv g_i^{n_2} \pmod{N}$ for all $i = 1, 2, \ldots, k$. The CA accepts if all the checks pass and rejects otherwise.
   If the bit is one then Alice sends the two integers k_1 and k_2. The CA then checks whether $u_i \equiv g_i h_i^{k_1} \pmod{N}$ and $v_i \equiv h_i g_i^{k_2} \pmod{N}$ for $i = 1, 2, \ldots, k$. The CA accepts if all the checks pass and rejects otherwise.

Fig. 2. Key certification protocol for g_i, h_i.

Theorem 2. *Let (N, g_i, h_i) be as above. For $N = p_1^{s_1} p_2^{s_2}$ define M to be the exponent of the group $\mathbb{Z}_N^*$ (i.e., $M = \lambda(N)$). Assume that the group generated by the g_i contains Q_N (and so, in particular, $(M/2)|\mathrm{lcm}(\mathrm{ord}(g_1), \ldots, \mathrm{ord}(g_k))$). The protocol Certification–2 has the following properties:*

Completeness: *If $h_i \equiv g_i^d \pmod{N}$ where $ed \equiv 1 \pmod{M/2}$ then the CA will accept Alice's proof.*

Soundness: *If some $h_i \not\equiv g_i^d \pmod{N}$ or if $ed \not\equiv 1 \pmod{M/2}$ then Alice, even computationally unbounded, cannot convince the CA to accept her proof with probability better than $1/2$.*

Zero-knowledge: *When Alice behaves correctly, the CA gains no information about Alice's private input apart from the validity of her proof.*

Proof. The proof of completeness is immediate from inspection of the protocol. We focus on the further two properties.

Soundness: If Alice can respond correctly to both choices of the bit b then she knows numbers e and d such that $h_i \equiv g_i^d \pmod{N}$ and $g_i \equiv h_i^e \pmod{N}$ for all $i = 1, 2, \ldots, k$. This proves that all the g_i and h_i are related by the same numbers.

We now show that the protocol implies $ed \equiv 1 \pmod{M}$. Suppose that there is some odd prime power $q^a | M$ such that $ed \equiv x \not\equiv 1 \pmod{q^a}$. By assumption, $q^a | \text{ord}(g_i)$ for some index i. Let $g = g_i^{(\text{ord}(g_i)/q^a)}$ and $h = h_i^{(\text{ord}(g_i)/q^a)}$. We have $g^x \equiv g^{ed} \equiv h^e \equiv g \pmod{N}$ which is a contradiction.

Zero-knowledge: This is immediate, a standard argument showing that transcripts of the protocol can be simulated. $\square$

The protocol must be repeated K_1 times to achieve a probability of successful cheating to be 2^{-K_1}. The flows of instances of the protocol may all be sent in parallel.

It would be very useful to have a more efficient protocol to prove the correctness of the g_i and h_i. It seems to be non-trivial to find such a protocol. The standard methods used when working in finite fields $\mathbb{F}_q^*$ do not immediately translate into secure protocols to solve our problem – they are vulnerable to attacks of the type described in Section 3.

5   Undeniable RSA Signatures

As usual with RSA it is not possible to allow any number $m \in \mathbb{Z}_N^*$ to be a valid message. Instead we need to use a cryptographically strong randomised padding scheme to provide an integrity check on messages. We return to this issue in Section 7, but for now we simply assume that this has been performed and that we want to provide a signature for some element $m \in \mathbb{Z}_N^*$.

Alice's signature on m is the usual RSA signature $s = m^d \pmod{N}$. There is one technicality as the signature confirmation protocol does not distinguish between signatures which differ by an element of order dividing 2. This is also the situation in [18]. The problem is easily solved allowing all four values s such that $s^{2e} \equiv m^2 \pmod{N}$ to be valid signatures.

5.1   Confirmation of an Undeniable Signature

Let (m, s) be an alleged signature pair. In Figure 3 we give a protocol (in the honest verifier case) for Alice to prove to a verifier Bob that the alleged pair is a genuine one. We assume that Alice has sent (m, s) and her certified public key information to Bob. There are two solutions to make the protocol robust against dishonest verifiers (i.e. perfect zero-knowledge) and we discuss them shortly.

The security of this protocol is discussed in Section 6. The protocol must be repeated $K_2/\log_2 B$ times (the flows of each instance may be executed in parallel) to achieve the desired probability of $1 - 2^{-K_2}$ that the signature is valid.

Protocol Confirm($N, g_1, \ldots, g_k, h_1, \ldots, h_k, m, s$).

1. Bob chooses random integers $r_0, r_1, \ldots, r_k$ such that all $1 < r_i < N$. Bob computes the challenge $C = s^{r_0} h_1^{r_1} \cdots h_k^{r_k} \pmod{N}$.
2. Bob sends to Alice the challenge C.
3. Alice sends to Bob the response $R = C^e \pmod{N}$.
4. Bob accepts if $R^2 \equiv (m^{r_0} g_1^{r_1} \cdots g_k^{r_k})^2 \pmod{N}$ and rejects otherwise.

Fig. 3. Undeniable signature confirmation protocol in the honest verifier case

We now discuss how to transform this protocol into a perfect zero-knowledge protocol (i.e. robust against dishonest verifiers). The first solution uses the standard technique: Instead of sending the response R, Alice publishes a commitment to it, and opens the commitment only once Bob has shown that the challenge C is of the correct form. We emphasise that the zero-knowledge property of the protocol only holds when the input is a valid message-signature pair (otherwise the protocol gives a message corresponding to a given signature). A signature confirmer must be careful to only execute the confirmation protocol on valid inputs (the denial protocol given in the next section should be used in other cases).

We note that, for undeniable signatures, it is usually preferable to use designated verifier proofs in protocols such as the one above. This can be easily achieved using the methods of Jakobsson, Sako and Impagliazzo [20].

There is another solution which provides security only against forgery of signatures. Instead of sending the response R, Alice sends $t = H(R^2 \pmod{N})$ where H is some cryptographically strong hash function. Bob can then check whether or not t is equal to $H((m^{r_0} g_1^{r_1} \ldots g_k^{r_k})^2 \pmod{N})$. This proof method does not allow Bob to use Alice as a signing oracle, which prevents Bob from being able to forge signatures. The main problem with this method is that Alice no longer knows which message signature pair (m, s) she is being requested to verify. This is an attack on an undeniable signature scheme since one intended feature of these schemes is that Alice should be aware which of her signatures are being verified. Nevertheless, in some contexts this proof technique might be of use.

5.2 Denial of an Undeniable Signature

Given a pair (m, s) which is not a valid signature for Alice on the message m (i.e. $s \not\equiv \xi m^d \pmod{N}$ where $\mathrm{ord}(\xi)|2$), it is important to provide a protocol allowing Alice to prove that this is the case.

One way to achieve this would be to compute the true signature $m^d \pmod{N}$ for m, send it to the verifier, and then prove its correctness using the signature confirmation protocol above. But this leaves the user open to a chosen-message attack, since she can be forced to publish valid signatures on any message m.

Instead, we proceed by allowing Alice to modify the alleged pair (m, s) to obtain a random related pair (m', s') and then perform the above process. To the eye of a seasoned cryptographer this still looks like a security flaw, but it is well-known that most undeniable signature schemes (e.g. [7], [18]) allow existential forgery using just the public key. These issues are handled using padding schemes on messages (further discussion is given in Section 7).

Our denial protocol runs as follows: Bob (or Alice and Bob running a joint protocol) chooses a random integer $1 < r < N$. Both parties can then compute the related elements $m' = m^r \pmod{N}$ and $s' = s^r \pmod{N}$. Alice publishes her correct signature s'' for the message m' and proves it is correct using the confirmation protocol above. The verifier can then determine whether $(s')^2 \equiv (s'')^2 \pmod{N}$.

The denial process requires only two more exponentiations than the confirmation protocol per participant, and the security is the same as that of the confirmation protocol. This is in contrast with Gennaro, Krawczyk and Rabin [18] where the denial protocol is much less efficient than the confirmation protocol. There is no loss of security from performing signature denial using our methods and so the protocol of [18] can be improved by using our approach. We note that Miyazaki [24] also gave an efficient denial protocol for this application based on the denial protocol of Chaum and van Antwerpen [7], although that protocol requires double the computation of ours.

Note that this denial method cannot be used in the classical case of undeniable signatures in a finite field $\mathbb{F}_q^*$ since the verifier can undo the transformation of exponentiation by r to recover a signature on a chosen message. We note that Jakobsson [21] has given an efficient denial protocol for Chaum's undeniable signatures. It is straightforward to adapt Jakobsson's ideas to the case of RSA-based undeniable signatures, but we obtain a denial protocol which takes at least twice the computation time as our method.

6   Security of the Confirmation Protocol

In this section we discuss the security of the confirmation and denial protocols. A discussion of security against existential forgery is given in the next section.

Theorem 3. *The confirmation protocol has the following properties:*

Completeness: *If (m, s) has been formed correctly then Bob will accept Alice's proof.*

Soundness: *If (m, s) is not valid, then Alice, even computationally unbounded, cannot convince Bob to accept her proof with probability better than $1/B$.*

Zero-knowledge: *When Alice behaves correctly, Bob gains no information about Alice's private input apart from the validity of her proof.*

Proof. The proof of completeness is immediate from inspection of the protocol. We focus on the further two properties.

Soundness: We assume that $N = p_1 p_2$ has the property that $p_i \equiv 3 \pmod 4$ and that all odd primes $l < B$ are coprime to $\varphi(N)$. We also assume that the g_i generate Q_N. This is certified with probability $1 - 2^{-K_1}$ by the key certification process.

Let (m, s') be an invalid signature prepared by a cheating Alice. Then we can write $s' = \alpha \xi m^d \pmod{N}$ where $\xi^2 \equiv 1 \pmod{N}$ and where α has odd order which is divisible only by powers of primes q where $q \geq B$. Let q be such a prime. Since the g_i generate Q_N, there exists at least one index I such that $q | \mathrm{ord}(g_I)$.

Given a challenge C of the correct form $s'^{r_0} h_1^{r_1} \cdots h_k^{r_k}$ a cheating Alice must construct a response R which will satisfy Bob's check. We now project all elements into the subgroup $\langle \alpha \rangle$ of q elements which is generated by α (if $q \| \varphi(N)$ then this is done by raising all elements to the power $\varphi(N)/q$). We continue to use the same notation for these elements, but the reader should be aware that they now only have order dividing q.

Expressing all elements in terms of powers of α we have $m = \alpha^{l_0}$ and $g_i = \alpha^{l_i}$. It follows that

$$\log_\alpha(C) \equiv r_0(1 + l_0 d) + \sum_{i=1}^{k} l_i r_i d \pmod{q} \tag{1}$$

where $d \not\equiv 0 \pmod{q}$ and $l_I \not\equiv 0 \pmod{q}$ (where I is as above). The response R (again, only considering the image in the subgroup of elements of order q) must satisfy

$$\log_\alpha(R) \equiv r_0 l_0 + \sum_{i=1}^{k} l_i r_i \pmod{q}.$$

Since Alice knows d, the ability to construct the right response R is therefore equivalent to knowledge of $r_0 \pmod{q}$. But this is information-theoretically hidden: Given any solution $(r_0, \ldots, r_k)$ to equation (1) and any integer $0 < x < q$, there is another solution $(r_0 + x, r_1, \ldots, r_I', \ldots, r_k)$ where

$$r_I' = (r_I l_I d - x(1 + l_0 d))/(l_I d) \pmod{q}.$$

Furthermore, since all values r_i may be reduced modulo $\varphi(N)$ the condition $1 < r_i < N$ is preserved. Hence, Alice has no better strategy than guessing the right power of α in her response R.

This argument applies to all primes $q | \mathrm{ord}(\alpha)$ but since Alice might choose α so that its order is the first prime power larger than B we can only conclude that Alice's cheating probability is $1/B$.

Zero-Knowledge: We analyse the protocol in the case where the commitment scheme is used. In this case Alice verifies the construction of C before giving Bob any information. Standard techniques show that the protocol can be simulated.

There is an interesting subtlety here though: If Alice's response R is actually of the form ξC^d then taking ratios gives an element ξ of order two, which is not simulatable. So only an "honest" run of the protocol is simulatable. In other words, if Alice at any time publishes signatures which are of the form ξm^d with $\xi \neq 1$ then she is potentially giving Bob useful information. $\square$

7   Security against Existential Forgery

We now turn to the problem of whether an adversary can construct a pair (m, s) which Alice is unable to deny.

As is well known, with standard RSA [27] it is easy to forge signatures: Given (N, e) one can choose any integer $s \in \mathbb{Z}_N^*$, define $m = s^e \pmod{N}$, and then s is a valid signature for the message m. Similarly, for our system it is very easy to construct random pairs (m, s) such that $s \equiv m^d \pmod{N}$ using the pairs g_i, h_i. Hence resistance to forgery must rely heavily on the padding scheme.

Methods to form secure RSA encryption/signature schemes are one of the most important and well-known parts of cryptography. The precise techniques used to achieve this are not relevant to the present paper. We simply assume that some padding scheme such as that developed by Bellare and Rogaway [1] is used.

The proof of security against existential forgery given by Gennaro, Krawczyk and Rabin in [18] applies directly to our situation. The proof shows that an adversary who is able to forge a signature for the undeniable signature scheme can be used to construct an adversary which forges signatures for the same RSA modulus and the same message padding scheme. The security result applies to any attack model, in particular, it applies to an adversary mounting an adaptive chosen ciphertext attack (CCA2). The details of the message padding scheme do not affect the proof. We refer to [18] for the details of this, and also a discussion of indistinguishability of signatures.

8   Discussion

As with the scheme of Gennaro, Krawczyk and Rabin [18] it is possible to add extra functionality such as confirmability and convertibility. We discuss some of these extensions in this section. We also discuss the efficiency of the scheme.

8.1   Confirmability

Confirmer signatures are an extension of undeniable signatures which were introduced by Chaum [11]. These systems allow a signer to delegate the tasks of signature confirmation and denial to another party.

The secret key of Alice includes the number e which is used for verifying signatures. Alice may give this number to a designated confirmer. This means that the task of confirming signatures can be performed either by Alice or by the confirmer. Assuming the hardness of the RSA problem, since the confirmer only knows e, they are unable to forge Alice's signature.

We note that Alice cannot fool the confirmer about the validity of a given signature. To the confirmer, Alice's signatures are standard RSA signatures (up to multiplication by an element of small order).

8.2   Convertibility

More generally, Alice could have a private key of the form (e, d, c) where $edc \equiv 1 \pmod{\varphi(N)}$ In this case, e could be public, d and c known to Alice, and c

known to a designated converter. The verification protocol in this case involves raising to the power ec rather than e and it is in all other respects identical to the one in Figure 3. An undeniable signature (m, s) can therefore be confirmed by the converter or the signer.

Any individual signature may be converted to a standard RSA signature by the transformation $(m, s) \to (m, s^c \pmod{N})$. This process may be performed by the converter or the signer. A proof analogous to that in Figure 3 must be used to show that this has been performed correctly. Once again, Alice cannot cheat against the confirmer.

Note that Alice may choose several different pairs (c_i, d_i) if she wants to use several converters with disjoint jurisdiction with the same public key (N, e).

Finally, by publishing c, all Alice's signatures become standard RSA signatures with respect to the exponent ec.

8.3   Efficiency

Here we discuss the performance of our scheme for general values of B, K_1 and K_2.

The one-off costs of public key certification are dominated by two factors. There are $\frac{K_1}{\log_2 3}$ iterations of the protocol of Figure 1, each iteration requiring one exponentiation by Alice and one by the CA. As we have discussed, when B becomes large (and the moduli special), these computations become costly. There are also K_1 iterations of the protocol of Figure 2, each iteration requiring $2k$ exponentiations (where k was defined as a function of K_1 and B earlier) for the two participants. These computations become *less* costly as B increases. The final public key is $2k + 1$ times as long as the modulus N.

Each of the $\frac{K_2}{\log_2 B}$ rounds of the signature confirmation protocol (in the zero-knowledge version) requires $k + 2$ exponentiations for Alice and $2(k + 1)$ for Bob. As we have discussed above, the denial protocol requires an additional pair of exponentiations in total.

For the typical choices of $K_1 = K_2 = 100$ and $B = 2^{10}$, we can take $k = 11$. Then key certification requires a few thousand exponentiations. For signature confirmation (and denial), Bob needs to do around 240 exponentiations and Alice around 10 (disregarding the cost of calculating and checking commitments). This is eminently practical.

When B is small then our signature confirmation is not as efficient as the scheme of [18]. On the other hand, the CA and the users can agree on how large B should be taken. Increasing the size of B reduces the size of k and improves the cost of signature confirmation at the expense of public key certification. For large B, say $B \geq 2^{20}$, one would use different certification techniques, e.g. the methods of [4]. The 'limiting case' is the case of [18], where signature confirmation is very efficient and public key certification is extremely inefficient.

We re-iterate that the scheme of [18] has improved efficiency when our signature denial protocol is used.

8.4   Completely General Moduli

We assumed that $N = p_1 p_2$ (or, more generally, $N = p_1^{s_1} p_2^{s_2}$) where $p_i \equiv 3 \pmod 4$. One can construct a scheme which does not require any condition on $p_i \pmod 4$ but this involves some extra techniques. The main difficulty in this case is that we no longer know which power of two divides the exponent of $\mathbb{Z}_N^*$.

First, we assume that $\varphi(N)$ has at least one prime factor of sufficiently large size. If this is not the case then the modulus is easily factored using the Pollard $P{-}1$ method. No RSA-based cryptosystem would be secure with such a modulus.

The public key certification proceeds as before with some choice of B. For completely general moduli choose $B = 3$ (in which case it is not necessary to execute the protocol of Figure 1).

Let $M = 2^{\lfloor \log_2(N) \rfloor}$. A signature on a message m is now defined to be any element of the set

$$\{\xi m^d \in \mathbb{Z}_N^* : \xi^M \equiv 1 \pmod N\}.$$

The signature confirmation protocol proceeds as in Figure 3 except the check is that

$$R^M \equiv (m^{r_0} g_1^{r_1} \cdots g_k^{r_k})^M \pmod N.$$

The probability of successful cheating by a dishonest signer in one instance of the protocol is still $1/B$. The proof of security against forgery of signatures is a slight modification of the one given in [18].

8.5   Another Approach

The trick used in the above subsection can be adapted to give a different solution to our original problem. Let $M = \prod_{2 \leq l < B} l^{\lfloor \log_l(N) \rfloor}$. Then a signature on message m could be any element such that $s^{eM} \equiv m^M \pmod N$. The advantage of this approach is that it can be used with a completely general modulus and no certification of the structure of the modulus is required.

There are two disadvantages of this approach. First, the size of the number M grows very quickly, and so only quite small values of B may realistically be used (not more than $B = 2^{10}$). Second, the size of candidate signatures could become extremely large. This has repercussions when analysing signature forgery, since for certain 'weak' moduli the probability of successful forgery might be relatively large compared to the desired security. Nevertheless, we feel it is worth mentioning this other approach.

9   Conclusion

We have presented an undeniable signature scheme for completely general RSA moduli. Our work provides a tunable family of schemes described by the parameters B, K_1 and K_2. These determine a set of trade-offs between efficiency of key certification (and generality of moduli) and the efficiency of signature confirmation/denial. For typical values, $B = 2^{10}$, $K_1 = K_2 = 100$, our scheme is completely practical. We also give a natural denial protocol for the RSA setting which is more efficient than the previous denial protocol of [18].

Acknowledgments

The authors would like to thank Ivan Damgård, Jan Camenisch and Markus Jakobsson for helpful comments. The first author would like to thank Hewlett-Packard Laboratories Bristol for support during the time this research was done.

References

1. Bellare, M. and Rogaway, P. The exact security of digital signatures–How to sign RSA and Rabin, in U. Maurer (ed.), EUROCRYPT '96, Springer LNCS 1070 (1996) 399–416.
2. Boyar, J., Chaum, C., Damgård, I. and Pedersen, T. Convertible undeniable signatures, in A.J. Menezes and S.A. Vanstone (eds.), CRYPTO '90, Springer LNCS 537, Springer (1991) 189–205.
3. Boyd, C. and Foo, E. Off-line fair payment protocols using convertible signatures, in K. Ohta et al (eds.), ASIACRYPT '98, Springer LNCS 1514 (1998) 271-285.
4. Camenisch J. and Michels, M. Proving in zero-knowledge that a number is the product of two safe primes, in J. Stern (ed.), EUROCRYPT '99, Springer LNCS 1592 (1999) 106–121.
5. Camenisch, J. and Michels, M. Confirmer signature schemes secure against adaptive adversaries, in B. Preneel (ed.), EUROCRYPT 2000, Springer LNCS 1870 (2000) 243-258.
6. Catalano, D., Gennaro, R. and Halevi, S. Computing inverses over a shared secret modulus, in B. Preneel (ed.), EUROCRYPT 2000, Springer LNCS 1807 (2000) 190–206.
7. Chaum, D. and van Antwerpen, H. Undeniable signatures, in G. Brassard (ed.), CRYPTO '89, Springer LNCS 435 (1990) 212-216.
8. Chaum, D. Zero-knowledge undeniable signatures, in I.B. Damgård (ed.), CRYPTO '90, Springer LNCS 473 (1991) 458-464.
9. Chaum, D., van Heijst, E., and Pfitzmann, B, Cryptographically strong undeniable signatures, unconditionally secure for the signer, in J. Feigenbaum (ed.), CRYPTO '91, Springer LNCS 576, (1992) 470–484.
10. Chaum, D. and Pedersen, T. P. Wallet databases with observers, in E. Brickell (ed.), CRYPTO '92, Springer LNCS 740 (1993) 89–105.
11. Chaum, D. Designated confirmer signatures, in A. de Santis (ed.), EUROCRYPT '94, Springer LNCS 950 (1995) 86–91.
12. Damgård, I. and Pedersen, T. New convertible undeniable signature schemes, in U. Maurer (ed.) EUROCRYPT '96, Springer LNCS 1070 (1996) 372–386.
13. Damgård, I. and Koprowski, M. Practical threshold RSA signatures without a trusted dealer, in B. Pfitzmann (ed.), EUROCRYPT 2001, Springer LNCS 2045 (2001) 152–165.
14. Desmedt, Y., Frankel Y. and Yung, M. Multi-receiver/multi-sender network security: efficient authenticated multicast/feedback, INFOCOM '92 (1992) 2045–2054.
15. Fouque, P.-A., Stern, J. Fully distributed threshold RSA under standard assumptions, in C. Boyd (ed.), ASIACRYPT 2001, Springer (2001).

16. Gennaro, R., Jerecki, S., Krawczyk, H. and Rabin, T. Robust and efficient sharing of RSA functions, in N. Koblitz (ed.), CRYPTO '96, Springer LNCS 1109 (1996) 157–172.

17. Gennaro, R., Miccianicio, D. and Rabin, T. An efficient non-interactive statistical zero-knowledge proof system for quasi-safe prime products, 5th ACM Conference on Computer and Communications Security, October 1998.

18. Gennaro, R., Krawczyk, H. and Rabin, T. RSA-based undeniable signatures, in W. Fumy (ed.), CRYPTO '97, Springer LNCS 1294 (1997) 132-149. Full version *Journal of Cryptology* (2000) 13:397–416.

19. van de Graaf, J. and Peralta, R. A simple and secure way to show that validity of your public key, in C. Pomerance (ed.), CRYPTO '87, Springer LNCS 293 (1988) 128–134.

20. Jakobsson, M., Sako, K. and Impagliazzo, R. Designated verifier proofs and their applications, in U. Maurer (ed.) EUROCRYPT '96, Springer LNCS 1070 (1996) 143–154.

21. Jakobsson, M. Efficient oblivious proofs of correct exponentiation, in B. Preneel (ed.), Communications and multimedia security, Kluwer (1999) 71–84.

22. Menezes, A.J., van Oorschot, P.C. and Vanstone, S.A. *Handbook of Applied Cryptography*, CRC Press, 1997.

23. Michels, M. and Stadler, M. Generic constructions for secure and efficient confirmer signature schemes, in K. Nyberg (ed.) EUROCRYPT '98, Springer LNCS 1403 (1998) 406–421.

24. Miyazaki, T. An improved scheme of the Gennaro-Krawczyk-Rabin undeniable signature system based on RSA, in D. Won (ed.), ICISC 2000, Springer LNCS 2015 (2001) 135–149.

25. Poupard, G. and Stern, J. Short proofs of knowledge for factoring, in PKC 2000, Springer LNCS 1751 (2000) 147–166.

26. Rabin., T. A simplified approach to threshold and proactive RSA, in H. Krawczyk (ed.), CRYPTO '98, Springer LNCS 1462 (1998) 89–104.

27. Rivest, R.L., Shamir, A. and Adleman, L.M. A method for obtaining digital signatures and public-key cryptosystems, *Communications of the ACM* v.21, n.2, 1978, pages 120–126.

28. Shoup, V. Practical threshold signatures, in B. Preneel (ed.), EUROCRYPT 2000, Springer LNCS 1807 (2000) 207–220.

29. Silverman, R. D. Fast generation of random, strong RSA primes, *CryptoBytes*, 3, No. 1, 1997, pp. 9–13.

Co-operatively Formed Group Signatures

Greg Maitland and Colin Boyd

Information Security Research Centre
Queensland University of Technology
Brisbane, Australia
{g.maitland,c.boyd}@qut.edu.au

Abstract. Group signatures and their applications have received considerable attention in the literature in recent times. Substantial gains have been made with respect to designing provably secure and efficient schemes. In practice, as with all signature schemes, deploying group signature schemes requires the group member's signing keys to be both physically and electronically secure from theft. Smartcards or similar devices are often offered as a solution to this problem.

We consider the possibility of co-operatively forming group signatures so as to balance the processing load between a modestly performed secure device and a much more powerful workstation. The constructions are based on the observation that several recent group signature schemes have adopted a structure which utilises two values in signature creation - a private signing key and a group membership certificate. We describe a co-operative group signature scheme based on a recently proposed scheme as well as a 'wallet with observer' variant.

1 Introduction

Unlike ordinary signatures, group signatures allow a group member to create anonymous (and unlinkable) signatures on behalf of a group. A single public key is required in order to verify signatures produced by any group member. Upon verifying a signature, the verifier does not learn the identity of the group member who created the signature. However, should the need arise, a group signature can be 'opened' by a trusted party and the identity of the member who created the signature will be revealed.

Specialized applications, such as voting and auctions, have been suggested as areas of application for group signatures [16]. Group signatures have also been integrated into electronic cash systems [14,19,17]. The anonymity and unlinkability afforded by group signatures suggest that they may have a role to play in these more specialized applications.

The practical deployment of group signatures schemes, like other signature schemes, must consider measures which secure a user's secrets from theft. The cryptographic security of a scheme is rendered irrelevant if a user's secrets are not both physically and electronically secure. Since workstations do not represent secure environments, it is necessary to consider the utilisation of some form of secure device in which a user's secrets can be stored and processed. Because of

B. Preneel (Ed.): CT-RSA 2002, LNCS 2271, pp. 218–235, 2002.
© Springer-Verlag Berlin Heidelberg 2002

their cost and portability, smartcards are typically considered suitable for this task.

While smartcards are cost effective, readily available and relatively tamper-resistant, they are constrained in terms of processing power and memory. In the future, the gulf between smartcards and workstations with respect to relative performance and storage capacity will increase, not decrease. It is therefore necessary to consider carefully the processing and storage demands that a signature scheme may make upon its host device.

Smartcards may also be used as active agents which are capable of enforcing various rules and regulations. Various electronic cash schemes [10,13,5] have suggested a 'wallet with observer' model where the smartcard observer actively prevents double-spending of coins. This is achieved by splitting the user's effective private key between the observer and a user-supplied computing device. The user can only produce valid signature with the co-operation of the tamper-resistant observer.

Main Contribution: We consider ways of co-operatively forming group signatures with a view to balancing the processing and storage demands between workstation and smartcard. We offer some general observations about the structure of recent group signature schemes and their suitability for the task at hand. In particular, we seek to exploit the natural occurrence of two secret values (private key and membership certificate) in these schemes. An example of co-operative group signature generation is provided using a group signature scheme by Ateniese et al. [2]. A further variation using the 'wallet with observer' paradigm is given which illustrates the flexibility of this approach.

Organisation of the Paper: In the next section we provide background information on group signatures schemes, specifically those targeted at 'large' groups. In section 3, we discuss the existing ways in which signatures are co-operatively formed and point out the possibility of co-operative signature formation with respect to a class of group signature schemes. Section 4 presents an example of such a scheme based around a recently proposed group signature scheme. A 'wallet with observer' variation of this same scheme is described in section 5.

2   Group Signature Background

In this section we present the concepts relevant to group signature schemes and explain the basic idea of Camenisch and Stadler [6] which underpins several recent group signature schemes [2,4,19,8].

Group signature schemes were first introduced by Chaum and van Heyst [9] in 1991. Unlike ordinary signatures, group signatures provide anonymity (and unlinkability) to the signer, i.e., a verifier can only tell that a member of a group signed. However, in exceptional cases, such as a legal dispute, any group signature can be 'opened' by a designated revocation manager to reveal unambiguously the identity of the signature's originator. At the same time, no coalition of entities (including the group manager) can create a group signature which a verifier will accept as valid but cannot be opened.

The following definitions are borrowed from Ateniese et al. [1]. A group signature scheme is a digital signature scheme comprised of the following procedures:

- SETUP: an algorithm for generating the initial group public key $\mathcal{Y}$ and a group private key $\mathcal{S}$.
- JOIN: a protocol between the group manager ($\mathcal{GM}$) and a user that results in the user becoming a new group member.
- SIGN: a protocol between a group member and a user whereby a group signature on a user-supplied message is computed by the group member.
- VERIFY: an algorithm for establishing the validity of a group signature given a group public key and a signed message.
- OPEN: an algorithm that, given a signed message and a group private key, allows the revocation manager ($\mathcal{RM}$) to determine the identity of the signer.

A group signature scheme must satisfy the following security properties:

- *Correctness:* Signatures formed by a group member using SIGN must be accepted by VERIFY.
- *Unforgeability:* Only group members are able to sign messages on behalf of the group.
- *Anonymity:* Given a valid signature of some message, identifying the actual signer is computationally hard for everyone but the revocation manager.
- *Unlinkability:* Deciding whether two different valid signatures were computed by the same group member is computationally hard (except for $\mathcal{RM}$).
- *Exculpability:* Neither a group member nor the group manager can sign on behalf of other group members. A closely related property is that of non-framing [12]. It captures the notion of a group member not being made responsible for a signature that the group member did not produce.
- *Traceability:* The group manager is always able to open a valid signature and identify the signer.
- *Coalition-resistance:* A colluding subset of group members (even if comprised of the entire group) cannot generate a valid signature that the group manager cannot link to one of the colluding group members.

The efficiency of a group signature scheme is typically based on the following parameters:

- The size of the group public key $\mathcal{Y}$.
- The size of a group signature.
- The efficiency of SIGN and VERIFY.
- The efficiency of SETUP, OPEN and JOIN.

With early group signature schemes, the length of the public key was proportional to the size of the group and therefore the running time of the verification algorithm depends on the number of group members. The scheme proposed by Camenisch and Stadler [6] was the first to offer:

- a fixed sized group public key and a group signature independent of the size of the group.
- the addition of new members at any time without requiring changes to the group public key.
- the computational complexity of SIGN and VERIFY is also independent of group size.

Camenisch and Stadler [6] proposed a general framework for constructing group signature schemes suitable for large groups and this approach has been followed by various recently proposed schemes [2,4,19,8]. An outline of this framework is as follows:

- SETUP: A group's public key consists of:
  - the public key of an ordinary digital signature scheme used by the group manager ($\mathcal{GM}$) to sign membership certificates. Let $Sig_{\mathcal{GM}}(m)$ denote the signature of this scheme on a message m.
  - the public key of a probabilistic encryption scheme used by the revocation manager ($\mathcal{RM}$). Let $Encr_{\mathcal{RM}}(r,m)$ denote the probabilistic encryption of a message m using a randomizer r and let $Decr_{\mathcal{RM}}(c)$ denote the decryption of ciphertext c.
- JOIN: A new group member $\mathcal{M}$ chooses a random private key $x_{\mathcal{M}}$ and computes a membership key $z \leftarrow f(x_{\mathcal{M}})$, where f is a suitable one-way function. The new member commits to z (for instance by signing it) and sends z and the commitment to the group manager $\mathcal{GM}$ who returns a membership certificate $V \leftarrow Sig_{\mathcal{GM}}(z)$. The group manager stores the association between the new member's identity and the new alias (the public key z) in the membership table.
- SIGN: To sign a message m on behalf of the group, a member probabilistically encrypts z using the public key of the revocation manager (let $c = Encr_{\mathcal{RM}}(r,z)$ denote this ciphertext for some randomiser r) and issues a signature of knowledge proving knowledge of values $\tilde{x}$ and $\tilde{V}$ such that $\tilde{V} = Sig_{\mathcal{GM}}(f(\tilde{x}))$ holds and that $f(\tilde{x})$ is encrypted in c.
- VERIFY: The verification of such a group-signature is done by checking the signature of knowledge.
- OPEN: The revocation manager can easily revoke the anonymity of a group signature by decrypting c to recover z and forwarding this value to the group manager. The group manager can find the member's real identity from the stored values in the membership table.

The challenge is to make suitable choices for the primitives used in this framework. In particular, one seeks to find suitable choices for the one-way function, the signature scheme and probabilistic encryption scheme that yield an efficient signature of knowledge for the values $\tilde{x}$ and $\tilde{V}$. Recently, particular attention has been paid to the task of designing membership certificates which are coalition resistant [1,3].

3   Co-operative Signature Generation

Ordinary signature schemes focus on the roles of two principle parties – the signer and the verifier. The signer creates a signature on a given message and the verifier determines if the signature is valid or not. Prior to signing, the signer chooses a private key and public key pair and the verifier is supplied with the public key for the purposes of signature verification. Because a signature is usually used as proof that the signer commits to the contents of the signed message, the verifier is interested in more than the fact that the signature is correctly constructed with respect to a given public key. The verifier is also interested in the association between the public key and the owner's identity. To this end, public key infrastructure (PKI) exists for the purpose of certifying the ownership of public keys. Therefore, in practice, a signature must be accompanied either explicitly or implicitly by a certified public key.

While two party signature schemes are common, examples of three party schemes also exist. Blind signature schemes [11] involve a signer and a recipient engaging in an interactive protocol so that the recipient may gain a signature on a message of choice and that this signature remains unknown to the signer. While the recipient may verify the resulting signature to ensure that the signer has acted honestly, the recipient of the blind signature is not the intended final destination for the signature. The recipient subsequently shows the blinded signature to a verifier for the purpose of proving to the verifier that they possess a signature formed by the signer. The exact meaning of this act depends on the application and is not of direct interest.

Another three party example can be found in some electronic cash schemes [10,13,5]. The 'wallet with observer' paradigm utilises a combination of a user-supplied computing device and an observer device charged with protecting the bank's interests. The customer's effective private key is split into two parts. The customer's computing device stores one part and the observer the other. Both devices must co-operate in order to form a valid customer signature.

The model for group signatures introduced by Camenisch and Stadler [6] is quite similar to the functional operation of ordinary signature schemes. The user generates a key pair, $(x_\mathcal{M}, z)$, and acquires a certificate, $\mathcal{V}$, on the public key of the pair from a trusted party (in this case the group manager). The difference is that the subsequent use of the key pair and certificate must be carried out in such a way so as not to reveal the identity of the signer (anonymity) or link together signatures created by the same user (unlinkability). Showing the public key or the certificate with each signature would compromise the anonymity and unlinkability required by the group signature scheme. The use of proofs of knowledge provides the means by which the public key and the certificate can be used in signature creation without being disclosed. This leads to the observation that there are effectively two secret values, $x_\mathcal{M}$ and $\mathcal{V}$, inherently present in this model.

As noted in section 2, the user must form a signature of knowledge of values $x_\mathcal{M}$ and $\mathcal{V}$ such that $\mathcal{V} = Sig_{g_\mathcal{M}}(f(x_\mathcal{M}))$ holds and that $f(x_\mathcal{M})$ has been encrypted under the revocation manager's public key. The notation suggests that the proof will be provided in some atomic and indivisible fashion. In fact, the two proposals in [6,7] demonstrate such a proof by providing two proofs of knowledge – one proof of knowledge of $x_\mathcal{M}$ such that $z = f(x_\mathcal{M})$ and another proof of knowledge of $\mathcal{V}$ such that $\mathcal{V} = Sig_{\mathcal{GM}}(z)$. The two proofs are not independent of one another. Values used in the construction of the two proofs are shared which links the two proofs together and allows the assertion of knowledge of both $x_\mathcal{M}$ and $\mathcal{V}$.

When the group signature scheme allows for the separation of the roles of $x_\mathcal{M}$ and $\mathcal{V}$ in forming the proof of knowledge, it is possible for two co-operating entities (one knowing $x_\mathcal{M}$ and the other knowing $\mathcal{V}$) to form a group signature if they share the required values during the signing process. In this situation, neither of the entities can independently create a group signature and each needs the co-operation of the other entity.

The above observation leads to an investigation of the usefulness of such co-operatively formed group signatures.

4 A Co-operative Group Signature Scheme

This section describes a co-operative group signature scheme based on a group signature scheme proposed by Ateniese, Camensich, Joye and Tsudik [2]. In order that a group member $\mathcal{M}_i$ create a signature on a message m of the type described by Ateniese et al. [2], knowledge of two undisclosed values is required – the group member's secret key x_i and the group membership certificate $\mathcal{V}_i$.

It is possible for a group member $\mathcal{M}_i$ and a recipient $\mathcal{R}$ to create a group signature co-operatively if $\mathcal{M}_i$ knows only the group member's secret key x_i and $\mathcal{R}$ knows only the group membership certificate $\mathcal{V}_i$. Both parties engage in an interactive variation of the steps described by Ateniese et al. [2]. The interaction is based on a proof of knowledge of discrete logs in a group of unknown order.

Although $\mathcal{R}$ knows the group membership certificate $\mathcal{V}_i$ and can easily calculate the group member's public key, $\mathcal{R}$ cannot learn the group member's secret key x_i. If this were possible, the group manager who issues $\mathcal{V}_i$ could also learn x_i and therefore sign on behalf of a group member. This would contradict the security properties already claimed for the original scheme.

4.1 The Group Signature Procedures

The SETUP, JOIN, VERIFY and OPEN procedures are as per the original scheme [2] where a more detailed discussion of these procedures can be found. For simplicity, we assume that the group manager is also the revocation manager.

SETUP: The initial phase involves the group manager ($\mathcal{GM}$) setting the group public $\mathcal{Y}$ and his secret key $\mathcal{S}$.

- Let $\epsilon > 1$, k, and ℓ_p be security parameters where the parameter ℓ_p sets the size of the modulus to use and the parameter ϵ controls the tightness of the statistical zero-knowledgeness.

- Let λ_1, λ_2, γ_1, and γ_2 denote lengths satisfying

$$\lambda_1 > \epsilon(\lambda_2 + k) + 2, \ \lambda_2 > 4\ell_p, \ \gamma_1 > \epsilon(\gamma_2 + k) + 2 \text{ and } \ \gamma_2 > \lambda_1 + 2.$$

The integral ranges Λ and Γ are defined as

$$\Lambda = [2^{\lambda_1} - 2^{\lambda_2} + 1, 2^{\lambda_1} + 2^{\lambda_2} - 1] \text{ and } \Gamma = [2^{\gamma_1} - 2^{\gamma_2} + 1, 2^{\gamma_1} + 2^{\gamma_2} - 1].$$

- Select $\mathcal{H}$ to be a collision-resistant hash function such that

$$\mathcal{H} : \{0,1\}^* \to \{0,1\}^k.$$

- Select random secret ℓ_p-bit primes p', q' such that $p = 2p' + 1$ and $q = 2q' + 1$ are prime. Set the RSA modulus used to sign group membership certificates to $n = pq$.
- Choose random generators a, a_0, g, h for the set of quadratic residues Q_n. Note that, by construction, the order of $\mathbb{Z}_n^*$ is $\varphi(n) = 4p'q'$ and therefore the order of Q_n is $p'q'$. Revealing the order of Q_n reveals $\varphi(n)$ and so compromises the security of RSA signatures formed using modulus n. Hence, the order of Q_n must remain unknown to all but the group manager.
- Choose a random element $x_R \in_R \mathbb{Z}_{p'q'}^*$ as the revocation private key and set $y_R = g^{x_R} \mod n$ as the revocation public key.
- The group public key is: $\mathcal{Y} = (n, a, a_0, y, g, h, \epsilon, k, \ell_p, \lambda_1, \lambda_2, \gamma_1, \gamma_2, \mathcal{H})$.
- The corresponding secret key (known only to $\mathcal{GM}$) is: $\mathcal{S} = (p', q', x_R)$.

JOIN: Each group member $\mathcal{M}_i$ joins the group by interacting with the group manager $\mathcal{GM}$ in order to acquire:

- A private key x_i known only to the group member $\mathcal{M}_i$ such that $x_i \in \Lambda$. The associated public key is $C_2 = a^{x_i} \mod n$ with $C_2 \in Q_n$.
- A membership certificate $[A_i, e_i]$ where e_i is a random prime chosen by $\mathcal{GM}$ such that $e_i \in_R \Gamma$ and A_i has been computed by the $\mathcal{GM}$ as $A_i := (C_2 a_0)^{1/e_i} \mod n$.

As part of JOIN, $\mathcal{GM}$ creates a new entry in the membership table and stores the new member's real identity along with $[A_i, e_i]$ in the new entry.

SIGN: A group member $\mathcal{M}_i$ can co-operatively create a group signature tuple $(c, s_1, s_2, s_3, s_4, T_1, T_2, T_3)$ on a message $m \in \{0,1\}^*$ by using knowledge of x_i and interacting with a recipient $\mathcal{R}$ who knows $[A_i, e_i]$ as shown in fig.1.

VERIFY: A verifier can verify a signature $(c, s_1, s_2, s_3, s_4, T_1, T_2, T_3)$ of the message m as follows:

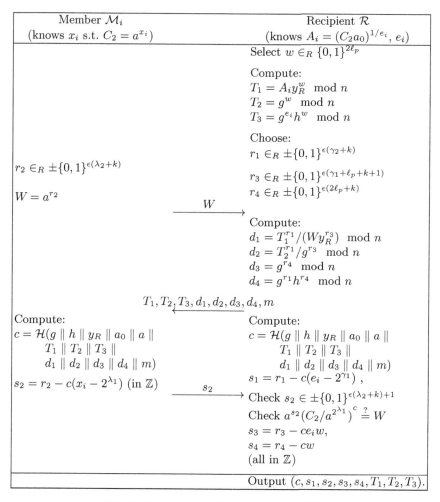

Member $\mathcal{M}_i$	Recipient $\mathcal{R}$
(knows x_i s.t. $C_2 = a^{x_i}$)	(knows $A_i = (C_2 a_0)^{1/e_i}$, e_i)

Select $w \in_R \{0,1\}^{2\ell_p}$

Compute:
$T_1 = A_i y_R^w \mod n$
$T_2 = g^w \mod n$
$T_3 = g^{e_i} h^w \mod n$

Choose:
$r_1 \in_R \pm\{0,1\}^{\epsilon(\gamma_2+k)}$

$r_2 \in_R \pm\{0,1\}^{\epsilon(\lambda_2+k)}$

$r_3 \in_R \pm\{0,1\}^{\epsilon(\gamma_1+\ell_p+k+1)}$

$W = a^{r_2}$

$r_4 \in_R \pm\{0,1\}^{\epsilon(2\ell_p+k)}$

$\xrightarrow{\qquad W \qquad}$

Compute:
$d_1 = T_1^{r_1}/(W y_R^{r_3}) \mod n$
$d_2 = T_2^{r_1}/g^{r_3} \mod n$
$d_3 = g^{r_4} \mod n$
$d_4 = g^{r_1} h^{r_4} \mod n$

$\xleftarrow{\quad T_1, T_2, T_3, d_1, d_2, d_3, d_4, m \quad}$

Compute:	Compute:
$c = \mathcal{H}(g \parallel h \parallel y_R \parallel a_0 \parallel a \parallel$	$c = \mathcal{H}(g \parallel h \parallel y_R \parallel a_0 \parallel a \parallel$
$\quad T_1 \parallel T_2 \parallel T_3 \parallel$	$\quad T_1 \parallel T_2 \parallel T_3 \parallel$
$\quad d_1 \parallel d_2 \parallel d_3 \parallel d_4 \parallel m)$	$\quad d_1 \parallel d_2 \parallel d_3 \parallel d_4 \parallel m)$
$s_2 = r_2 - c(x_i - 2^{\lambda_1})$ (in $\mathbb{Z}$)	$s_1 = r_1 - c(e_i - 2^{\gamma_1})$,

$\xrightarrow{\qquad s_2 \qquad}$

Check $s_2 \in \pm\{0,1\}^{\epsilon(\lambda_2+k)+1}$
Check $a^{s_2}(C_2/a^{2^{\lambda_1}})^c \stackrel{?}{=} W$
$s_3 = r_3 - c e_i w,$
$s_4 = r_4 - cw$
(all in $\mathbb{Z}$)

Output $(c, s_1, s_2, s_3, s_4, T_1, T_2, T_3)$.

Fig. 1. Co-operative Group Signature SIGN

1. Compute:

$$c' = \mathcal{H}\,(\,g \parallel h \parallel y_R \parallel a_0 \parallel a \parallel T_1 \parallel T_2 \parallel T_3 \parallel$$
$$a_0{}^c T_1^{s_1-c2^{\gamma_1}}/(a^{s_2-c2^{\lambda_1}} y_R^{s_3}) \mod n \parallel$$
$$T_2^{s_1-c2^{\gamma_1}}/g^{s_3} \mod n \parallel$$
$$T_2{}^c g^{s_4} \mod n \parallel$$
$$T_3{}^c g^{s_1-c2^{\gamma_1}} h^{s_4} \mod n \parallel m)$$

2. Accept the signature if and only if $c = c'$ and
   - $s_1 \in \pm\{0,1\}^{\epsilon(\gamma_2+k)+1}$
   - $s_2 \in \pm\{0,1\}^{\epsilon(\lambda_2+k)+1}$
   - $s_3 \in \pm\{0,1\}^{\epsilon(\gamma_1+2\ell_p+k+1)+1}$
   - $s_4 \in \pm\{0,1\}^{\epsilon(2\ell_p+k)+1}$

OPEN: In the event that the actual signer must be subsequently identified (e.g., in case of a dispute) $\mathcal{GM}$ executes the following procedure:

1. Check the signature's validity via the VERIFY procedure.
2. Recover A_i (and thus the identity of P_i) as $A_i = T_1/T_2^{x_R} \mod n$.
3. Prove that $log_g y_R = log_{T_2}(T_1/A_i \mod n)$

4.2   Security

As the interactive protocol derived from the original group signature scheme is an honest-verifier zero-knowledge proof of knowledge [2], a verifier cannot learn any useful information from the co-operatively formed signatures produced by our scheme if the hash function $\mathcal{H}$ acts as a random oracle. The major concern is whether or not the recipient can learn information about the group member's private key by participating in the interactive co-operative group signature protocol.

Definition 1 (Strong-RSA Problem). *Let $n = pq$ be an RSA-like modulus and let G be a cyclic subgroup of $\mathbb{Z}_n^*$ of order $|G|$. Let $\lceil \log_2 |G| \rceil = \ell_G$. Given n and $z \in G$, the Strong-RSA Problem consists of finding $u \in G$ and $e \in \mathbb{Z}_{>1}$ satisfying $z \equiv u^e \mod n$.*

Assumption 1 (Strong-RSA Assumption). *A probabilistic polynomial-time algorithm $\mathcal{K}$ exists which on input of a security parameter ℓ_G outputs a pair (n, z) such that, for all probabilistic polynomial-time algorithms $\mathcal{P}$, the probability that $\mathcal{P}$ can solve the Strong-RSA Problem is negligible.*

Theorem 1. *Under the Strong-RSA assumption, the interactive protocol between the group member and the recipient is an honest-verifier statistical zero-knowledge proof of knowledge of the member's private key for any $\epsilon > 1$.*

If the recipient is allowed to choose the challenge c, we cannot make a definitive statement about the security of the scheme as the recipient may not act honestly. If we assume that the hash function $\mathcal{H}$ acts as a random oracle, then, by requiring that the member $\mathcal{M}_i$ calculates the challenge as the output from the hash function $\mathcal{H}$, we force the challenge to be randomly chosen and hence we can claim that the interactive protocol is a statistical zero-knowledge proof of knowledge. The proof is outlined in appendix A.

4.3   Practical Deployment

There is a real risk that a group member's signing secrets could be stolen by an attacker if they were stored on an insecure desktop workstation. The signing secrets need to be stored in a secure portable device such as a smartcard. While the group signature scheme of Ateniese et al. [2] is significantly more efficient than previous proposals, implementing the entire scheme within currently available smartcards is likely to result in performance which is unacceptable to those well used to the high performance of current desktop workstations. In particular, the smartcard environment suffers from the inability to perform pre-computations when not in contact with a reader.

The co-operative scheme described above is well suited to the scenario of a high performance workstation with attached smartcard reader. The group member's private key can be stored securely on a smartcard and the corresponding membership certificate can be stored on the workstation. The two devices can co-operatively form group signatures. Theft of the membership certificate provides no more information about the member's private key than was already available to the group manager. Since the group manager cannot forge a group member's signature, theft of the membership certificate alone does not compromise security. The division of labour is particularly well suited to an implementation of the group member's role by a modestly performed smartcard. Only one exponentiation is required by the smartcard compared to around 10 exponentiations if the original group signature scheme were implemented entirely by the smartcard.

5   A 'Wallet with Observer' Scheme

The 'wallet with observer' model was first proposed by Chaum and Pedersen [10] and has been used by various electronic cash schemes [13,5] in order to proactively protect the bank's interests. In this model, the bank-issued smartcard observer actively prevents the double-spending of coins.

Maitland and Boyd [15] have proposed a fair cash scheme based on the group signature described by Ateniese et al [2]. A natural extension of this fair cash scheme would be the introduction of prior restraint of double-spending. Since a new member interacts with the group manager to obtain a group membership certificate, it is natural for the group manager to consider issuing a tamper-resistant observer which contains this membership certificate. In this scenario, the group member could only produce a valid signature with the co-operation of the tamper-resistant observer.

The scheme described in section 4 can be rearranged into the 'wallet with observer' model through a minor adjustment to the JOIN procedure and a re-structuring of the SIGN procedure.

5.1   The Restructured Procedures

Except for JOIN and SIGN, the other procedures are as per those described in Sect. 4.1.

JOIN: The JOIN procedure follows the steps used by the JOIN presented in Sect. 4.1. At the conclusion of the JOIN procedure, the group manager $\mathcal{GM}$ issues the new group member $\mathcal{M}_i$ with a tamper-resistant observer $\mathcal{O}_i$ in which the group manager has embedded the new group member's certificate $[A_i, e_i]$.

SIGN: A group member $\mathcal{M}_i$ can co-operatively create group signatures on messages $m \in \{0,1\}^*$ by using knowledge of x_i and interacting with an observer $\mathcal{O}_i$ who knows $[A_i, e_i]$ as follows:

1. $\mathcal{O}_i$:
   - Generate a random value $w \in_R \{0,1\}^{2\ell_p}$ and compute:

   $$T_1 = A_i y_R^w \mod n; \quad T_2 = g^w \mod n; \quad T_3 = g^{e_i} h^w \mod n.$$

   where the pair (T_1, T_2) represent an ElGamal encryption of A_i using the revocation manager's public key y_R.
   - Randomly choose: $r_1 \in_R \pm\{0,1\}^{\epsilon(\gamma_2+k)}$,
     $$r_3 \in_R \pm\{0,1\}^{\epsilon(\gamma_1+\ell_p+k+1)},$$
     $$r_4 \in_R \pm\{0,1\}^{\epsilon(2\ell_p+k)}$$
   - Compute: $\bar{d}_1 = T_1^{r_1}/y_R^{r_3} \mod n; d_2 = T_2^{r_1}/g^{r_3} \mod n;$
     $\qquad\quad d_3 = g^{r_4} \qquad \mod n; d_4 = g^{r_1} h^{r_4} \mod n.$
   - Send $(T_1, T_2, T_3, \bar{d}_1, d_2, d_3, d_4,)$ to $\mathcal{M}_i$.
2. $\mathcal{M}_i$:
   - Randomly choose $r_2 \in_R \pm\{0,1\}^{\epsilon(\lambda_2+k)}$ and
   - Compute $d_1 = \bar{d}_1/a^{r_2} \mod n$ and
   - $c = \mathcal{H}(g \parallel h \parallel y_R \parallel a_0 \parallel a \parallel T_1 \parallel T_2 \parallel T_3 \parallel d_1 \parallel d_2 \parallel d_3 \parallel d_4 \parallel m)$.
   - Send d_1 to $\mathcal{O}_i$.
3. $\mathcal{O}_i$: Compute
   - $c = \mathcal{H}(g \parallel h \parallel y_R \parallel a_0 \parallel a \parallel T_1 \parallel T_2 \parallel T_3 \parallel d_1 \parallel d_2 \parallel d_3 \parallel d_4 \parallel m)$.
   - $s_1 = r_1 - c(e_i - 2^{\gamma_1})$, $s_3 = r_3 - ce_i w$, $s_4 = r_4 - cw$ (all in $\mathbb{Z}$)
   - Send (s_1, s_3, s_4) to $\mathcal{M}_i$.
4. $\mathcal{M}_i$: Compute
   - $s_2 = r_2 - c(x_i - 2^{\lambda_1})$, (in $\mathbb{Z}$)
   - Use **VERIFY** to confirm that $(c, s_1, s_2, s_3, s_4, T_1, T_2, T_3)$ is a valid signature on m.
   - Output $(c, s_1, s_2, s_3, s_4, T_1, T_2, T_3)$.

5.2 Security

We are interested in whether the group member $\mathcal{M}_i$ can learn information about the membership certificate stored in the observer $\mathcal{O}_i$ from the protocol interactions with the observer.

Theorem 2. *Under the Strong-RSA assumption, the interactive protocol between the observer and the group member is an honest-verifier statistical zero-knowledge proof of knowledge of a membership certificate corresponding to the membership secret key.*

If we assume that the hash function $\mathcal{H}$ acts as a random oracle, then, by requiring the observer $\mathcal{O}_i$ to calculate the challenge as the output from the hash function $\mathcal{H}$, we force the challenge to be randomly chosen and hence we can claim that the interactive protocol is a statistical zero-knowledge proof of knowledge. A proof of theorem 2 is given in appendix A.

5.3   Practical Deployment

The use of an observer containing the member's certificate fits in well with the natural flow of the JOIN procedure. The last step in a successful JOIN procedure sees the group manager providing the new group member with the membership certificate necessary to form group signatures. As has already been raised, there is a real risk that a group member's signing secrets could be stolen by an attacker when stored on an insecure workstation. If a secure computing device capable of accepting a smartcard (e.g. palmtop or mobile phone) is available, the smartcard observer can be placed in this device and maintained in an active state. In this situation, the pre-computation of values is possible and the division of labour afforded by the above scheme becomes viable.

The original work on 'wallets with observers' [10] raised the possibility of an inflow/outflow of information between the observer and its issuer through the use of subliminal channels. The protocol, as presented, needs to be protected against such information leakage; otherwise, the anonymity of the group member may be compromised. To this end, the observer's choice of random values used to generate the commitments needs to be salted by the group member.

6   Conclusions and Further Work

We have presented a method for co-operatively forming group signatures that applies to a number of the recently proposed schemes. We have demonstrated that the methods have practical application in the real deployment of group signature schemes.

It is intended to investigate the practical deployment of these methods in the context of an electronic cash scheme. The 'wallet with observer' model is susceptible to information outflow through a subliminal channel and will be the subject of further work in order to protect against this possibility.

The notions introduced in this paper can also be applied to group blind signatures. The scheme proposed by Ramzan and Lysyanskaya [14] is based on the group signature scheme of Camenisch and Stadler [6] and uses two connected proofs of knowledge. It therefore lends itself to our construction. If the blind signature recipient is provided with the group member's certificate, the proof of knowledge of the certificate can be formed more efficiently by the recipient. Only the proof of knowledge of the private key need be blinded. In this way, the recipient can be assured that the group signature was formed by the member concerned and not diverted to another member.

The group blind signatures proposed by Ramzan and Lysyanskaya [14,18] are the only ones published in the literature. A group blind signature scheme with provable security and more efficient construction is desirable. The notions presented in this paper are worthy of further investigation with a view to designing an alternative group blind signature scheme.

Acknowledgments

We would like to thank the anonymous referees for their insightful observations and suggestions. This research is part of an ARC SPIRT project undertaken jointly by Queensland University of Technology and Telstra (Australia).

References

1. G. Ateniese and G. Tsudik. Some open issues and new directions in group signatures. In Matthew Franklin, editor, *Financial cryptography: Third International Conference, FC '99, Anguilla, British West Indies, February 22–25, 1999: proceedings*, volume 1648 of *Lecture Notes in Computer Science*, pages 196–211, Berlin, Germany / Heidelberg, Germany / London, UK / etc., 1999. Springer-Verlag.
2. Giuseppe Ateniese, Jan Camenisch, Marc Joye, and Gene Tsudik. A practical and provably secure coalition-resistant group signature scheme. In Mihir Bellare, editor, *Advances in Cryptology—CRYPTO 2000*, volume 1880 of *Lecture Notes in Computer Science*, pages 255–270. Springer-Verlag, 20–24 August 2000.
3. Giuseppe Ateniese, Marc Joye, and Gene Tsudik. On the difficulty of coalition-resistance in group signature schemes. In *Second Conference Security in Communication Networks (SCN '99)*, 16–17 September 1999.
4. Giuseppe Ateniese and Gene Tsudik. Group signatures á la carte. In *Proceedings of the Tenth Annual ACM-SIAM Symposium on Discrete Algorithms (SODA)*, pages 848–849, N.Y., January 17–19 1999. ACM-SIAM.
5. Stefan Brands. Untraceable off-line cash in wallets with observers. In Douglas R. Stinson, editor, *Advances in Cryptology—CRYPTO '93*, volume 773 of *Lecture Notes in Computer Science*, pages 302–318. Springer-Verlag, 22–26 August 1993.
6. J. L. Camenisch and M. A. Stadler. Efficient group signature schemes for large groups. In Jr. Burton S. Kaliski, editor, *Advances in Cryptology—CRYPTO '97*, volume 1294 of *Lecture Notes in Computer Science*, pages 410–424. Springer-Verlag, 17–21 August 1997.
7. Jan Camenisch. *Group Signature Schemes and Payment Systems Based on the Discrete Logarithm Problem*. PhD thesis, ETH, 1998. Reprinted as Vol. 2 of ETH-Series in Information Security an Cryptography, editied by Ueli Maurer, Hartung-Gorre Verlag, Konstanz, ISBN 3-89649-286-1.
8. Jan Camenisch and Markus Michels. A group signature scheme based on an RSA-variant. BRICS Report Series RS-98-27, University of Aarhus, November 1998.
9. D. Chaum and E. van Heyst. Group signatures. In D. W. Davies, editor, *Advances in Cryptology—EUROCRYPT 91*, volume 547 of *Lecture Notes in Computer Science*, pages 257–265. Springer-Verlag, 8–11 April 1991.
10. D. Chaum and T. Pryds Pedersen. Wallet databases with observers. In Ernest F. Brickell, editor, *Advances in Cryptology—CRYPTO '92*, volume 740 of *Lecture Notes in Computer Science*, pages 89–105. Springer-Verlag, 1993, 16–20 August 1992.
11. David Chaum. Blind signatures for untraceable payments. In David Chaum, Ronald L. Rivest, and Alan T. Sherman, editors, *Advances in Cryptology: Proceedings of Crypto 82*, pages 199–203. Plenum Press, New York and London, 1983, 23–25 August 1982.
12. L. Chen and T. P. Pedersen. New group signature schemes. In Alfredo De Santis, editor, *Advances in Cryptology—EUROCRYPT 94*, volume 950 of *Lecture Notes in Computer Science*, pages 171–181. Springer-Verlag, 1995, 9–12 May 1994.

13. Ronald Cramer and Torben Pedersen. Improved privacy in wallets with observers. In Tor Helleseth, editor, *Advances in Cryptology—EUROCRYPT 93*, volume 765 of *Lecture Notes in Computer Science*, pages 329–343. Springer-Verlag, 1994, 23–27 May 1993.

14. A. Lysyanskaya and Z. Ramzan. Group blind digital signatures: A scalable solution to electronic cash. In R. Hirschfeld, editor, *Financial Cryptography: Second International Conference, FC '98*, volume 1465 of *Lecture Notes in Computer Science*, pages 184–197. Springer-Verlag, February 1998.

15. Greg Maitland and Colin Boyd. Fair electronic cash based on a group signature scheme. In *Third International Conference on Information and Communications Security (ICICS 2001)*. Springer-Verlag, November 2001. To appear.

16. Toru Nakanishi, Toru Fujiwara, and Hajime Watanabe. A linkable group signature and its application to secret voting. *PSJ Transactions*, 40(7):3085–3096, 1999.

17. Toru Nakanishi and Yuji Sugiyama. Unlinkable divisible electronic cash. In E. Okamoto and J. Pieprzyk, editors, *The Third International Workshop on Information Security (ISW2000)*, volume 1975 of *Lecture Notes in Computer Science*, pages 121–134, 2000.

18. Zulfikar Ramzan. Group blind digital signatures: Theory and applications. Master's thesis, Department of Electrical Engineering and Computer Science, MIT, 1999.

19. Jacques Traoré. Group signatures and their relevance to privacy-protecting off-line electronic cash systems. In J. Pieprzyk, R. Safavi-Naini, and J. Seberry, editors, *Australasian Conference on Information Security and Privacy (ACISP'99)*, volume 1587 of *Lecture Notes in Computer Science*, pages 228–243. Springer-Verlag, 1999.

A   Security Proofs

Proof of Theorem 1.

Honest-verifier Zero Knowledge: To prove that the protocol is statistical honest-verifier zero-knowledge for any $\epsilon > 1$, we have to show that an honest recipient, i.e., one who chooses the challenge c uniformly random from $\{0,1\}^k$, can simulate a protocol-conversation that is statistically indistinguishable from a protocol-conversation with the group member.

The recipient can construct a simulator S as follows:
- randomly choose c' from $\{0,1\}^k$ according to the uniform distribution,
- randomly choose s'_2 from $\pm\{0,1\}^{\epsilon(\lambda_2+k)}$ according to the uniform distribution,
- compute $W' = a^{s'_2}\left(C_2/a^{2^{\lambda_1}}\right)^{c'}$.

(W', c', s'_2) provides a simulated view of the protocol as seen by the recipient. Note that each value in the range occurs with probability $1/(2^{\epsilon(\lambda_2+k+1)} - 1)$. That is, the probability distribution $P_S(s'_2)$ according to which the simulator S chooses s'_2 is as follows:

$$P_S(s'_2) = \begin{cases} 0 & \text{for } s'_2 < -(2^{\epsilon(\lambda_2+k)} - 1) \\ \frac{1}{2^{\epsilon(\lambda_2+k+1)}-1} & \text{for } -(2^{\epsilon(\lambda_2+k)} - 1) \leq s'_2 \leq (2^{\epsilon(\lambda_2+k)} - 1) \\ 0 & \text{for } s'_2 > (2^{\epsilon(\lambda_2+k)} - 1) \end{cases}$$

To prove that these values are statistical indistinguishable from a view of a protocol run with the group member, it suffices to consider the probability distribution $P_{\mathcal{M}}(s_2)$ of the response s_2 of the group member $\mathcal{M}$ and the probability distribution $P_{\mathcal{S}}(s_2')$ according to which the simulator $\mathcal{S}$ chooses s_2'.

The probability distribution of s_2 can be evaluated as follows.

– The group member chooses the private key x_i randomly from Λ according to any distribution

$$u_i = x_i - 2^{\lambda_1} \in \{-2^{\lambda_2}+1,\ldots,0,\ldots,2^{\lambda_2}-1\}.$$

– The group member chooses r uniformly at random from $\pm\{0,1\}^{\epsilon(\lambda_2+k)}$

$$r_2 \in \{-(2^{\epsilon(\lambda_2+k)}-1),\ldots,0,\ldots,(2^{\epsilon(\lambda_2+k)}-1)\}.$$

– The recipient chooses c from $\{0,1\}^k$ according to any distribution.

$$c \in \{0,1,\ldots,2^k-1\}.$$

Therefore, in general,

$$(c\cdot u_i) \in \{-(2^{\lambda_2}-1)(2^k-1),\ldots,0,\ldots,(2^{\lambda_2}-1)(2^k-1)\}$$

and, since $s_2 = r_2 - cu_i$,

$$s_2 \in \{-(2^{\epsilon(\lambda_2+k)}-1)-(2^{\lambda_2}-1)(2^k-1),\ldots,0,\ldots,(2^{\epsilon(\lambda_2+k)}-1)+(2^{\lambda_2}-1)(2^k-1)\}$$

For a given group member, the private key x_i is fixed and hence $u_i = x_i - 2^{\lambda_1}$ is fixed. Therefore, $(c\cdot u_i)$ takes on (2^k-1) possible values with $(c\cdot u_i) \in \{0,u_i,2u_i,\ldots,(2^k-1)u_i\}$ if $u_i > 0$ and, if $u_i < 0$, $(c\cdot u_i) \in \{(2^k-1)u_i,\ldots,2u_i,u_i,0\}$. Without loss of generality, assume $u_i > 0$. The case $u_i < 0$ produces a distribution which is a mirror image of its positive counterpart.

The probability that the group member $\mathcal{M}$ selects r_2 is

$$p_{\mathcal{M}}(r_2) = \begin{cases} \frac{1}{2^{\epsilon(\lambda_2+k+1)}-1} & \text{for } -(2^{\epsilon(\lambda_2+k)}-1) \le r_2 \le (2^{\epsilon(\lambda_2+k)}-1) \\ 0 & \text{otherwise} \end{cases}$$

Let $p_{\mathcal{R}}(c)$ be the probability that the recipient $\mathcal{R}$ selects the challenge c. (Therefore, $\sum_{c=0}^{2^k-1} p_{\mathcal{R}}(c) = 1$.) Because the selection processes for r_2 and c are independent, the probability of a valid pair $(r_2, c\cdot u_i)$ being selected is

$$p_{\mathcal{M}}(r_2) \times p_{\mathcal{R}}(c) = \frac{p_{\mathcal{R}}(c)}{(2^{\epsilon(\lambda_2+k+1)}-1)}.$$

Let $N_{s_2} = \{(r_2,c)|\, s_2 = r_2 - cu_i\}$. The probability that s_2 is output by the interactive protocol is

$$P_{\mathcal{M}}(s_2) = \sum_{(r_2,c)\in N_{s_2}} \frac{p_{\mathcal{R}}(c)}{(2^{\epsilon(\lambda_2+k+1)}-1)}$$

$$= \frac{1}{(2^{\epsilon(\lambda_2+k+1)}-1)} \sum_{(r_2,c)\in N_{s_2}} p_{\mathcal{R}}(c) \le \frac{1}{(2^{\epsilon(\lambda_2+k+1)}-1)}$$

For $-(2^{\epsilon(\lambda_2+k)} - 1) \le s_2 \le 2^{\epsilon(\lambda_2+k)} - (2^k - 1)(2^{\lambda_2} - 1)$, each s_2 can be calculated in 2^k ways using all possible values of c. That is, the output is s_2 when $(r_2, cu_i) \in \{(s_2, 0), (s_2 + u_i, u_i), (s_2 + 2u_i, 2u_i), \ldots, (s_2 + (2^k - 1)u_i, (2^k - 1)u_i)\}$. The probability of s_2 being output is exactly

$$P_{\mathcal{M}}(s_2) = \frac{1}{(2^{\epsilon(\lambda_2+k+1)} - 1)} \sum_{c=0}^{2^k-1} p_{\mathcal{R}}(c) = \frac{1}{(2^{\epsilon(\lambda_2+k+1)} - 1)}$$

For the left tail $-(2^{\epsilon(\lambda_2+k)} - 1) - (2^{\lambda_2} - 1)(2^k - 1) \le s_2 < -(2^{\epsilon(\lambda_2+k)} - 1)$, s_2 can be calculated in strictly less than 2^k ways and hence occurs with probability less than $\frac{1}{(2^{\epsilon(\lambda_2+k+1)} - 1)}$.

For the right tail $(2^{\epsilon(\lambda_2+k)} - 1) < s_2 \le (2^{\epsilon(\lambda_2+k)} - 1) + (2^{\lambda_2} - 1)(2^k - 1)$, s_2 can be calculated in at most 2^k ways and hence occurs with probability less than or equal to $\frac{1}{(2^{\epsilon(\lambda_2+k+1)} - 1)}$.

Thus we have

$$\sum_{\alpha \in \mathbb{Z}} |P_{\mathcal{M}}(\alpha) - P_{\mathcal{S}}(\alpha)| = \sum_{\alpha \in \text{Left Tail}} |P_{\mathcal{M}}(\alpha) - P_{\mathcal{S}}(\alpha)| +$$

$$\sum_{\alpha \in \text{Right Tail}} |P_{\mathcal{M}}(\alpha) - P_{\mathcal{S}}(\alpha)|$$

$$= \sum_{\alpha \in \text{Left Tail}} |P_{\mathcal{M}}(\alpha) - 0| +$$

$$\sum_{\alpha \in \text{Right Tail}} \left| P_{\mathcal{M}}(\alpha) - \frac{1}{(2^{\epsilon(\lambda_2+k+1)} - 1)} \right|$$

$$\le \frac{(2^{\lambda_2} - 1)(2^k - 1)}{(2^{\epsilon(\lambda_2+k+1)} - 1)} + \frac{(2^{\lambda_2} - 1)(2^k - 1)}{(2^{\epsilon(\lambda_2+k+1)} - 1)}$$

$$\le \frac{2 \times 2^{\lambda_2} \times 2^k}{(2^{\epsilon(\lambda_2+k+1)} - 1)} = \frac{2^{\lambda_2+k+1}}{(2^{\epsilon(\lambda_2+k+1)} - 1)}$$

$$= \frac{1}{(2^{(\epsilon-1)(\lambda_2+k+1)} - 2^{-(\lambda_2+k+1)})}$$

Thus, as the number of active bits in the private key x_i (namely λ_2 bits) increases, the difference between the two distributions is driven to zero for $\epsilon > 1$ faster than any inverse polynomial function of λ_2.

Proof of Knowledge: If the group member follows the protocol, then

$$W = a^{r_2} = a^{s_2 + c(x_i - 2^{\lambda_1})} = a^{s_2} \left(a^{x_i} / a^{2^{\lambda_1}} \right)^c = a^{s_2} \left(C_2 / a^{2^{\lambda_1}} \right)^c \mod n$$

with $s_2 \in \pm \{0, 1\}^{\epsilon(\lambda_2+k)+1}$. Hence, the recipient will accept and the protocol is complete.

Using the usual construction for a knowledge extractor, it is sufficient two show that the knowledge extractor can compute x_i once two accepting triples with the same commitment have been found. Let (W, c, s_2) and (W, c', s_2') be two such accepting triples. Since $W = a^{r_2} = a^{s_2}(C_2/a^{\lambda_1})^c = a^{s'_2}(C_2/a^{\lambda_1})^{c'}$ holds, we have $(C_2/a^{\lambda_1})^{c-c'} = a^{s'-s}$. Let $d = \gcd(c - c', s' - s)$. Using the extended Euclidean algorithm, we obtain values u and v such that $u\frac{c-c'}{d} + v\frac{s'-s}{d} = 1$ and hence we have

$$a = a^{u\frac{c-c'}{d}+v\frac{s'-s}{d}} = \left(a^u(C_2/a^{\lambda_1})^v\right)^{\frac{c-c'}{d}}$$

By construction, $d \le c - c'$. If $d < c - c'$ then $a^u(C_2/a^{\lambda_1})^v$ is a $\frac{c-c'}{d}$-root of a. Since this contradicts the Strong-RSA assumption, we must have $d = c - c'$. Hence $c - c'$ divides $s' - s$, and we can compute the integer

$$x_i = \frac{s'_2 - s_2}{c - c'} + 2^{\lambda_1}$$

such that $a^{x_i} = C_2$. $\square$

Proof of Theorem 2.

That the interactive protocol is honest-verifier statistical zero-knowledge follows from the proof given for the original scheme [2]. We restrict our attention the proof of knowledge part. We have to show that the knowledge extractor is able to recover the group certificate $[A_i, e_i]$ once it has found two accepting tuples.

Let (c, s_1, s_3, s_4) and (c', s'_1, s'_3, s'_4) be tuples that are derived from the same commitments $(T_1, T_2, T_3, d_3, d_4)$. Without loss of generality, assume that $c' > c$.

Extracting A_i using T_1, T_2 and d_3: Recall $T_1 = A_i y^w \mod n$ and $T_2 = g^w \mod n$ for a randomly chosen $w \in_R \{0,1\}^{2\ell_p}$, $d_3 = g^{r_4} \mod n$ for a randomly chosen $r_4 \in_R \pm\{0,1\}^{\epsilon(2\ell_p+k)}$ and the response is $s_4 = r_4 - cw$.

$$d_3 = g^{r_4} = g^{s_4+cw} = g^{s_4}T_2^c = g^{s'_4+c'w} = g^{s'_4}T_2^{c'} \quad \mod n$$

$$g^{s_4}T_2^c = g^{s'_4}T_2^{c'} \quad \mod n$$

$$g^{s_4-s'_4} = T_2^{c'-c} \quad \mod n \tag{1}$$

$$g^{(s_4-s'_4)/(c'-c)} = T_2 \quad \mod n \tag{2}$$

Let $\delta_4 = \gcd(s_4 - s'_4, c' - c)$. Therefore, by the extended Euclidean algorithm, there exists α_4, β_4 such that $\alpha_4(s_4 - s'_4) + \beta_4(c' - c) = \delta_4$ and so, using equation (1),

$$g = g^{[\alpha_4(s_4-s'_4)+\beta_4(c'-c)]/\delta_4} = \left(g^{\alpha_4(s_4-s'_4)}g^{\beta_4(c'-c)}\right)^{1/\delta_4} = \left(T_2^{\alpha_4(c'-c)}g^{\beta_4(c'-c)}\right)^{\frac{1}{\delta_4}}$$

$$= \left(T_2^{\alpha_4}g^{\beta_4}\right)^{\frac{(c'-c)}{\delta_4}} = \left(T_2^{\alpha_4}g^{\beta_4}\right)^k \quad \mod n \text{ with integer } k = \frac{(c'-c)}{\delta_4}$$

Since $\delta_4 = \gcd(s_4 - s'_4, c' - c)$ and $c' > c$, δ_4 divides $(c' - c)$ and $\delta_4 \le (c' - c) \implies 1 \le \frac{c'-c}{\delta_4} = k$. If $k > 1$, then $T_2^{\alpha_4}g^{\beta_4}$ is a kth root of g which contradicts the Strong-RSA assumption. Therefore $k = 1$ and $\delta_4 = (c' - c)$.

However, δ_4 also divides $s_4 - s_4'$ and so $s_4 - s_4' = \phi_4(c' - c)$ for some integer ϕ_4. Because $s_4 = r_4 - cw$ and $s_4' = r_4 - c'w$,

$$s_4 - s_4' = (c' - c)w$$
$$\frac{s_4 - s_4'}{c' - c} = w$$
$$\phi_4 = w$$

and so $A_i = T_1/y^w = T_1/y^{\phi_4} \mod n$.

Extracting e_i using ϕ_4, T_3 and d_4: Note that $\phi_4 = \frac{s_4 - s_4'}{c' - c}$ from above.

$$d_4 = g^{r_1}h^{r_4} = g^{s_1 + c(e_i - 2^{\gamma_1})}h^{s_4 + cw} = g^{s_1}h^{r_4}g^{c(e_i - 2^{\gamma_1})}h^{cw}$$
$$= g^{s_1}h^{s_4}(g^{-2^{\gamma_1}}g^{e_i}h^w)^c = g^{s_1}h^{s_4}(g^{-2^{\gamma_1}}T_3)^c \mod n$$
$$d_4 = g^{r_1}h^{r_4} = g^{s_1' + c'(e_i - 2^{\gamma_1})}h^{s_4' + c'w} = g^{s_1'}h^{s_4'}(g^{-2^{\gamma_1}}T_3)^{c'} \mod n$$
$$g^{s_1}h^{s_4}(g^{-2^{\gamma_1}}T_3)^c = g^{s_1'}h^{s_4'}(g^{-2^{\gamma_1}}T_3)^{c'} \mod n$$
$$g^{s_1 - s_1'} = h^{s_4' - s_4}(g^{-2^{\gamma_1}}T_3)^{c' - c} = (h^{\phi_4}g^{-2^{\gamma_1}}T_3)^{c' - c} \mod n$$

Let $\delta_1 = \gcd(s_1 - s_1', c' - c)$. Therefore, by the extended Euclidean algorithm, there exists α_1, β_1 such that $\alpha_1(s_1 - s_1') + \beta_1(c' - c) = \delta_1$ and so, using equation (1),

$$g = g^{[\alpha_1(s_1 - s_1') + \beta_1(c' - c)]/\delta_1} = \left(g^{\alpha_1(s_1 - s_1')}g^{\beta_1(c' - c)}\right)^{1/\delta_1}$$
$$= \left((h^{\phi_4}g^{-2^{\gamma_1}}T_3)^{\alpha_1(c' - c)}g^{\beta_1(c' - c)}\right)^{1/\delta_1} = \left((h^{\phi_4}g^{-2^{\gamma_1}}T_3)^{\alpha_1}g^{\beta_1}\right)^{\frac{(c' - c)}{\delta_1}}$$
$$= \left((h^{\phi_4}g^{-2^{\gamma_1}}T_3)^{\alpha_1}g^{\beta_1}\right)^j \text{ with integer } j = \frac{(c' - c)}{\delta_1}.$$

Since $\delta_1 = \gcd(s_1 - s_1', c' - c)$ and $c' > c$, δ_1 divides $(c' - c)$ and $\delta_1 \leq (c' - c) \implies 1 \leq \frac{c' - c}{\delta_1} = j$. If $j > 1$, then $(h^{\phi_4}g^{-2^{\gamma_1}}T_3)^{\alpha_1}g^{\beta_1}$ is a jth root of g which contradicts the Strong-RSA assumption. Therefore $j = 1$ and $\delta_1 = (c' - c)$. However, $\delta_1 = (c' - c)$ also divides $(s_1 - s_1')$ and so there is an integer $\phi_1 = \frac{s_1 - s_1'}{c' - c}$. Because $s_1 = r_1 - c(e_i - 2^{\gamma_1})$ and $s_1' = r_1 - c'(e_i - 2^{\gamma_1})$,

$$s_1 - s_1' = (c' - c)(e_i - 2^{\gamma_1})$$
$$\frac{s_1 - s_1'}{c' - c} = (e_i - 2^{\gamma_1})$$

So $e_i = \frac{s_1 - s_1'}{c' - c} + 2^{\gamma_1} = \phi_1 + 2^{\gamma_1}$.

$\square$

Transitive Signature Schemes

Silvio Micali and Ronald L. Rivest

Laboratory for Computer Science, Massachusetts Institute of Technology,
Cambridge, MA 02139
rivest@mit.edu

Abstract. We introduce and provide the first example of a transitive digital signature scheme. Informally, this is a way to digitally sign vertices and edges of a dynamically growing, transitively closed, graph G so as to guarantee the following properties:

– Given the signatures of edges (u, v) and (v, w), anyone can easily derive the digital signature of the edge (u, w).
– It is computationaly hard for any adversary to forge the digital signature of any new vertex or other edge of G, even if he can request the legitimate signer to digitally sign any number of G's vertices and edges of his choice in an adaptive fashion (i.e., even if he can choose which vertices and edges the legitimate signer should sign next after he sees the legitimate signatures of the ones requested so far).

Keywords: public-key cryptography, digital signatures, graphs, transitive closure.

1 Introduction

Sometimes cryptosystems have (or can be designed to have) algebraic properties that make them exceptionally useful for certain applications.

For example, cryptosystems with appropriate homomorphisms can be used for "computing with encrypted data" [14,8,9,15].

Similarly, blind signatures [4,5] utilize similar homomorphic properties.

Of course, algebraic properties of cryptosystems are often undesirable; they may yield avenues for attacking the cryptosystem such as undesirable "malleability" properties [7].

We propose here a new property for signature schemes that may have interesting applications. We call signature schemes with this property "transitive" signature schemes, because the signature scheme is compatible with computing the transitive closure of the graph being signed.

Subsequent to our work, Johnson et al. [11] have investigated related generalizations under the rubric of "homomorphic signature schemes."

Graphs are commonly used to represent a binary relation on a finite set. A graph $G = (V, E)$ has a finite set V of vertices and a finite set $E \subseteq V \times V$ of edges. We write an edge from u to v as the ordered pair (u, v) in any case, whether the graph is directed or undirected. Cormen et al. [6] describe graph representations and algorithms.

B. Preneel (Ed.): CT-RSA 2002, LNCS 2271, pp. 236–243, 2002.
© Springer-Verlag Berlin Heidelberg 2002

Many graphs are naturally *transitive*: that is, there is an edge from u to v whenever there is a path from u to v.

For a directed example, consider a graph representing a military chain-of-command. Here vertices represent personnel and a directed edge (u, v) from u to v means that u commands (or controls) v. Clearly, if u commands v and v commands w, then u commands w.

For an undirected example, consider a graph representing a set of administrative domains. The vertices represent computers and an undirected edge means that u and v are in the same administrative domain. Again, it is clear that if u and v are in the same administrative domain, and if v and w are in the same administrative domain, then u and w are in the same administrative domain. Transitive (and reflexive) undirected graphs represent equivalence relations.

We are interested in situations where someone (let's call her Alice) wishes to publish a transitive graph in an authenticated (i.e., signed) manner. Signing the graph allows others to know that they are working with the authentic graph (or with authentic components of the graph).

Of course, Alice could just sign a representation of the entire graph as a single signed message. This approach may, however, be awkward in practice, particularly if the graph changes frequently or the components are large.

Regarding the efficiency of representation, we observe that a graph with n vertices may have $O(n^2)$ edges. It may be more efficient for Alice to sign a smaller graph, with the explicit understanding that the intended graph is the *transitive closure* of the signed graph. In this way she never needs to sign more than $O(n)$ edges (in the undirected case).

The *transitive closure* $G^* = (V^*, E^*)$ of a graph $G = (V, E)$ is defined to have $V^* = V$ and to have an edge (u, v) in E^* if and only if there is a path (of length zero or greater) from u to v in G. (This is more properly called the reflexive transitive closure, but we stick with standard usage.)

For further efficiency, we now restrict our attention to schemes wherein Alice signs the vertices and edges of the graph G individually. This allows G to grow dynamically: vertices and edges may be added later on without Alice having to re-sign everything done before. (We assume that vertices and edges are never deleted.)

Alice thus has two signature schemes for signing the components of the graph G: one signature scheme for signing vertices, and one for signing edges. We denote her signature of the vertice v as $\sigma(v)$ and her signature of the edge (u, v) as $\sigma(u, v)$, with the understanding that the underlying signature schemes may be different.

Because we are focussing on the situation where the graph Alice intends to sign is transitive, she need only sign a subset of the edges, as long as the transitive closure of that subset is equal to the intended graph. For example, in the chain-of-command example, she need only sign edges representing the relationship between an individual and his immediate superior; other edges can be inferred from these. Or, for another example, in the administrative domain example, she need only sign enough edges to form a spanning tree within each administrative domain.

An observer who then sees a signed edge $\sigma(u,v)$ and another signed edge $\sigma(v,w)$ can then infer that (u,w) is also in the graph being signed by Alice. It is as if Alice had actually signed the edge (u,w) directly. And then if the observer sees another edge (w,x), he can infer that (u,x) is in the graph as well.

In some applications, however, it may be necessary for a party to actually *prove* that Alice signed (either explicitly or implicitly by transitivity) an edge. For example, a party x may need to prove that he is in the chain of command underneath party u. Or, party x may need to prove that he is in the same administrative domain as party u. How can this be done, if the edge (u,x) was not explicitly signed, but is only inferrable by transitivity?

A straightforward approach to proving that (u,x) is in the graph is to produce a "proof" consisting of a sequence of signed edges forming a path from u to x (with their signatures). In this example, a proof that (u,x) is in the graph might consist of the sequence:

$$(u,v),(v,w),(w,x) \tag{1}$$

and their corresponding signatures:

$$\sigma(u,v),\sigma(v,w),\sigma(w,x) \ . \tag{2}$$

(Actually, the proof includes Alice's signatures on each of the vertices as well:

$$\sigma(u),\sigma(v),\sigma(w),\sigma(x) \ .) \tag{3}$$

This sequence of signed edges forms a path from u to x, thus proving that the edge (u,x) is in the graph being signed by Alice.

The problem with using such "path-proofs" is that they may become cumbersome if the path is long, and may introduce unnecessary detail and information in the proof. (Why should a soldier need to mention each of his superior officers if all he is trying to prove is that he is in the U.S. Army?)

We are thus led to wonder whether some signature schemes might be compatible with a "path compression" operator that produces an inferred signature that is indistinguishable from one that might have been produced by Alice.

More specifically, let us informally define a *"transitive signature scheme"* to be a scheme for signing the vertices and edges of a graph such that if someone sees Alice's signatures on vertices u, v, and w and also sees Alice's signatures on edges (u,v) and (v,w), then that someone can easily compute a valid signature on the edge (u,w) that is indistinguishable from a signature on that edge that Alice would have produced herself. See Figure 1.

With a transitive signature scheme Alice only needs to sign a minimum number of edges that has the same transitive closure as her intended graph; an observer can infer her signatures on the remaining (inferred) edges.

We note for the record that this minimum subset of edges having the same transitive closure is called the *transitive reduction* of a graph and can be computed efficiently in both the undirected and directed cases [1].

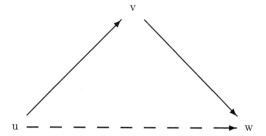

Fig. 1. With a transitive signature scheme anyone can compute a signature on edge (u, w) given signatures on edges (u, v) and (v, w).

With a transitive signature scheme, anyone can produce a short "proof" that a given edge is in the graph. Even if Alice didn't sign that edge explicitly, the proof is a signature that might as well been produced by Alice. Most conveniently, one does not need to have the verifier understand chains or sequences of signed edges (as one typically has to do for certificate chains in an analagous situation, for example [12, Section 13.6.2]).

Having providing some motivation, we can then ask: do transitive signature schemes exist?

We provide a partial answer below.

2 An Undirected Transitive Signature Scheme

In this section we describe a transitive signature scheme for working on undirected graphs (which we dub a *undirected transitive signature scheme*) and prove it secure. It is based on the difficulty of the discrete logarithm problem.

User setup: Each user selects a public-key signature scheme for signing vertices. The vertex signature scheme should have the usual security properties, but need not have any special algebraic properties. For example, the scheme proposed by Goldwasser, Micali, and Rivest [10] is satisfactory here as a vertex signature scheme. The user selects a public-key/private-key pair, and publishes the public-key.

For use in signing edges, each user selects and publishes the following parameters:

- large primes p and q such that q divides $p - 1$, and
- generators g and h of the subgroup G_q of order q of $\mathbf{Z}_p^*$, such that the base-g logarithm of h modulo p is infeasible for others to compute.

Creating a new vertex: When a user Alice wishes to create a new vertex and add it to the graph, she does the following:

- Let n denote the number of vertices previously created. Increment n by 1.
- Select two values x_n and y_n independently at random from $\mathbf{Z}_q$.
- Compute v_n as $g^{x_n} h^{y_n} \pmod{p}$.

- Compose, sign, and publish a statement of the form, "The n-th vertex of the graph is represented by the value v_n." (The message is signed using the vertex-signing public-key signature scheme. The value v_n is given explicitly in the message. The values x_n and y_n are kept secret by Alice.)

Signing the edge (i, j): To sign the edge between the i-th vertex and the j-th vertex, Alice computes and publishes the quadruple:

$$(i, j, \alpha_{ij}, \beta_{ij})$$

where

$$\alpha_{ij} = x_i - x_j \pmod{q}$$
$$\beta_{ij} = y_i - y_j \pmod{q} .$$

We note that the vertex-signing procedure is very similar to the information-theoretically secure commitment scheme of Pedersen [13] (or equivalently, of Chaum, van Heijst, and Pfitzmann [3]).

Verifying an edge signature: Anyone can verify the signature on an edge by checking that

$$v_i = v_j g^{\alpha_{ij}} h^{\beta_{ij}} \pmod{q} . \tag{4}$$

Composing edge signatures: Given a signature $(i, j, \alpha_{ij}, \beta_{ij})$ of edge (i, j) and a signature $(j, k, \alpha_{jk}, \beta_{jk})$ of edge (j, k), anyone can compute the signature

$$(i, k, \alpha_{ik}, \beta_{ik})$$

on edge (i, k) as:

$$\alpha_{ik} = \alpha_{ij} - \alpha_{jk} \pmod{q}$$
$$= x_i - x_k \pmod{q}$$
$$\beta_{ik} = \beta_{ik} - \beta_{ik}$$
$$= y_i - y_k \pmod{q} .$$

This signature is identical to what Alice would produce when signing the edge (i, k).

3   Security

As with any signature scheme, proving security means proving that an adversary will not be able to forge new signatures having seen some previous legitimate signatures. Of course, when the signature scheme has the kind of algebraic property considered in this paper, the adversary is intentionally given for free the ability to "forge" signatures on new edges, as long as they are in the transitive closure of previously signed edges. Since the adversary is being explicitly given this capability for free, it is considered a feature (and not a defect) that the adversary can compute these signatures. It is thus the ability of the adversary to compute signatures on edges *outside* the transitive closure of previously signed edges that should be considered as "forgery" and a break of the signature scheme; our proof shows that forging signatures outside of the transitive closure of previously signed edges is provably hard.

There is another subtle issue regarding the definition of security, regarding how the previously signed edges are determined. The simplest sort of "static" adversary would be asked to forge a signature on a "new" edge, given as input a graph with a given set of edge signatures.

A more powerful adversary, which our scheme is capable of defeating, would be given the following capabilities, which he could exercise at will, until he is ready to attempt a forgery. The adversary can make the following requests of the signer.

- **Initialize.** Ask the signer to discard any previously signed vertices and edges, and begin with a "clean slate"; ask the signer to make up a new name N for a new graph to be constructed.
- **Add a new vertex i.** Ask the signer to sign a statement saying that "vertex i is now part of the graph N."
- **Add a new edge (i, j).** Ask the signer to sign a statement saying that "edge (i, j) is now part of the graph N."

Once the adversary is done with his requests, he is challenged to forge an edge signature on any edge of his choice, as long as the edge is not in the transitive closure of the edges previously signed.

It is important to emphasize that the adversary is *adaptive* in the sense that each such request need be formulated by the adversary only after he has seen the response to all previous requests. The adversary does not need to commit to all of his requests in advance; he makes them up as he goes along.

(We note that if the adversary is static rather than adaptive, in the sense that he must commit to *all* of requests before seeing the responses to any of them, then other simpler schemes will work. For example, it suffices for the signer to publicly assign a random number x_i to each vertex i, and to sign edge (i, j) by giving an RSA signature on the value x_i/x_j, that is, $(x_i/x_j)^d \pmod{n}$. The multiplicative property of RSA ensure the desired transitive property. However, we don't know how to prove this scheme is secure against an adaptive adversary, as the usual reduction techniques seem to require knowing ahead of time (when the x_i values are committed to) what the final connected components will be, so that the x_i values can be appropriately set up for the reduction, and there are too many possible arrangements of the connected components to guess it correctly with a inverse polynomial chance of success.)

In our scheme, starting a new graph means making up a new set of system parameters for that graph, and publishing the details of the corresponding vertex and edge signature schemes. Each vertex and edge signature is accomplished as described in the previous section.

Theorem 1. *The proposed undirected transitive signature scheme is secure in the sense that an adversary can not forge a signature for an edge not in the transitive closure of edges already seen, even if he can adaptively request new vertices to be created and signed and request edges to be signed, assuming that computing discrete logarithms is hard.*

Proof sketch: A straightforward reduction from the discrete logarithm in G_q, the subgroup of prime order q modulo p. Given an instance (g, y, p, q) of the discrete logarithm problem (where g and y are in G_q and the goal is to compute $\log_g(y) \bmod p$) we create a transitive signature scheme with $g = g$ and $h = y$. It is easy to see how a simulator can respond successfully to requests for signatures of edges. However, the usual linear algebra (see Pedersen [13]) can be used to show that it is hard for an adversary to forge signatures since it would imply being able to compute $\log_g(h)$. One needs to argue that the adversary learns nothing about the true x_i and y_i values from the signatures he observes. The adversary's forgery on an edge is thus extremely unlikely to be equal to the simulator's signature for that edge. Given two signatures on the same edge, $\log_g(h)$ can be computed. (A more complete proof will be given in the complete version of this paper.)   □

4   Remarks and Discussion

The problem of finding a *directed* transitive signature scheme remains a very interesting open problem. We have not been able to make much progress on this problem.

We note that there is some similarity between our problem and the problem of "atomic proxy cryptography" due to Blaze et al. [2]. However, we have not been able to come up with a provably secure undirected transitive signature scheme for a scheme modelled on their ideas without losing too much efficiency in the proof.

Another interesting open problem is the following: Given Alice's signature on message M and on (in some special way) Bob's public key, Bob can then "cut himself out of the middle" and produce Alice's signature on M and on (in the same special way) Carol's public key. This would be very useful for collapsing chains of delegation or certification.

Acknowledgments

We would like to thank Shafi Goldwasser for Mihir Bellare for many interesting discussions, encouragement, and suggestions. (In particular, Mihir Bellare sent us a proof that the RSA-based scheme give works against a static adversary.)

References

1. A. V. Aho, M. R. Garey, and J. D. Ullman. The transitive reduction of a directed graph. *SIAM J. Comput.*, 1:131–137, 1972.
2. Matt Blaze, Gerrit Bleumer, and Martin Strauss. Divertible protocols and atomic proxy cryptography. In Kaisa Nyberg, editor, *Proceedings EUROCRYPT '98*, pages 127–144. Springer, 1998.
3. D. Chaum, E. van Heijst, and B. Pfitzmann. Cryptographically strong undeniable signatures, unconditionally secure for the signer. In J. Feigenbaum, editor, *Proceedings CRYPTO '91*, pages 470–484. Springer, 1992. Lecture Notes in Computer Science No. 576.

4. David Chaum. Blind signatures for untraceable payments. In R. L. Rivest, A. Sherman, and D. Chaum, editors, *Proceedings CRYPTO 82*, pages 199–203, New York, 1983. Plenum Press.

5. David Chaum. Security without identification: Transaction systems to make big brother obsolete. *Communications of the ACM*, 28(10):1030–1044, Oct 1985.

6. Thomas H. Cormen, Charles E. Leiserson, and Ronald L. Rivest. *Introduction to Algorithms.* MIT Press/McGraw-Hill, 1990.

7. D. Dolev, C. Dwork, and M. Naor. Non-malleable cryptography. In *Proc. STOC '91*, pages 542–552. ACM, 1991.

8. Joan Feigenbaum. Encrypting problem instances: Or...can you take advantage of someone without having to trust him? In H. C. Williams, editor, *Proceedings CRYPTO 85*, pages 477–488. Springer, 1986. Lecture Notes in Computer Science No. 218.

9. Joan Feigenbaum and Michael Merritt. Open questions, talk abstracts, and summary of discussions. In *DIMACS Series in Discrete Mathematics and Theoretical Computer Science*, volume 2, pages 1–45, 1991.

10. Shafi Goldwasser, Silvio Micali, and Ronald L. Rivest. A digital signature scheme secure against adaptive chosen-message attacks. *SIAM*, 17(2):281–308, April 1988.

11. Robert Johnson, David Molnar, Dawn Song, and David Wagner. Homomorphic signature schemes. In *Topics in Cryptology—CT-RSA 2002*, pages 244–262. Springer, 2002. Lecture Notes in Computer Science No. 2271 (This Volume).

12. Alfred J. Menezes, Paul C. van Oorschot, and Scott A. Vanstone. *Handbook of Applied Cryptography.* CRC Press, 1997.

13. T.P. Pedersen. Non-interactive and information-theoretic secure verifiable secret sharing. In J. Feigenbaum, editor, *Proceedings CRYPTO '91*, pages 129–140. Springer, 1992. Lecture Notes in Computer Science No. 576.

14. Ronald L. Rivest, Leonard Adleman, and Michael L. Dertouzos. On data banks and privacy homomorphisms. In R. DeMillo, D. Dobkin, A. Jones, and R. Lipton, editors, *Foundations of Secure Computation*, pages 169–180. Academic Press, 1978.

15. Tomas Sander, Adam Young, and Moti Yung. Non-interactive cryptocomputing for NC^1. In *Proceedings 40th IEEE Symposium on Foundations of Computer Science*, pages 554–566, New York, 1999. IEEE.

Homomorphic Signature Schemes

Robert Johnson[1], David Molnar[2], Dawn Song[1,*], and David Wagner[1]

[1] University of California at Berkeley
[2] ShieldIP

Abstract. Privacy homomorphisms, encryption schemes that are also homomorphisms relative to some binary operation, have been studied for some time, but one may also consider the analogous problem of homomorphic signature schemes. In this paper we introduce basic definitions of security for homomorphic signature systems, motivate the inquiry with example applications, and describe several schemes that are homomorphic with respect to useful binary operations. In particular, we describe a scheme that allows a signature holder to construct the signature on an arbitrarily redacted submessage of the originally signed message. We present another scheme for signing sets that is homomorphic with respect to both union and taking subsets. Finally, we show that any signature scheme that is homomorphic with respect to integer addition must be insecure.

1 Introduction

A cryptosystem $f : G \to R$ defined on a group $(G, \cdot)$ is said to be homomorphic if f forms a (group) homomorphism. That is, given $f(x)$ and $f(y)$ for some unknown $x, y \in G$, anyone can compute $f(x \cdot y)$ without any need for the private key. Somewhat surprisingly, this property has a wide range of applications, including secure voting protocols [8] and multiparty computation [26].

In a series of talks, Rivest suggested the investigation of homomorphic *signature* schemes. For instance, the RSA signature scheme is a group homomorphism, as $m_1^d \cdot m_2^d = (m_1 \cdot m_2)^d$. This property was previously considered to be undesirable and much energy has been spent on eliminating it [5]. The question is whether this property can be put to positive use instead. More generally, Rivest asked whether homomorphic signature schemes with positive applications can be found.

Our goal is to shed light on this question. In the process, we give a formal definition of what it means to be a secure homomorphic signature scheme (see Section 3). Then, we construct a redactable signature scheme where, given a signature $\mathsf{Sig}(x)$, anyone can compute a signature $\mathsf{Sig}(w)$ on any redaction w of x obtained by rubbing out some positions of x (Section 4). We give proofs of

* This research was supported in part by the Defense Advanced Research Projects Agency under DARPA contract N6601-99-28913 (under supervision of the Space and Naval Warfare Systems Center San Diego) and by the National Science foundation under grants FD99-79852 and CCR-0093337.

B. Preneel (Ed.): CT-RSA 2002, LNCS 2271, pp. 244–262, 2002.
© Springer-Verlag Berlin Heidelberg 2002

security for this scheme (Appendix A). We present a scheme for signing sets that allows anyone to compute the signature on the union of two signed sets, and the signature on any subset of a signed set, and corresponding proofs of security (Section 5).

We also consider additively homomorphic signature schemes $\mathsf{Sig} : \mathbb{Z}/m\mathbb{Z} \to R$. We argue that all such schemes are insecure: they unavoidably possess properties that are likely to render them useless in practice (Section 6). The problematic properties of additive signature schemes are general and completely independent of the way the scheme is implemented. In response, we define and consider *semigroup-homomorphic* signature schemes, which would permit us to avoid these bad properties. We pose as an open problem to find a signature scheme that is semigroup-homomorphic but not group-homomorphic.

2   Related Work

The notion of homomorphic signature schemes was first given by Rivest in a series of talks on "two new signature schemes" [24]. The first of these signature schemes, due to Micali and Rivest, is a "transitive signature scheme" for undirected graphs [19]. In this scheme, given a signature on two graph edges $\mathsf{Sig}((x, y))$, $\mathsf{Sig}((y, z))$, a valid signature $\mathsf{Sig}((x, z))$ on any edge in their transitive closure can be computed without access to the secret key. This scheme works only for undirected graphs; given signatures on the transitive closure edge $\mathsf{Sig}((x, z))$ and the edge $\mathsf{Sig}((x, y))$, a signature on the "intermediate" edge $\mathsf{Sig}((y, z))$ can be computed It was left as an open problem to find a similar scheme for *directed* graphs.

The second signature scheme, due to Rivest, Rabin, and Chari, allows "prefix aggregation" [24]. Given two signatures $\mathsf{Sig}(p\|0)$ and $\mathsf{Sig}(p\|1)$ on the two messages obtained from p by appending a zero and one bit, their scheme allows computation of a signature $\mathsf{Sig}(p)$ on p without access to the secret key. The scheme as presented has a property similar to the transitive graph signature scheme: the signature $\mathsf{Sig}(p\|1)$ can be easily computed. from $\mathsf{Sig}(p)$ and $\mathsf{Sig}(p\|0)$. It was left as an open problem to find a similar scheme that does not have this property. Rivest also posed the open problem of finding a "concatenation signature scheme," in which given two signatures $\mathsf{Sig}(x)$ and $\mathsf{Sig}(y)$ a signature $\mathsf{Sig}(x\|y)$ on their concatenation can be computed.

Rivest also suggested investigating what other "signature algebras" can be constructed. In Section 4 we give a construction for "redactable signatures." Then in Section 5 we show that RSA accumulators can be used to construct signatures homomorphic with respect to the union and subset operations.

Homomorphic signature schemes are intriguing in part because homomorphic cryptosystems have proved to be so useful. Rivest, Adleman, and Dertouzos noted applications of "privacy homomorphisms" to computing on encrypted data soon after the introduction of RSA [25]. Peralta and Boyar showed that an XOR-homomorphic bit commitment could be exploited to yield more efficient zero-knowledge proofs of circuit satisfiability [23]. Feigenbaum and Merritt noted that a "cryptosystem which is a ring homomorphism on $\mathbb{Z}/2\mathbb{Z}$ could be used to implement completely non-interactive secure circuit evaluation" and called such cryptosystems "algebraic" [14]. Benaloh gave a secure election scheme based on a homomorphic encryption scheme [12]. Cramer and Damgard use homomorphic bit commitments to drastically simplify zero-knowledge proofs [13]. Many other examples exist in which homomorphic properties are used to construct cryptographic protocols.

The initial promise of privacy homomorphisms was tempered by a string of negative results. Ahituv, Lapid, and Neumann showed that any cryptosystem that is XOR-homomorphic on $GF(2^{64})$ is insecure under chosen ciphertext attack [1]. Boneh and Lipton showed that any deterministic cryptosystem that is a field homomorphism must fall victim to a subexponential attack [10]. They further conjectured that any field-homomorphic cryptosystem, which they called "completely malleable," would prove to be insecure. Brickell and Yacobi broke a number of candidate constructions of privacy homomorphisms [11]. These negative results have their analogue in our results of Section 6 showing the triviality of signature schemes that are group homomorphisms on $(\mathbb{Z}, +)$.

Besides RSA, several other homomorphic cryptosystems are currently known. Goldwasser-Micali encryption takes the form of a group homomorphism $\mathbb{Z}/2\mathbb{Z} \to (\mathbb{Z}/n\mathbb{Z})^*$ [17], and others have proposed a number of other public-key encryption schemes that have various useful homomorphic properties [15,8,22,20]. Of particular interest is Sander, Young, and Yung's slick construction of an encryption algorithm that is both AND- and XOR-homomorphic [26]; they note that this is the first cryptosystem homomorphic over a semigroup.

Redactable signature schemes are related in both spirit and construction to the "incremental cryptography" of Goldwasser, Goldreich, and Bellare [3]. Our notion of privacy for redactable signatures has a parallel in their notion of privacy for incremental signatures [4].

3   Definitions

We define the notion of a homomorphic signature scheme as follows. A specification of a signature scheme includes a message space $\mathcal{M}$, a set of private keys $\mathcal{K}$, a set of public keys $\mathcal{K}'$, a (possibly randomized) signature algorithm $\mathsf{Sig} : \mathcal{K} \times \mathcal{M} \to \mathcal{Y}$, and a verification algorithm $\mathsf{Vrfy} : \mathcal{K}' \times \mathcal{M} \times \mathcal{Y} \to \{\mathsf{ok}, \mathsf{bad}\}$ so that $\mathsf{Vrfy}(k', x, \mathsf{Sig}(k, x)) = \mathsf{ok}$ for all $x \in \mathcal{M}$ when (k, k') is a matching private key and public key. As a notational matter, for conciseness we often omit the private and public keys, writing $\mathsf{Sig}(x)$ instead of $\mathsf{Sig}(k, x)$ and $\mathsf{Vrfy}(x, s)$ instead of $\mathsf{Vrfy}(k', x, s)$ when this abbreviation will not cause confusion. Also, for a binary operation $\odot : \mathcal{M} \times \mathcal{M} \to \mathcal{M}$ and a set $S \subseteq \mathcal{M}$, we let $\mathrm{span}_\odot(S)$ denote the least set T with $S \subseteq T$ and $x \odot y \in T$ for all $x, y \in T$.

Definition 1. *Fix a signature scheme* $\mathsf{Sig} : \mathcal{K} \times \mathcal{M} \to \mathcal{Y}$, $\mathsf{Vrfy} : \mathcal{K}' \times \mathcal{M} \times \mathcal{Y} \to \{\mathsf{bad}, \mathsf{ok}\}$ *and a binary operation* $\odot : \mathcal{M} \times \mathcal{M} \to \mathcal{M}$. *We say that* Sig *is homomorphic with respect to* $\odot$ *if it comes with an efficient family of binary operations* $\otimes_{k'} : \mathcal{Y} \times \mathcal{Y} \to \mathcal{Y}$ *so that* $y \otimes_{k'} y' = \mathsf{Sig}(x \odot x')$ *for all* x, x', y, y' *satisfying* $\mathsf{Vrfy}(x, y) = \mathsf{Vrfy}(x', y') = \mathsf{ok}$.

As an example, if $(G, \times_G)$ and $(R, \times_R)$ are two groups and we have a signature scheme $\mathsf{Sig} : \mathcal{K} \times G \to R$ that is also a group homomorphism from G to R for each $k \in \mathcal{K}$, then this will qualify as a signature scheme that is homomorphic with respect to $\times_G$, since we may take $y \otimes_{k'} y' = y \times_R y'$.

This definition requires that signatures derived via $\otimes_{k'}$ be indistinguishable from signatures generated by the private key holder, which we refer to as history-independence. This precludes trivial schemes that, for example, allow the ordered pair (y, y') to serve as a signature on $x \odot x'$. Although the definition above is for deterministic signature schemes, extending it to probabilistic schemes is straightforward; one can simply require that the distribution of $y \otimes_{k'} y'$ be indistinguishable from that of $\mathsf{Sig}(x \odot x')$.

For homomorphic signature schemes, we need a new definition of security. The standard notion of security against existential forgeries is too strong: no homomorphic signature scheme could ever satisfy it, because given two signatures on messages x and x', one can generate a signature on the new message $x'' = x \odot x'$ without asking the signer for a signature on x'' explicitly.

Fortunately, there is a natural extension. Suppose the adversary has obtained signatures on queries $x_1, \ldots, x_q$. Such an adversary can deduce signatures on $x_i \odot x_j$, $(x_i \odot x_j) \odot x_k$, and so on; in other words, no message in $\mathrm{span}_\odot(x_1, \ldots, x_q)$ seems to be safe. Therefore, we will require that no adversary be able to deduce a valid signature on anything outside $\mathrm{span}_\odot(x_1, \ldots, x_q)$.

Definition 2. *We say that a homomorphic signature scheme Sig is (t, q, ϵ)-secure against existential forgeries with respect to $\odot$ if every adversary A making at most q chosen-message queries and running in time at most t has advantage $Adv\,A \leq \epsilon$. The advantage of an adversary A is defined as the probability that, after queries on the messages $x_1, \ldots, x_q$, A outputs a valid signature $\langle x', y' \rangle$ on some message $x' \notin \mathrm{span}_\odot(x_1, \ldots, x_q)$. In other words, $Adv\,A = \Pr[A^{\mathsf{Sig}(k,\cdot)} = \langle x', y' \rangle \wedge \mathsf{Vrfy}(x', y') = \mathsf{ok} \wedge x' \notin \mathrm{span}_\odot(x_1, \ldots, x_q)]$.*

In some cases, we might want an additional guarantee that the operation $\odot$ does not allow the adversary to create too many new signatures. Consider the additive signature schemes broken later in Section 6: They are good candidates to satisfy Definition 2, yet knowledge of a single signature could allow an attacker to sign every other possible message. Therefore, we introduce the concept of security against random forgeries:

Definition 3. *The signature scheme $\mathsf{Sig} : \mathcal{K} \times \mathcal{M} \to \mathcal{Y}$ is said to be (t, q, ϵ)-secure against random forgeries if every adversary A making at most q chosen-message queries and running in time at most t has advantage $Adv\,A \leq \epsilon$. The advantage of an adversary is the probability that it can output a valid signature on a new message x' chosen by the referee uniformly at random from $\mathcal{M}$ and independently of all the adversary's chosen-message queries. In other words, $Adv\,A = \Pr[\mathsf{Vrfy}(x', A^{\mathsf{Sig}(k,\cdot), c_{x'}}) = \mathsf{ok}]$ where $c_{x'}$ is the constant function that always returns x' and where the adversary is not allowed to make any further queries to its signing oracle after looking at the value x'.*

For example, one might reasonably conjecture that textbook RSA signatures are secure against random forgeries.

The security of a homomorphic signature scheme seems to depend on two related notions: the size of $\text{span}_\odot(x_1, \ldots, x_q)$ for small sets of messages, and the difficulty of the decomposition problem, i.e. given $x \in \text{span}_\odot(x_1, \ldots, x_q)$, find an explicit decomposition of x in terms of the x_i's and the $\odot$ operation. If decomposition is hard, then the scheme may be secure even if the spans of small sets of messages are quite large, as is the case with RSA. The scheme may also be secure if the decomposition problem is easy, but most sets of messages only span a small portion of the message space. Redactable signatures are just such a scheme. The trouble occurs when a scheme admits both easy decomposition and large spans. Indeed, this is exactly what renders additive schemes so vulnerable to random forgeries.

One can easily generalize the above notions to operations that take an arbitrary number of operands, and to schemes that are simultaneously homomorphic with respect to more than one operation.

We can also consider signature schemes that are homomorphic with respect to a number of interesting mathematical structures. If $(G, \times_G)$ is a group, then we say that $\mathsf{Sig} : G \to \mathcal{Y}$ is a group-homomorphic signature scheme if it is homomorphic with respect to the binary operator $\times_G$ as well as the unary inversion operation $x \mapsto x^{-1}$. If $(M, \times_M)$ is a semigroup[1], then we say that $\mathsf{Sig} : M \to \mathcal{Y}$ is a semigroup-homomorphic signature scheme if it is homomorphic with respect to the binary operator $\times_M$. If $(R, \times_R, +_R)$ is a ring, we say that $\mathsf{Sig} : R \to \mathcal{Y}$ is a ring-homomorphic signature scheme if it is homomorphic with respect to both $\times_R$ and $+_R$. Boneh et al. have shown that every[2] field-homomorphic signature scheme $\mathsf{Sig} : \mathbb{F} \to \mathcal{Y}$ can be broken in subexponential time [10].

4   Redactable Signatures

The problem. Redactable signatures are intended to model a situation where a censor can delete certain substrings of a signed document without destroying the ability of the recipient to verify the integrity of the resulting (redacted) document. In particular, we allow the censor to replace arbitrary bit positions in the document with a special symbol ♯ representing the location of the deletions. (Our construction can be readily generalized to any alphabet, so that the signer can limit redactions to whole words, sentences, etc., but for simplicity we describe only the case of bit strings here.)

Redactable signatures might have several applications. They permit deletion of arbitrary substrings of a document, and thus might be useful in proxy cryp-

---

[1] Recall that a *semigroup* is a set M together with an associative binary operator $\times_M$ with the property that M is closed under $\times_M$.

[2] Boneh et al. noted that their attack applies to all *deterministic* field-homomorphic encryption schemes, but not to randomized encryption schemes. We observe that their result also applies to all field-homomorphic signature schemes, whether the signature scheme is deterministic or randomized.

Alice spends 60 hours a week trying to find ways to add value to our bottom line, and never have I known her to shirk her duties. Alice is a true asset to our company, and I cannot think of one person better suited to your requirements.	Alice spends 60 hours a week trying to find ways to shirk her duties, and I can think of one person better suited to your requirements.	Alice spends 60 hours a week trying to find ways to ♯♯♯♯♯♯♯♯♯♯♯♯♯♯♯♯♯♯♯♯♯♯♯♯♯ ♯♯♯♯♯♯♯♯♯♯♯♯♯♯♯♯♯♯♯♯♯♯♯♯♯ ♯♯♯ shirk her duties ♯♯♯♯♯♯ ♯♯♯♯♯♯♯♯♯♯♯♯♯♯♯♯♯♯♯♯♯♯♯♯♯ ♯♯, and I can♯♯♯ think of one person better suited to your requirements.

Fig. 1. An illustration of the need to disclose the location of redactions. The left shows a sample document that one might sign with a redactable signature scheme. The middle shows one substring that might be obtained if the location of redactions is not made explicit in the result; note how the meaning of the original document has been violated. The right shows the corresponding redaction obtained if deletions are disclosed explicitly; note how the attempted trickery is revealed by this countermeasure. Because of this potential for this sort of semantic attack when the location of deletions are not made explicit, as illustrated here, all our constructions follow the model shown on the right of making deletions explicit.

tography or incremental cryptography. As their name suggests, they also permit redaction and censorship of signed records with an untrusted censor: the censor cannot forge documents, except those obtained by redaction of a legitimately signed document.

Semantic attacks and our formulation of redaction. Making the problem statement more precise exposes a subtle trap here. A natural first attempt at a definition might require that, given a signature $\mathsf{Sig}(x)$ on x, one can obtain a signature $\mathsf{Sig}(w)$ on any substring w of x that is obtained by deleting some of the symbols in x. However, this formulation of the problem has a serious limitation: such a scheme will conceal the presence of deletions, which introduces the risk of semantic attacks. Without indication of where the redaction has occurred, an attacker might legally be able to truncate the end of one sentence, delete the beginning of the next, and slice them together to get a message the sender would not have authorized. For an example, see Figure 1. To defeat semantic attacks, we instead use a formulation where the presence of redacted segments are made explicit.

We define the concept more formally as follows. Let us take $\Sigma = \{0, 1, \sharp\}$. We define a partial order $\preceq$ on Σ so that $\sharp \preceq 0$, $\sharp \preceq 1$, and $a \preceq a$ for each $a \in \Sigma$ (and no other non-trivial relations hold). This induces a partial order $\preceq$ on Σ^* by pointwise comparison, namely, $w_1 \cdots w_n \preceq x_1 \cdots x_n$ holds if $w_i \preceq x_i$ for each i. We say that the document w is a *redaction* of x if $w \preceq x$, or in other words, if w can be obtained from x by replacing some positions of x with the $\sharp$ symbol. The signature scheme $\mathsf{Sig} : \Sigma^* \to R$ is called *redactable* if we can derive from a signature $\mathsf{Sig}(x)$ on x a signature $\mathsf{Sig}(w)$ on any redaction $w \preceq x$ we like, without the help of the signer.

Note that, in this model, we do not attempt to hide the location of the censored portions of the documents. Thus, a signature $\mathsf{Sig}(w)$ on a redaction w of x may legitimately reveal which portions of x were deleted, and specifically, the presence and location of redacted segments is protected from tampering by the signature. However, it does not reveal the previous contents of the redacted portions of the document. We expect that this privacy property may be important to many applications.

Redactable signatures may be viewed as an instance of a homomorphic signature scheme endowed with operations $D_i : \Sigma^* \to \Sigma^*$ that replace the i-th bit position with a $\sharp$ symbol. The requirement that signed documents be redactable is equivalent to requiring that our signature scheme be homomorphic with respect to these unary operators.

A trivial construction. There is one obvious way to build redactable signatures. We fix a traditional signature scheme Sig_0; no special homomorphic properties are required from Sig_0. We assume for simplicity that our base signature scheme permits message recovery, but this assumption is not essential. In the trivial construction, to sign a message x of length n, we generate a fresh key pair (s, v) for some secure conventional signature scheme, and then the signature on x is

$$\mathsf{Sig}(x) = \langle \mathsf{Sig}_0(n, v), s(1, x_1), \ldots, s(n, x_n) \rangle.$$

Verification is straightforward.

The key pair essentially serves as a document ID. Without it, an attacker could replace the ith component of any signed message with the ith component of any other signed message, which we do not wish to allow.

To redact a signed message, we simply erase the appropriate segments of the message and the corresponding positions in the signature. For instance, redacting the first symbol in $\mathsf{Sig}(x)$ yields a signature $\langle \mathsf{Sig}_0(n, v), s(2, x_2), \ldots, s(n, x_n) \rangle$. This scheme supports a privacy property: redacted signatures do not reveal the redacted parts of the original message. We can see that they reveal the locations that have been redacted, much like typical redaction of paper documents does, but this is all that is leaked.

However, this trivial scheme has a serious limitation: the signature $\mathsf{Sig}(x)$ is very long. If s produces m-bit signatures, then the construction above yields signatures of length $mn+O(1)$, which is likely to be large compared to the length of the message, n. Therefore, the challenge is to find a scheme with much shorter signatures.

Our construction. We describe how to build a secure redactable signature scheme out of any traditional signature scheme. The main idea is to combine Merkle hash trees [18] with the GGM tree construction for increasing the expansion factor of a pseudorandom generator [21]. We first generate a pseudorandom value at each leaf by working down the tree (using GGM), then we compute a hash at the root by working up the tree (using Merkle's tree hashing).

We describe in detail how to sign a message x of length n. First, we arrange the symbols of the message at the leftmost leaves of a binary tree, with each leaf

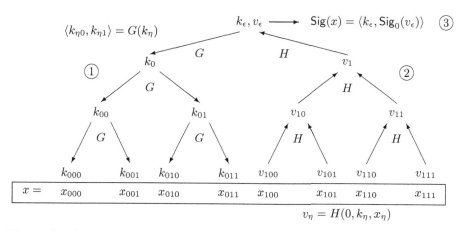

Fig. 2. A redactable signature on the message $x = \langle x_{000}, x_{001}, \ldots, x_{111} \rangle$. The message is signed in three phases: first, we generate k-values by recursing down the tree with the PRG G; then, we generate v-values by recursing up the tree with the hash H; finally, we sign v_ϵ using our conventional signature scheme. To avoid cluttering the diagram, we show the downward phase only on the left branch of the tree, and the upwards phase only on the right branch of the tree, but the full scheme requires we traverse both branches in both directions.

at depth $\lceil \lg n \rceil$. We identify nodes of the tree with elements of $\{0, 1\}^*$, so that a node η has children $\eta0$ and $\eta1$, and the empty string ϵ denotes the root node. Let $G : \mathcal{K} \to \mathcal{K} \times \mathcal{K}$ be a length-doubling pseudorandom generator, let H be a cryptographic hash function, and pick $k_\epsilon \in \mathcal{K}$ uniformly at random. We use a three-phase algorithm (see Figure 2, which shows phase one on the left and phase two on the right):

Expansion: We use the GGM tree construction to associate a key k_η to each node η. In other words, we define a key for each node of the tree by the recursive relation $\langle k_{\eta0}, k_{\eta1} \rangle = G(k_\eta)$.

Hashing: We then compute a hash value v_ℓ for each leaf a as $v_\ell = H(0, k_\ell, x_\ell)$ and apply the Merkle hash construction, i.e., we recursively compute $v_\eta = H(1, v_{\eta0}, v_{\eta1})$ (or, if only a left child exists, $H(1, v_{\eta0})$).

Signing: Finally, we sign the root of the hash tree using our base signature scheme Sig_0, and compute the signature on x as $\mathsf{Sig}(x) = \langle k_\epsilon, \mathsf{Sig}_0(v_\epsilon) \rangle$.

Verification is straightforward, since given k_ϵ and x we can compute v_ϵ and check the signature.

Next, we describe how to redact a signed message. To erase position ℓ from a signed message x, we need to reveal v_ℓ. At first sight, revealing $v_\ell = H(0, k_\ell, x_\ell)$ would appear to be very dangerous, since given k_ϵ we can compute k_ℓ and then iterate over all possible values of x_ℓ to learn the value of the erased symbol. This would violate the secrecy property.

To restore secrecy, we take advantage of the GGM tree. When we erase x_ℓ, we will also erase k_ϵ, and reveal only what is needed to verify the signature without

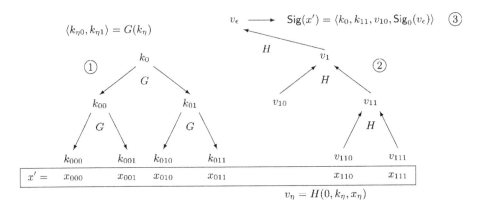

$$v_\eta = H(0, k_\eta, x_\eta)$$

Fig. 3. A redaction of the signature on x, with the two bits x_{100} and x_{101} deleted. Like in Fig. 2, we show only part of the signature process, to avoid cluttering the diagram.

leaking k_ℓ. The notion we need is as follows: define the *co-nodes* associated to a leaf ℓ to be the siblings of the nodes along the path from ℓ to the root. We reveal k_η at each co-node η associated to ℓ, as this is exactly what is needed to check the resulting signature. Thus, redacting x_ℓ from $\mathsf{Sig}(x)$ would yield $\langle v_\ell, \{k_\eta : \eta \text{ a co-node of } \ell\}, \mathsf{Sig}_0(v_\epsilon)\rangle$. Note that there are at most $\lg n$ co-nodes associated to any single leaf.

We can repeat the above procedure to further redact an already-redacted signature, if we like. In the case of redactions in consecutive parts of the document, we compact the signature by using the tree structure: if the signature contains simultaneously $v_{\eta 0}$ and $v_{\eta 1}$ for some η, then we replace these two quantities with the single value v_η, and we similarly replace $k_{\eta 0}, k_{\eta 1}$ with k_η. This both shortens the signature and hides the order in which redactions were performed, ensuring a sort of path-independence property. Thus, though signatures will grow in length, the length may grow only slowly if there is some locality to the sequence of redactions. At worst, each consecutive segment of erasures of length n' from a message of length n adds $O(\lg n')$ hash values and $O(\lg n')$ key values to the signature.

This signature scheme produces signatures that are relatively short. If Sig_0 yields m-bit signatures and if we use m'-bit hash values and keys, then an unredacted signature is only $m + m'$ bits long. This is a constant in n, the length of the message. After erasing s segments, each of length at most n', the signature will be $m + O(sm' \lg(nn'))$ bits long. In general, signatures could be as long as $m + O(nm')$ bits after many redactions, but we can expect that in practice the number of erased segments will typically not be too large and hence signatures should often be quite short. Therefore, this is a considerable improvement over the trivial construction outlined earlier.

This signature scheme also has an additional homomorphic property. For a fixed message x, we can form the lattice of substrings $L_x = \{w \in \Sigma^* : w \preceq x\}$, where the join $v \sqcup v'$ is the unique least $w \in L_x$ satisfying $v \preceq w$ and $v' \preceq w$, and the meet is defined dually. We note that given two redactions $\mathsf{Sig}(v)$ and $\mathsf{Sig}(v')$ of a common signature on a message x, we can compute signatures $\mathsf{Sig}(v \sqcup v')$ and $\mathsf{Sig}(v \sqcap v')$ on their join and meet. In other words, our scheme is also homomorphic with respect to the binary operations $\sqcup$ and $\sqcap$. We imagine that this might be useful in some applications.

Security analysis. We show in Appendix A that this construction is secure against existential forgeries under reasonable cryptographic assumptions[3]. See the appendix for details.

5   Set-Homomorphic Signatures

We now turn to operations on sets and derive a scheme which simultaneously supports the union and subset operations. More precisely, the scheme allows anyone possessing sets U_1, U_2 and $\mathsf{Sig}(U_1), \mathsf{Sig}(U_2)$ to compute $\mathsf{Sig}(U_1 \cup U_2)$ and $\mathsf{Sig}(U)$ for any $U \subseteq U_1$. Note that these two operations, union and subset, can be combined to create many others, such as set difference, symmetric difference, and intersection, so this scheme seems particularly powerful. As an example application, at the end of this section we will construct an alternate redactable signature scheme which has many advantages over the one presented in Section 4.

The construction is based on the accumulator technique [9]. To begin, let h be a public hash function so that, for all x, $h(x)$ is uniformly distributed over the odd primes less than n [16]. If we extend h to sets via $h(U) = \prod_{u \in U} h(u)$, then $h(U_1 \cup U_2) = \mathrm{lcm}(h(U_1), h(U_2))$, assuming there are no h-collisions between elements of U_1 and U_2. Each user creates a rigid RSA modulus n and $v \in (Z/nZ)^*$ selected uniformly at random. Recall that an RSA modulus $n = pq$ is rigid if $\frac{p-1}{2}$ and $\frac{q-1}{2}$ are prime and $|p| = |q|$ [9]. By choosing n this way, we ensure that $\gcd(h(x), \varphi(n)) = 1$ for almost all x. To sign a set U, one computes $\mathsf{Sig}(U) = v^{\frac{1}{h(U)}} \bmod n$. Note that since $h(U)$ is relatively prime to $\varphi(n)$, there is exactly one solution y to the equation $y^{h(U)} = v$. To verify a signature y on a set U, one need only check that $y^{h(U)} = v \bmod n$.

Given U_1, U_2 and signatures $v^{\frac{1}{h(U_1)}}, v^{\frac{1}{h(U_2)}}$, computing the signature on $U_1 \cup U_2$ is easy. First use the Euclidean algorithm to find a and b such that $a \cdot h(U_1) + b \cdot h(U_2) = \gcd(h(U_1), h(U_2))$. Then $(v^{\frac{1}{h(U_1)}})^b (v^{\frac{1}{h(U_2)}})^a = v^{\frac{1}{\mathrm{lcm}(h(U_1), h(U_2))}}$, which is the desired signature. If $U \subseteq U_1$ then $\mathsf{Sig}(U_1 \setminus U) = \mathsf{Sig}(U_1)^{h(U)}$, so computing subset signatures is also straightforward.

As with the redactable signature problem, we'd like to show this scheme is resistant to forgeries and that it satisfies the history-independence requirement for homomorphic signature schemes. The latter is clearly satisfied since the signature on a set U' is independent of how we obtain that signature: starting with a signature $\mathsf{Sig}(U_1)$ on U_1 and then removing the elements in U_2 to get $\mathsf{Sig}(U_1)^{h(U_2)}$ yields exactly the same result as asking for the signature on $U_1 \setminus U_2$

---

[3] We do not address security against random forgeries, as it is not clear what is the right distribution on the message space.

directly, i.e., $\mathsf{Sig}(U_1)^{h(U_2)} = \mathsf{Sig}(U_1 \setminus U_2)$. A similar property holds for the union operation, and so this scheme is history-independent.

The resistance of this scheme to forgeries requires a more detailed security analysis, but basically it rests on the difficulty of computing kth roots modulo an RSA modulus and on the randomness of the hash function.

Before we give the proof, we should state clearly the parameterization of the difficulty of the RSA problem. We assume that RSA behaves as a good trapdoor permutation, as others have suggested before [6,7]. This assumption appears to be weaker than the so-called Strong RSA assumption [2].

Definition 4. *Let $H_k = \{pq : p$ and q are safe primes, $p \neq q$, and $|p| = |q| = k\}$. We say RSA is a (t, ϵ_r)-secure trapdoor permutation if for any adversary A with running time less than t, we have $\Pr[A(n, e, x^e) = x] < \epsilon_r$, where the probability is taken over the choice of $n \in H_k$, $e \in (Z/\varphi(n)Z)^*$, $x \in (Z/nZ)^*$, and the coin tosses of A.*

The following theorem relates the security of the set-homomorphic signature scheme to the security of RSA, and shows essentially that if the signature scheme is insecure, then an attacker could exploit that weakness to break RSA without too much more work. Note that the theorem guarantees security even when the adversary can make a number of adaptively chosen signature queries, which seems to be an extension of previous results.

Theorem 1. *Let h be a random oracle as above. Assume that RSA is a (t, ϵ_r)-secure one-way function. Then the set-homomorphic signature scheme Sig defined above is (t', q_h, ϵ)-secure against existential forgery with respect to subset and union operations given that the total number of hash oracle queries performed is less than q_h, where $\epsilon \approx q_h \epsilon_r \log n + q_h^2 \log n/n$ and $t' \approx t$ [4].*

Proof. (sketch) We give a proof by contradiction. Assume the above theorem is false, meaning there exists an adversary A which can (t', ϵ)-break the set-homomorphic signature scheme Sig. Assume after A performs a number of random oracle queries and obtains signatures on sets $x_1, \ldots, x_{q_s}$, A outputs a forgery on x^* where x^* is not in the span of $x_1, \ldots, x_{q_s}$ under subset and union operations. This means there exists an element $y \in x^*$ such that $y \notin x_i$, for all $1 \leq i \leq q_s$, and A knows u such that $u^{h(y)} = v$. There are two cases:

- We have $h(y) \neq h(y')$ for all $y' \in \cup_{1 \leq i \leq q_s} x_i$.
- We have $h(y) = h(y')$ for some $y' \in \cup_{1 \leq i \leq q_s} x_i$. This happens with probability at most $q_h^2 \log n/n$ (by the prime number theorem and the birthday paradox).

If the first case happens with probability higher than $q_h \epsilon_r \log n$, we show as following that we can construct an adversary $B(n, e, z)$ using A which can (t, ϵ_r)-break RSA. In particular, if e is prime, B runs the following game using A. (Otherwise, the simulation fails; this happens with probability at most about $1/\log n$, by the prime number theorem.)

---

[4] The total number of hash oracle queries include the ones made by A and the ones made by the signing procedure.

- B first selects uniformly randomly q_h primes $p_1, \ldots, p_{q_h}$ not equal to e. Then B sets $f = \prod_{1 \le i \le q_h} p_i$ and $v = z^f \mod n$. Due to the choice of n, we have $\gcd(f, \varphi(n)) = 1$ with overwhelming probability, and thus the map $z \mapsto z^f \mod n$ is bijective on $(Z/nZ)^*$. This means that v is uniformly distributed on $(Z/nZ)^*$, since z is, so we can feed v as an input to A and it will have the right distribution. B also selects a random integer $k \in \{1, \ldots, q_h\}$.
- B constructs the hash oracle as follows. Given a hash oracle query on element w, if w has been queried before, then returns the previous corresponding answer. For the j-th unique hash oracle query w_j, if $j \ne k$, then return the prime p_j; otherwise, return e.
- B constructs the signing oracle that A queries as follows. Given a signing query on a set of m elements $U = \{a_1, \ldots, a_m\}$, B queries the hash oracle to obtain $h(a_1), \ldots, h(a_m)$. If for some $1 \le i \le m$, $h(a_i) = e$, then abort; otherwise, return $b = z^{f/(\prod_{1 \le i \le m} h(a_i))}$. It is easy to see that b is a valid signature on the set U.
- If at the end of the game, A outputs an existential forgery on a set of elements that includes an element y such that $h(y) = e$, then B can learn from A's forgery a value u satisfying $u^{h(y)} = v = z^f$. Because e is prime, $\gcd(e, f) = 1$, and we can compute α and β such that $\alpha e + \beta f = 1$ using the Eulidean algorithm. Thus $z = (u^\beta z^\alpha)^e$, and B outputs $u^\beta z^\alpha$ as the e-th root of z. Otherwise, abort the game.

If A outputs an existential forgery and e is prime, the probability that B succeeds is at least $1/q_h$. So if there exists an adversary A that (t', ϵ)-breaks the set-homomorphic signature scheme, we can construct an adversary B that (t, ϵ_r)-breaks RSA.

The set-union scheme suggests another solution to the redactable document signature problem. To sign the document $x = (x_1, \ldots, x_k)$, first select a random unique document identifier, k_x, and then sign the set of triples $D = \{(k_x, i, x_i)\}$, say $y = \mathsf{Sig}(D)$. The complete redactable signature is (k_x, y), and verification is straightforward. To compute the signature on the message with word i redacted, simply compute $y' = \mathsf{Sig}(D \setminus \{(k_x, i, x_i)\}) = y^{h(k_x, i, x_i)}$, and the new signature is (k_x, y'). Note that this scheme also reveals the locations of the redactions, preventing the semantic attacks described in Section 4. Not only is this scheme much simpler to implement and easier to understand than our prior redactable scheme, the signatures produced in this scheme are of constant length.

6   Additive Signature Schemes

We describe next a number of schemes that we have studied in our search for an additive signature scheme. All of them turn out to be insecure, and in interesting ways. More precisely, the schemes we examine all have an undesirable property that is likely to make them useless in practice: given signatures on a small set of known messages, we can forge signatures on all other possible messages. Thus, such schemes are insecure against random forgeries as described in Definition 3.

Constructions built from multiplicative signature schemes. If we have a multi-plicative signature scheme $\mathsf{Sig}_\times : G \to R$ as well as a group homomorphism $\varphi : \mathbb{Z}/m\mathbb{Z} \to G$ from the additive group $\mathbb{Z}/m\mathbb{Z}$ to the multiplicative group G, then a natural candidate for an additive signature scheme $\mathsf{Sig}_+ : \mathbb{Z}/m\mathbb{Z} \to R$ is $\mathsf{Sig}_+ = \mathsf{Sig}_\times \circ \varphi$. For instance, we can instantiate $\mathsf{Sig}_\times$ with RSA. In this case, $G = (\mathbb{Z}/n\mathbb{Z})^*$, and every homomorphism from $\mathbb{Z}/m\mathbb{Z}$ to $(\mathbb{Z}/n\mathbb{Z})^*$ takes the form $\varphi(x) = g^x \pmod{n}$ for some $g \in G$, so our construction takes the form $\mathsf{Sig}_+(x) = g^{xd} \pmod{n}$.

To see why any such scheme must be insecure against random forgeries, suppose $\mathsf{Sig}_+ : \mathbb{Z} \to G$ is a group-homomorphic signature scheme. Then Sig_+ is entirely determined by $\mathsf{Sig}_+(1)$. Thus, if we can recover a few messages $m_1, \ldots, m_k$ such that $\gcd(m_1, \ldots, m_k) = 1$, then since there exist $a_1, \ldots, a_k \in \mathbb{Z}$ such that $\sum_i a_i m_i = 1$, we can compute $\mathsf{Sig}_+(1) = \sum_i a_i \mathsf{Sig}_+(m_i)$. Given this information, we can compute $\mathsf{Sig}_+(m) = m\mathsf{Sig}_+(1)$ for any m. This attack easily extends to group homomorphic signatures $\mathsf{Sig}_+ : \mathbb{Z}/n\mathbb{Z} \to G$ by lifting to $\mathbb{Z}$, applying the Euclidean algorithm, and projecting back down to $\mathbb{Z}/n\mathbb{Z}$. Since any small set of messages will likely have gcd 1, this scheme is vulnerable to random forgeries after only a few messages have been signed.

These weaknesses are very general. In particular, they apply to almost any instance of the construction $\mathsf{Sig}_+ = \mathsf{Sig}_\times \circ \varphi$. As another important example, every additive signature scheme that forms a group homomorphism is insecure against random forgeries.

It is perhaps counterintuitive that multiplicative signature schemes are plentiful while fully-additive schemes do not exist. The explanation seems to be the fact that $\mathbb{Z}/m\mathbb{Z}$ has a Euclidean algorithm, but $(\mathbb{Z}/n\mathbb{Z})^*$ does not. In other words, it is not that the span is larger in $\mathbb{Z}/m\mathbb{Z}$, but that the decomposition problem is easy in $\mathbb{Z}/m\mathbb{Z}$ but hard in $(\mathbb{Z}/n\mathbb{Z})^*$.

One might imagine that the problem is the need to be homomorphic with respect to both addition as well as negation, and thus one idea might be to look for a scheme that respects only the addition operator but not the negation. In other words, given $\mathsf{Sig}_+(x), \mathsf{Sig}_+(y)$, we still want to be able to compute $\mathsf{Sig}_+(x + y)$, but it should be hard to compute $\mathsf{Sig}_+(-x)$ from $\mathsf{Sig}_+(x)$. Note that the problems with additive signatures arise because one can find $a, a' \in \mathbb{Z}$ so that $ax + a'x' = 1$ and then compute $\mathsf{Sig}_+(1) = \mathsf{Sig}_+(ax) \times \mathsf{Sig}_+(a'x')$; yet one of a, a' will necessarily be negative. If we can somehow ensure that computing $\mathsf{Sig}_+(-x)$ from $\mathsf{Sig}_+(x)$ is hard, then the Euclidean algorithm will no longer apply, and the above attacks will fail. This suggests that we may want to look for an additive signature scheme without inverses (a semigroup-homomorphism), as such a scheme would resist the attacks described so far.

Further challenges. Yet even an additive signature scheme without inverses still has some properties that might not be expected. In particular, there is the problem that signatures can always be forged on all large enough messages, given signatures on two messages m, m'. This is because the equation

$$am + a'm' = x \qquad (\text{in } \mathbb{Z})$$

typically has solutions with $a, a' \geq 0$ when $x \geq \text{lcm}(m, m')$, and in this case a signature on x can be obtained from signatures on m, m'.

More generally, if we have signatures $\mathsf{Sig}(m_1), \ldots, \mathsf{Sig}(m_k)$, then we can forge a signature $\mathsf{Sig}(x)$ whenever we can write x in the form $x = a_1 m_1 + \cdots + a_k m_k$ for some known $a_1, \ldots, a_k \in \mathbb{Z}_{\geq 0}$. This is a subset sum problem, and the issue is that if the m_i are small enough, the subset sum problem is easy to solve. So our only hope is to choose additive signatures without inverses, only sign messages large enough that subset sum is hard (require $m_i \geq \ell$ for some lower bound ℓ), and refuse to accept unusually large messages (enforce $m_i \ll \ell^2$) in the verification algorithm.

Additive schemes in higher dimensions. More generally, we could look for an additive scheme $\mathsf{Sig}_+ : (\mathbb{Z}/m\mathbb{Z})^d \to R$ in dimension $d > 1$. Unfortunately, this does not seem to offer much opportunity to design a secure signature scheme, either.

Although increasing the dimension does make more work for the attacker, a slight extension of the previous remarks still applies.

Observation 1 *In any additive signature scheme on the lattice $L = (\mathbb{Z}/m\mathbb{Z})^d$, if one can obtain signatures $\mathsf{Sig}(x_1), \ldots, \mathsf{Sig}(x_d)$, where $x_1, \ldots, x_d$ are a basis for L, then one can succeed at any random forgery.*

Knowing the signatures on a basis is useless if computing the representation of a given message in that basis is hard. If we could compute the signatures of the standard basis elements given the signatures on the elements of another basis, then committing random forgeries would be easy. Note that the elements of the standard basis are the shortest vectors in the lattice $\mathbb{Z}^d$, so lattice reduction techniques may be used to discover representations of the standard basis elements in terms of messages with known signatures. The theoretical bounds on the length of the shortest vector returned by LLL are not very tight, but in practice it can find a representation of the standard basis of $(\mathbb{Z}/m\mathbb{Z})^d$ with only $(1 + \epsilon)d$ input vectors. Thus, we only need to collect $(1 + \epsilon)d$ signed messages before we can commit random forgeries with ease.

Therefore, it seems to be a challenging open problem to find a secure additive signature scheme.

7   Open Problems

Set-homomorphic signatures. We may look for a set-homomorphic signature scheme that is homomorphic with respect to operations other than union and subset. For example, consider the following construction: Let $\mathsf{Sig}_\times : G \to R$ be an arbitrary multiplicative signature scheme on some group G. For a set $U = \{x_1, \ldots, x_k\}$, define the hash function $f_h(U) = h(x_1) \cdots h(x_k)$, where $h : \mathcal{X} \to G$ is a cryptographic hash function on elements in $\mathcal{X}$. Then $\mathsf{Sig} = \mathsf{Sig}_\times \circ f_h$ is a signature scheme with the property that with signatures, $\mathsf{Sig}(U)$ and $\mathsf{Sig}(V)$, on two disjoint sets U and V, one can compute the signature on their union,

$\mathsf{Sig}(U \cup V) = \mathsf{Sig}(U) \times \mathsf{Sig}(V)$. If $U \subseteq V$, one can also compute the signature on their difference, $\mathsf{Sig}(V \setminus U) = \mathsf{Sig}(V) \times \mathsf{Sig}(U)^{-1}$.

We gave another example of a set-homomorphic scheme, based on RSA accumulators, in Section 5. One interesting question is whether we can design a signature scheme that is homomorphic only with respect to the union operation.

Concatenable signatures. Let $\mathsf{Sig} : \{0,1\}^* \to R$ be a signature scheme. We say that it is a concatenable signature scheme if, given $\mathsf{Sig}(x), \mathsf{Sig}(y)$, one can compute (without help from the signer) a signature $\mathsf{Sig}(x\|y)$ on the concatenation of x and y. In other words, a concatenable signature scheme should be homomorphic with respect to the semigroup $(\{0,1\}^*, \|)$ of bit-strings with the concatenation operator $\|$. Rivest has asked [24]: can we design a concatenable signature scheme?

Semigroup signatures without inverses. Giving an example of a secure semigroup-homomorphic signature scheme seems to be an intriguing open problem that is suggested by this work. We have pointed out instances of this problem several times throughout this paper. Micali and Rivest asked whether there exists a transitive signature scheme on directed graphs [19], and this domain has a semigroup structure. One might seek a redactable signature scheme supporting only redaction but not the join operation. An additive scheme that doesn't respect subtraction might have a chance of being secure. A set-homomorphic scheme that allows only the union operation (but not subsetting) would have a semigroup-homomorphic property, as would a concatenable signature scheme. More generally, we suspect that any scheme that is semigroup-homomorphic but not group-homomorphic might yield insights into these open problems.

8   Conclusions

Homomorphic signature schemes present a promising new direction for research. Since such schemes necessarily do not satisfy traditional definitions of security, we have proposed new definitions of security for these new schemes. We have shown that a variety of homomorphic signature schemes can be designed. We also examined limits on their existence, showing that, for example, no additively group-homomorphic scheme can ever be secure against random forgeries. Perhaps most importantly, we have suggested several open problems that, if solved, might provide useful new schemes supporting a variety of applications.

Acknowledgements

We thank Eric Bach and Mika R.S. Kojo for early observations on weaknesses in additive signature schemes. Adrian Perrig originally suggested the idea of using Merkle hash trees to support redaction, which was a key step in the development of the scheme presented in Section 4. We also gratefully acknowledge many helpful comments from Ron Rivest and the anonymous referees.

References

1. Niv Ahituv, Yeheskel Lapid, and Seev Neumann. Processing encrypted data. *Communications of the ACM*, 30(9):777–780, 1987.

2. Niko Baric and Birgit Pfitzmann. Collision-free accumulators and fail-stop signature schemes without trees. In *Advances in Cryptology–EUROCRYPT '97*, volume 1233 of *Lecture Notes in Computer Science*, pages 480–494. Springer-Verlag, 1997.

3. M. Bellare, O. Goldreich, and S. Goldwasser. Incremental cryptography: the case of hashing and signing. In Yvo Desmedt, editor, *Advances in Cryptology–CRYPTO '94*, pages 216–233, Berlin, 1994. Springer-Verlag. Lecture Notes in Computer Science Volume 839.

4. M. Bellare, O. Goldreich, and S. Goldwasser. Incremental cryptography with application to virus protection. In *FOCS 1995*, Berlin, 1995. Springer-Verlag.

5. M. Bellare and P. Rogaway. The exact security of digital signatures–how to sign with RSA and Rabin. In Ueli Maurer, editor, *Advances in Cryptology–EUROCRYPT '96*, pages 399–416, Berlin, 1996. Springer-Verlag. Lecture Notes in Computer Science Volume 1070.

6. Mihir Bellare and Phillip Rogaway. Random oracles are practical: A paradigm for designing efficient protocols. In *First ACM Conference on Computer and Communications Security*, pages 62–73, Fairfax, 1993.

7. Mihir Bellare and Phillip Rogaway. The exact security of digital signatures–how to sign with RSA and Rabin. In Ueli Maurer, editor, *Advances in Cryptology–EUROCRYPT 96*, volume 1070 of *Lecture Notes in Computer Science*. Springer-Verlag, 1996.

8. J Benaloh. Dense probabilistic encryption. In *Selected Areas in Cryptography*, 1994.

9. J.C. Benaloh and M. de Mare. One-way accumulators: A decentralized alternative to digital signatures. In *EUROCRYPT'93*, 1993.

10. D. Boneh and R. J. Lipton. Algorithms for black-box fields and their application to cryptography. In Neal Koblitz, editor, *Advances in Cryptology–CRYPTO '96*, pages 283–297, Berlin, 1996. Springer-Verlag. Lecture Notes in Computer Science Volume 1109.

11. E. F. Brickell and Y. Yacobi. On privacy homomorphisms. In David Chaum and Wyn L. Price, editors, *Advances in Cryptology–EUROCRYPT '87*, pages 117–126, Berlin, 1987. Springer-Verlag. Lecture Notes in Computer Science Volume 304.

12. J. Cohen and M. Fischer. A robust and verifiable cryptographically secure election scheme. In *26th Symposium on the Foundations of Computer Science*, 1985.

13. Cramer and Damgard. Zero knowledge proofs for finite field arithmetic – or, can zero knowledge be for free? In *Advances in Cryptology–CRYPTO '98*, Berlin, 1998. Springer-Verlag.

14. J. Feigenbaum and Merritt. Open questions, talk abstracts, and summary of discussions. In *DIMACS Series in Discrete Mathematics and Theoretical Computer Science*, pages 1–45, 1991.

15. E. Fujisaki, T. Okamoto, and Uchiyama. EPOC : Efficient probabilistic encryption. In *Submission to IEEE P1363*, 1998.

16. Rosario Gennaro, Shai Halevi, and Tal Rabin. Secure hash-and-sign signatures without the random oracle. In *Advances in Cryptology–EUROCRYPT'99*, pages 123–139. Springer-Verlag, 1999. Lecture Notes in Computer Science Volume 1592.

17. S. Goldwasser and S. Micali. Probabilistic encryption. *Journal of Computer and System Sciences*, 28(2):270–299, April 1984.

18. Ralph Merkle. Protocols for public key cryptosystems. In *Proceedings of the IEEE Symposium on Research in Security and Privacy*, Oakland, CA, April 1980. IEEE Computer Society Press.

19. S. Micali and R. Rivest. Transitive signature schemes. In *Topics in Cryptology–CT-RSA 2002*, pages 236–243. Springer-Verlag, 2002. Lecture Notes in Computer Science Volume 2271 (This Volume).

20. D. Naccache and J. Stern. A new public key cryptosystem based on higher residues. In *5th ACM Symposium on Computer and Communications Security*, 1998.

21. Goldreich Oded, Shafi Goldwasser, and Silvio Micali. How to construct random functions. *Journal of the ACM*, 33(4):792–807, October 1986.

22. P Paillier. Public-key cryptosystems based on composite degree residuosity classes. In *Advances in Cryptology–EUROCRYPT '99*, volume 1592 of *LNCS*, 1999.

23. R Peralta and J. Boyar. Short discreet proofs. In *Journal of Cryptology*, 2000.

24. R. Rivest. Two new signature schemes. Presented at Cambridge seminar; see http://www.cl.cam.ac.uk/Research/Security/seminars/2000/ rivest-tss.pdf, 2001.

25. R. Rivest, L. Adleman, and M.L. Dertouzos. On data banks and privacy homomorphisms. In *Foundations of Secure Computation*, pages 169–178. Academic Press, 1978.

26. T. Sander, A Young, and M Yung. Non-interactive cryptocomputing in NC^1. In *FOCS '99*, 1999.

A   Security Analysis for the Tree Redaction Scheme

The combination of redaction and randomization in the tree redaction scheme introduces one tricky aspect into the definition of security. The adversary might choose a message $x \in \{0,1\}^*$, and though the censor might be unwilling to reveal $\mathsf{Sig}(x)$, the adversary might be able to gain access to a redacted signature $\mathsf{Sig}(w)$ on some message $w \preceq x$ of the attacker's choice. In this case our security analysis should ensure that the attacker cannot use this partial knowledge to obtain, e.g., a signature $\mathsf{Sig}(x)$ on the full message x. Therefore, in our model we give the adversary access to two different oracles, denoted R and S. Invoking $R(x)$ denotes registration of a chosen message $x \in \{0,1\}^*$, and the registered messages are stored in a list $x^1, x^2, \ldots, x^q$. Calling $S(i, w)$ returns the signature $\mathsf{Sig}(w)$ on the redacted message $w \in \Sigma^*$, if $w \preceq x^i$; otherwise, if w is not a valid redaction of the message x^i specified in the i-th call to R, the computation aborts. Let W^i denote the join of all values w that appear in some oracle query of the form $S(i, w)$. If an adversary can output a signature on a message that is not a redaction of W^i for some i, we say that the adversary has found an existential forgery.

Theorem 2. *Let G be a (t, ϵ_G)-secure pseudorandom generator, H a (t, p_H)-collision-resistant hash function, and Sig_0 a conventional signature scheme that is (t, q, p_S)-secure against existential forgery. Suppose that $F_k(x) = H(0, k, x)$ is*

a (t, q, ϵ_H)-secure pseudorandom function. Then the redactable signature scheme
Sig defined above for messages of length at most n is (t', q, p')-secure against
existential forgery with respect to redaction, where $p' = q\lceil \lg n \rceil \epsilon_G + nq\epsilon_H + p_S + p_H + 2nq/2^{m'}$ and $t' \approx t$.

Proof. (sketch) Consider any adversary that attempts to break the resulting
modified scheme by exhibiting existential forgeries, i.e., by finding a valid signature on a message w^* that is not a redaction of the join of its previous queries.
We let x^i denote the i-th message registered with R, k^i_η, v^i_η denote the key and
hash value at node η, and we introduce the notation u^i_η for the input to the hash
function at η so that $v^i_\eta = H(u^i_\eta)$. For example, k^i_ϵ denotes the key randomly
chosen for use with x^i and its redactions, and v^i_ϵ denotes the root of the hash
tree on the signature for message x^i.

Suppose the adversary forges a signature $(\{v^*_\epsilon\}, \{k^*_\eta\}, \mathsf{Sig}\,(v^*_\epsilon))$ on w^*. If $v^*_\epsilon \neq v^i_\epsilon$ for all i, then we have found an existential forgery of Sig_0, which by assumption
happens with probability at most p_S. Therefore, we assume that for some i we
have $v^*_\epsilon = v^i_\epsilon$. Let T^* denote the tree corresponding to w^*, i.e., the leaf nodes
corresponding to unredacted symbols in w^* along with all their ancestors.

Thanks to the properties of Merkle tree hashing, we see that T^* must be a
sub-tree of the tree corresponding to x^i. (Otherwise, there is some node ℓ that
is a leaf node in the tree for x^i but is an internal node in T^*, but then we have
a hash collision $H(u^*_\ell) = H(u^i_\ell)$ since u^*_ℓ starts with a 1 and u^i_ℓ starts with a 0,
which contradicts the collision-resistance of H.) Similarly, the leaves of T^* must
form leaves in the tree for x^i, and the internal nodes in T^* form internal nodes
in the tree for x^i.

Moreover, the hash pre-images u_η must satisfy $u^*_\eta = u^i_\eta$ for each $\eta \in T^*$. This
tells us that $v^*_\eta = v^i_\eta$ for all $\eta \in T^*$. It also tells us that $k^*_\ell = k^i_\ell$ and $w^*_\ell = x^i_\ell$ for
each leaf node ℓ. This shows that w^* is a redaction of x^i.

The only case left to worry about is that possibly w^* includes some symbol
not present in W^i (recall that W^i denotes the join of the redactions of x^i that
were queried under oracle S). In this case, there is some leaf node $\ell \in T^*$ so that
$w^*_\ell = x^i_\ell$ but ℓ is not found in the tree corresponding to W^i.

In this case, the forged signature must reveal k_η where η is some ancestor
of ℓ. But now we can note that k_η was never disclosed by any of the oracle
queries (nor was any of the key values at η's ancestors). We argue that this
would constitute a break of G.

We can imagine replacing the key k_η and the keys at each of its descendant
nodes with truly random values, chosen independently of everything else. If any
adversary can recognize this substitution with advantage better than $q\lceil \lg n \rceil \epsilon_G$,
then according to the proof of security for the GGM tree construction [21], they
can distinguish G from random with advantage better than ϵ_G; thus we can
assume that this substitution makes no noticeable difference.

Now the only values related to k_η that are disclosed is $v^i_\ell = H(0, k_\ell, x^i_\ell)$
for leaves ℓ that are descendants of k_η. But, since $H(0, k, \cdot)$ is a good PRF,
we can in turn imagine replacing each such v^i_ℓ by truly random values, chosen

independently of everything else, and no attacker can recognize this substitution with advantage better than $nq\epsilon_H$ without breaking the PRF assumption.

Since the adversary must present the value k_η in the forged signature, this means that the adversary has guessed the value of k_η^*, even though k_η^* has never been disclosed and was chosen independently of everything else. We can bound the probability that this happens by $1/2^{m'}$ per node, and summing over all the nodes gives at most $2nq/2^{m'}$.

Adding up each of these distinguishing probabilities yields the claimed bound.

GEM: A Generic Chosen-Ciphertext Secure Encryption Method

Jean-Sébastien Coron[1], Helena Handschuh[1], Marc Joye[2], Pascal Paillier[1],
David Pointcheval[3], and Christophe Tymen[1,3]

[1] Gemplus Card International
34 rue Guynemer, 92447 Issy-les-Moulineaux, France
{jean-sebastien.coron, helena.handschuh, pascal.paillier,
christophe.tymen}@gemplus.com
[2] Gemplus Card International
Parc d'Activités de Gémenos, B.P. 100, 13881 Gémenos Cedex, France
marc.joye@gemplus.com, http://www.geocities.com/MarcJoye/
[3] École Normale Supérieure, Computer Science Department
45 rue d'Ulm, 75230 Paris Cedex 05, France
david.pointcheval@ens.fr, http://www.di.ens.fr/~pointche/

Abstract. This paper proposes an efficient and provably secure transform to encrypt a message with any asymmetric one-way cryptosystem. The resulting scheme achieves adaptive chosen-ciphertext security in the random oracle model.

Compared to previous known generic constructions (Bellare, Rogaway, Fujisaki, Okamoto, and Pointcheval), our embedding reduces the encryption size and/or speeds up the decryption process. It applies to numerous cryptosystems, including (to name a few) ElGamal, RSA, Okamoto-Uchiyama and Paillier systems.

Keywords. Public-key encryption, hybrid encryption, chosen-ciphertext security, random oracle model, generic conversion, block ciphers, stream ciphers.

1 Introduction

A major contribution of cryptography is *information privacy*: through encryption, parties can securely exchange data over an insecure channel. Loosely speaking this means that unauthorized recipients can learn nothing useful about the exchanged data.

Designing a "good" encryption scheme is a very challenging task. There are basically two criteria to compare the performances of encryption schemes: *efficiency* and *security*. Security is measured as the ability to resist attacks in a given adversarial model [1,8]. The standard security notion is IND-CCA2 *security*, *i.e.*, indistinguishability under adaptive chosen-ciphertext attacks (cf. Section 2). Usually, an (asymmetric) encryption scheme is proven secure by exhibiting a *reduction*: if an adversary can break the IND-CCA2 security then the same adversary can solve a related problem assumed to be infeasible.

B. Preneel (Ed.): CT-RSA 2002, LNCS 2271, pp. 263–276, 2002.
© Springer-Verlag Berlin Heidelberg 2002

This paper is aimed at simplifying the security proof by providing a *Generic Encryption Method* (GEM) to convert *any* asymmetric one-way cryptosystem into a *provably secure* encryption scheme. Hence, when a new asymmetric one-way function is identified, one can easily design a secure encryption scheme. Moreover, the conversion we propose is very efficient (computationally and memory-wise): the converted scheme has roughly the same cost as that of the one-way cryptosystem it is built from.

1.1 Previous Work

In [3], Bellare and Rogaway described OAEP, a generic conversion to transform a *"partial-domain" one-way trapdoor permutation* into an IND-CCA2 secure encryption scheme in the random oracle model [2,7]. Later, Fujisaki and Okamoto [5] presented a way to transform, in the random oracle model, any *chosen-plaintext* (IND-CPA) secure encryption scheme into an IND-CCA2 one. They improved their results in [6] where they gave a generic method to convert a *one-way* (OW-CPA) cryptosystem into an IND-CCA2 secure encryption scheme in the random oracle model. A similar result was independently discovered by Pointcheval [13]. More recently, Okamoto and Pointcheval [12] proposed a more efficient generic conversion, called REACT, to convert any one-way cryptosystem secure under *plaintext-checking attacks* (OW-PCA) into an IND-CCA2 encryption scheme. Contrary to [5,6,13], re-encryption is unnecessary in the decryption process to ensure IND-CCA2 security.

1.2 Our Results

This paper presents GEM, a generic IND-CCA2 conversion. The converted scheme, $\mathbb{E}_{pk}$, built from any OW-PCA asymmetric encryption $\mathcal{E}_{pk}$ and any length-preserving IND-secure symmetric encryption scheme E_K, is secure in the sense of IND-CCA2, in the random oracle model. As discussed in Section 2, the security levels we require for $\mathcal{E}_{pk}$ and E_K are *very* weak and the security level we obtain for $\mathbb{E}_{pk}$ is very high.

1.3 Organization of the Paper

The rest of this paper is organized as follows. In Section 2, we review the security notions for encryption, in both the symmetric and the asymmetric settings. Section 3 is the core of the paper. We present our new padding to convert, in the random oracle model, any asymmetric one-way cryptosystem into an encryption scheme that is secure in the *strongest* sense. We prove the security of our construction in Section 4 by providing and proving a *concrete* reduction algorithm. Finally, we illustrate the merits of our conversion in Section 5.

2   Security Notions

In this section, we recall the definition of an encryption scheme and discuss some related security notions. A good reference to the subject is [1].

2.1 Encryption Schemes

Definition 1. *An encryption scheme consists of three algorithms* $(\mathcal{K}, \mathcal{E}_{pk}, \mathcal{D}_{sk})$.

1. *On input a security parameter k, the key generation algorithm $\mathcal{K}(1^k)$ outputs a random matching pair (pk, sk) of encryption/decryption keys. For the symmetric case, we assume wlog that the encryption and decryption keys are identical, $K := pk = sk$. The key pk is public and the keys sk and K are secret.*
2. *The encryption algorithm $\mathcal{E}_{pk}(m, r)$ outputs a ciphertext c corresponding to a plaintext $m \in \mathsf{MSPC} \subseteq \{0, 1\}^*$, using the random coins $r \in \Omega$. When the process is deterministic, we simply write $\mathcal{E}_{pk}(m)$. In the symmetric case, we note E_K instead of $\mathcal{E}_{pk}$.*
3. *The decryption algorithm $\mathcal{D}_{sk}(c)$ outputs the plaintext m associated to the ciphertext c or a notification $\perp$ that c is not a valid ciphertext. In the symmetric case, we use the notations D_K.*

Furthermore, we require that for all $m \in \mathsf{MSPC}$ and $r \in \Omega$, $\mathcal{D}_{sk}(\mathcal{E}_{pk}(m, r)) = m$.

The converted scheme we propose in this paper is a combination of an asymmetric encryption scheme and a length-preserving symmetric encryption scheme. We assume very weak security properties from those two schemes. Namely, we require that the asymmetric scheme is OW-PCA and that the symmetric scheme is IND, as defined below.

2.2 Security Requirements

An attacker is said *passive* if, in addition to the ciphertext, s/he only obtains some auxiliary information s, which may depend on the potential plaintexts (but not on the key) [9, § 1.5] and other public information (e.g., the security parameter k and the public key pk). Note that in the asymmetric case an attacker can always construct valid pairs of plaintext/ciphertext from the public encryption key pk.

A minimal security requirement for an encryption scheme is *one-wayness* (OW). This captures the property that an adversary cannot recover the *whole* plaintext from a given ciphertext. In some cases, partial information about a plaintext may have disastrous consequences. This is captured by the notion of *semantic security* or the equivalent notion of *indistinguishability* [10]. Basically, indistinguishability means that the only strategy for an adversary to distinguish the encryptions of any two plaintexts is by guessing at random.

In the asymmetric case, suppose that the attacker has access to an oracle telling whether a pair (m, c) of plaintext/ciphertext is valid; *i.e.*, whether $m = \mathcal{D}_{sk}(c)$ holds. Following [12], such an attack scenario is referred to as the *plaintext-checking attack* (PCA). From the pair of adversarial goal (OW) and adversarial model (PCA), we derive the security notion of OW-PCA.

Definition 2. *An asymmetric encryption scheme is* OW-PCA *if no attacker with access to a plaintext-checking oracle* $\mathcal{O}^{\text{PCA}}$ *can recover the whole plaintext corresponding to a ciphertext with non-negligible probability. More formally, an asymmetric encryption scheme is* (τ, q, ε)-*secure in the sense of* OW-PCA *if for any adversary* $\mathcal{A}$ *which runs in time at most* τ, *makes at most* q *queries to* $\mathcal{O}^{\text{PCA}}$, *its success* ε *satisfies*

$$\Pr_{\substack{m \leftarrow \{0,1\}^* \\ r \leftarrow \Omega}} \left[\begin{array}{c} (sk, pk) \leftarrow \mathcal{K}(1^k), c \leftarrow \mathcal{E}_{pk}(m, r) : \\ \mathcal{A}^{\mathcal{O}^{\text{PCA}}}(c, s) = m \end{array} \right] \leq \varepsilon$$

where the probability is also taken over the random choices of $\mathcal{A}$.

For the symmetric case, we consider a passive attacker who tries to break the indistinguishability property of the encryption scheme. The attacker $\mathcal{A} = (\mathcal{A}_1, \mathcal{A}_2)$ runs in two stages. In the first stage (or *find stage*), on input k, $\mathcal{A}_1$ outputs a pair of messages (m_0, m_1) and some auxiliary information s. Next, in the second stage (or *guess stage*), given the encryption c_b of either m_0 or m_1 and the auxiliary information s, $\mathcal{A}_2$ tells if the challenge ciphertext c_b encrypts m_0 or m_1.

Definition 3. *A symmetric encryption scheme is* IND *if no attacker can distinguish the encryptions of two equal-length plaintexts with probability non-negligibly greater than* $1/2$. *More formally, a symmetric encryption scheme is* (τ, ν)-*secure in the sense of* IND *if for any adversary* $\mathcal{A} = (\mathcal{A}_1, \mathcal{A}_2)$ *which runs in time at most* τ, *its advantage* ν *satisfies*

$$\Pr_{b \leftarrow \{0,1\}} \left[\begin{array}{c} K \leftarrow \mathcal{K}(1^k), (m_0, m_1, s) \leftarrow \mathcal{A}_1(k), \\ c_b \leftarrow E_K(m_b) : \mathcal{A}_2(m_0, m_1, c_b, s) = b \end{array} \right] \leq \frac{1 + \nu}{2}$$

where the probability is also taken over the random choices of $\mathcal{A}$.

In contrast, for our converted encryption scheme we require the *highest* security level, namely IND-CCA2 security. The notion of IND-CCA2 for an asymmetric encryption scheme considers an *active* attacker who tries to break the system by probing with chosen-ciphertext messages. Such an attack can be non-adaptive (CCA1) [11] or adaptive (CCA2) [14]. In a CCA2 scenario, the adversary may run a second chosen ciphertext attack upon receiving the challenge ciphertext c_b (the only restriction being not to probe on c_b).

Definition 4. *An asymmetric encryption scheme is* IND-CCA2 *if no attacker with access to a decryption oracle* $\mathcal{O}^{\mathcal{D}_{sk}}$ *can distinguish the encryptions of two equal-length plaintexts with probability non-negligibly greater than* $1/2$. *More formally, an asymmetric encryption scheme is* (τ, q, ε)-*secure in the sense of* IND-CCA2 *if for any adversary* $\mathcal{A} = (\mathcal{A}_1, \mathcal{A}_2)$ *which runs in time at most* τ, *makes at most* q *queries to* $\mathcal{O}^{\mathcal{D}_{sk}}$, *its advantage* ε *satisfies*

$$\Pr_{\substack{b \leftarrow \{0,1\} \\ r \leftarrow \Omega}} \left[\begin{array}{c} (sk, pk) \leftarrow \mathcal{K}(1^k), \\ (m_0, m_1, s) \leftarrow \mathcal{A}_1^{\mathcal{O}^{\mathcal{D}_{sk}}}(k, pk), \\ c_b \leftarrow \mathcal{E}_{pk}(m_b, r) : \\ \mathcal{A}_2^{\mathcal{O}^{\mathcal{D}_{sk}}}(m_0, m_1, c_b, s) = b \end{array} \right] \leq \frac{1 + \varepsilon}{2}$$

where the probability is also taken over the random choices of $\mathcal{A}$, and $\mathcal{A}_2$ is not allowed to query on c_b.

3   GEM: Generic Encryption Method

A very appealing way to encrypt a message consists in using a *hybrid encryption* mode. A random session key R is first encrypted with an asymmetric cryptosystem. Then the message is encrypted under that session key with a symmetric cryptosystem. Although seemingly sound, this scheme does not achieve IND-CCA2 security under weak security assumptions for the two underlying cryptosystems. This section shows how to modify the above paradigm for attaining the IND-CCA2 security level.

3.1   REACT Transform

The authors of REACT imagined to append a checksum to the previous construction and prove the IND-CCA2 security of the resulting scheme in the random oracle model [12]. Briefly, REACT works as follows. A plaintext m is transformed into the ciphertext (c_1, c_2, c_3) given by

$$\text{REACT}(m) = \underbrace{\mathcal{E}_{pk}(R, u)}_{=c_1} \,\|\, \underbrace{\text{E}_K(m)}_{=c_2} \,\|\, \underbrace{\text{H}(R, m, c_1, c_2)}_{=c_3}$$

where u is a random, $K = \text{G}(R)$, and G, H are hash functions.

Building on this, we propose a new generic encryption method. Our method is aimed at shortening the whole ciphertext by incorporating the checksum (*i.e.*, c_3) into c_1 while maintaining the IND-CCA2 security level, in the random oracle model.

3.2   New Method

Let $\mathcal{E}_{pk}$ and E_K denote an asymmetric and a length-preserving symmetric encryption algorithms, respectively, and let F, G, H denote hash functions. Let also $\mathcal{D}_{sk}$ and D_K denote the decryption algorithms corresponding to $\mathcal{E}_{pk}$ and E_K, respectively. For convenience, for any element x defined over a domain A, we write $\sharp x$ for $|\{x \in A\}|$. So, for example, $\sharp m$ represents the cardinality of the message space, *i.e.*, the number of different plaintexts.

Encryption

Input: Plaintext m, random $\rho = r\|u$.
Output: Ciphertext (c_1, c_2) given by
$$\mathbb{E}_{pk}(m, \rho) = \underbrace{\mathcal{E}_{pk}(w, u)}_{=c_1} \,\|\, \underbrace{\text{E}_K(m)}_{=c_2}$$
where $s = \text{F}(m, r)$, $w = s \,\|\, r \oplus \text{H}(s)$,
and $K = \text{G}(w, c_1)$.

Decryption

Input: Ciphertext (c_1, c_2).
Output: Plaintext $\dot{m}$ or symbol $\perp$ according to

$$\mathbb{D}_{sk}(c_1 \| c_2) = \begin{cases} \dot{m} & \text{if } \dot{s} = \mathrm{F}(\dot{m}, \dot{r}) \\ \perp & \text{otherwise} \end{cases}$$

where $\dot{w} := \dot{s} \| \dot{t} = \mathcal{D}_{sk}(c_1)$, $\dot{K} = \mathrm{G}(\dot{w}, c_1)$,
$\dot{m} = \mathrm{D}_{\dot{K}}(c_2)$, and $\dot{r} = \dot{t} \oplus \mathrm{H}(\dot{s})$.

---

4   Security Analysis

We now prove the security of our conversion. We show that if the hybrid encryption scheme $\mathbb{E}_{pk}$ can be broken under an adaptive chosen-ciphertext attack then either the length-preserving symmetric encryption scheme E_K or the asymmetric encryption scheme $\mathcal{E}_{pk}$ underlying our construction is *highly* insecure, namely the IND-security of E_K or the OW-PCA security of $\mathcal{E}_{pk}$ gets broken.

Theorem 1. *Suppose that there exists an adversary that breaks, in the random oracle model, the* IND-CCA2 *security of our converted scheme $\mathbb{E}_{pk}$ within a time bound τ, after at most $q_\mathrm{F}, q_\mathrm{G}, q_\mathrm{H}, q_{\mathbb{D}_{sk}}$ queries to hash functions $\mathrm{F}, \mathrm{G}, \mathrm{H}$ and decryption oracle $\mathcal{O}^{\mathbb{D}_{sk}}$, respectively, and with an advantage ε. Then, for all $0 < \nu < \varepsilon$, there exists*

- *an adversary that breaks the* IND *security of E_K within a time bound τ and an advantage ν; or*
- *an adversary with access to a plaintext-checking oracle $\mathcal{O}^{\mathrm{PCA}}$ (responding in time bounded by τ_{PCA}) that breaks the* OW-PCA *security of $\mathcal{E}_{pk}$ within a time bound*

$$\tau' = \tau + (q_\mathrm{F}\, q_\mathrm{H} + q_\mathrm{G} + q_{\mathbb{D}_{sk}}(q_\mathrm{F} + q_\mathrm{G}))\,(\tau_{\mathrm{PCA}} + O(1)),$$

after at most

$$q_{\mathrm{PCA}} \le q_\mathrm{F}\, q_\mathrm{H} + q_\mathrm{G} + q_{\mathbb{D}_{sk}}(q_\mathrm{F} + q_\mathrm{G})$$

queries to $\mathcal{O}^{\mathrm{PCA}}$, and with success probability

$$\varepsilon' \ge \frac{\varepsilon - \nu}{2} - \frac{q_\mathrm{F}}{\sharp r} - q_{\mathbb{D}_{sk}}\left(\frac{1}{\sharp s} + q_\mathrm{F}\left(\frac{1}{\sharp r} + \frac{1}{\sharp s}\right) + \nu + \frac{1}{\sharp m}\right) \ .$$

From this, we immediately obtain:

Corollary 1. *For any* OW-PCA *asymmetric encryption $\mathcal{E}_{pk}$ and any length-preserving* IND-*secure symmetric encryption scheme E_K, our converted scheme $\mathbb{E}_{pk}$ is* IND-CCA2 *secure in the random oracle model.* $\square$

To prove Theorem 1, we suppose that there exists an adversary $\mathcal{A} = (\mathcal{A}_1, \mathcal{A}_2)$ able to break the INC-CCA2 security of $\mathbb{E}_{pk}$. We further suppose that E_K is (τ, ν)-secure in the sense of IND. From $\mathcal{A}$, we then exhibit an adversary $\mathcal{B}$ (*i.e.*, a reduction algorithm) that inverts $\mathcal{E}_{pk}$ using a plaintext-checking oracle, and thus breaks the OW-PCA security of $\mathcal{E}_{pk}$.

4.1   A Useful Lemma

The assumption that E_K is length-preserving and (τ, ν)-IND secure implies the following lemma.

Lemma 1. *Assume that E_K is a length-preserving (τ, ν)-IND symmetric encryption scheme, where τ denotes the time needed for evaluating $\mathrm{E}_K(\cdot)$. Then given a pair (m, c) of plaintext/ciphertext, we have*

$$\Pr_K[\mathrm{E}_K(m) = c] \leq \nu + \frac{1}{\sharp m} \ .$$

Proof. Given the pair (m, c), we consider the following distinguisher $\mathcal{A}$. $\mathcal{A}$ randomly chooses a bit $d \in \{0, 1\}$, sets $m_d = m$ and $m_{\neg d}$ to a random value m'. The pair $(m_d, m_{\neg d})$ is then sent to the encryption oracle which returns $c_b = \mathrm{E}_K(m_b)$ for a random key K and a random $b \in \{0, 1\}$. $\mathcal{A}$ then checks if $c_b = c$, returns d if the equality holds and $\neg d$ otherwise. Letting ε the advantage of $\mathcal{A}$, we have

$$\begin{aligned}
\varepsilon &= 2 \Pr_{m', K, d, b}[\mathcal{A} \text{ returns } b] - 1 \\
&= 2 \Pr_{m', K, d, b}[(c_b = c) \wedge (d = b)] + \\
&\quad\ 2 \Pr_{m', K, d, b}[(c_b \neq c) \wedge (\neg d = b)] - 1 \\
&= 2 \Pr_{m', K, d, b}[(\mathrm{E}_K(m) = c) \wedge (d = b)] + \\
&\quad\ 2 \Pr_{m', K, d, b}[(\mathrm{E}_K(m') \neq c) \wedge (\neg d = b)] - 1 \\
&= \Pr_K[\mathrm{E}_K(m) = c] + \Pr_{m', K}[\mathrm{E}_K(m') \neq c] - 1 \\
&= \Pr_K[\mathrm{E}_K(m) = c] - \Pr_{m', K}[\mathrm{E}_K(m') = c] \leq \nu \ .
\end{aligned}$$

The proof follows by noting that as E_K is length-preserving, it permutes the set of messages for any key K and so $\Pr_{m', K}[\mathrm{E}_K(m') = c] = 1/\sharp m$. □

4.2   Description of the Reduction Algorithm

$\mathcal{B}$ is given a challenge encryption $y = \mathcal{E}_{pk}(\widetilde{w}, *)$, an oracle $\mathcal{O}^{\mathrm{PCA}}$ which answers plaintext-checking requests on $\mathcal{E}_{pk}$, and an adversary $\mathcal{A} = (\mathcal{A}_1, \mathcal{A}_2)$ that breaks the IND-CCA2 security of $\mathbb{E}_{pk}$. $\mathcal{B}$'s goal is to retrieve all the bits of $\widetilde{w}$. Wlog, we assume that $\mathcal{O}^{\mathrm{PCA}}$ responds to any of $\mathcal{B}$'s requests with no error and within a time bounded by τ_{PCA}.

Throughout, the following notations are used. For any predicate $\mathrm{R}(x)$, $\mathrm{R}(*)$ stands for $\exists x$ s.t. $\mathrm{R}(x)$. If $\mathcal{O}$ is an oracle to which $\mathcal{A}$ has access, we denote by *query* $\mapsto$ *response* the correspondence $\mathcal{O}$ establishes between $\mathcal{A}$'s request *query* and the value *response* returned to $\mathcal{A}$. $\mathrm{HIST}\,[\mathcal{O}]$ stands for the set of correspondences established by $\mathcal{O}$ as time goes on: $\mathrm{HIST}\,[\mathcal{O}]$ can be seen as an history tape which gets updated each time $\mathcal{A}$ makes a query to $\mathcal{O}$. We denote by $q_{\mathcal{O}}$ the number of calls $\mathcal{A}$ made to $\mathcal{O}$ during the simulation.

Overview of $\mathcal{B}$. At the beginning, $\mathcal{B}$ chooses a random value $\widetilde{K}$. $\mathcal{B}$ then runs $\mathcal{A}$ and provides a simulation for F, G, H and $\mathbb{D}_{sk}$. $\mathcal{A} = (\mathcal{A}_1, \mathcal{A}_2)$ runs in two stages. At the end of the first stage (find stage), $\mathcal{A}_1$ outputs a pair (m_0, m_1). $\mathcal{B}$ then randomly chooses $b \in \{0, 1\}$, computes $\widetilde{c}_2 = \mathrm{E}_{\widetilde{K}}(m_b)$ and builds $(y, \widetilde{c}_2)$. This challenge is provided to $\mathcal{A}_2$, which outputs some bit at the end of the second stage (guess stage). Once finished, $\mathcal{B}$ checks whether some $\widetilde{w}$ has been defined during the game. If so, $\widetilde{w}$ is returned as the inverse of $\mathcal{E}_{pk}$ on y; otherwise a failure answer is returned. The detailed description of the simulation follows.

Wlog, we assume that $\mathcal{A}$ keeps track of all the queries throughout the game so that $\mathcal{A}$ never has to make the same query twice to the same oracle.

Simulation of F. For each new query (m, r),

(Event E_1) if processing guess stage and $m = m_b$ and there exists $s \mapsto h \in$ Hist [H] such that $y = \mathcal{E}_{pk}(s\|r \oplus h, *)$ then F sets $\widetilde{w} := s\|r \oplus h$, returns s and updates its history,

(no event) else F outputs a random value and updates its history.

Simulation of G. For each new query (w, c_1),

(Event E_2) if $c_1 = y$ and $y = \mathcal{E}_{pk}(w, *)$ then G sets $\widetilde{w} := w$, returns $\widetilde{K}$ and updates its history,

(Event E_3) else if $c_1 \neq y$ and $y = \mathcal{E}_{pk}(w, *)$ then G sets $\widetilde{w} := w$, returns a random value and updates its history,

(no event) else G outputs a random value and updates its history.

Simulation of H. For each new query s, H outputs a random value and updates its history.

Simulation of $\mathbb{D}_{sk}$ (Plaintext Extractor). For each new query (c_1, c_2), $\mathbb{D}_{sk}$ first checks (this verification step only stands while the guess stage $\mathcal{A}_2$ is running) that $(c_1, c_2) \neq (y, \widetilde{c}_2)$ since if this equality holds, the query must be rejected as $\mathcal{A}$ attempts to decrypt its own challenge ciphertext. Then, $\mathbb{D}_{sk}$ tries to find the only (if any) message m matching the query. To achieve this, $\mathbb{D}_{sk}$ invokes the simulation of G, H and F provided by $\mathcal{B}$ as follows.

- Find the unique pair (r, s) such that $(*, r) \mapsto s \in$ Hist [F], $s \mapsto h \in$ Hist [H] and $c_1 = \mathcal{E}_{pk}(s\|r \oplus h, *)$. If such a pair exists,
  - query G to get $K = G(s\|r \oplus h, c_1)$,
  - letting $m = \mathrm{D}_K(c_2)$, query F to check if $F(m, r) = s$. If the equality holds, return m; otherwise reject the query (Event RJ_1).
- If the search for (r, s) is unsuccessful, check if there exists w with $(w, c_1) \mapsto K \in$ Hist [G] and $c_1 = \mathcal{E}_{pk}(w, *)$. If such an w exists,
  - define s and t by $w = s\|t$, and query H to get $h = H(s)$,
  - letting $m = \mathrm{D}_K(c_2)$, query F to check if $F(m, t \oplus h) = s$. If the equality holds, return m; otherwise reject the query (Event RJ_2).
- If the search for w is unsuccessful, reject the query (Event RJ_3).

4.3   Soundness of $\mathcal{B}$

Unless otherwise mentioned, all probabilities are taken over the random choices of $\mathcal{A}$ and $\mathcal{B}$.

Simulation of Random Oracles. The plaintext $\widetilde{w}$ uniquely defines $\widetilde{s}$ and $\widetilde{t}$ such that $\widetilde{w} = \widetilde{s}\|\widetilde{t}$. We note $\widetilde{r}$ the random variable $\widetilde{t} \oplus H(\widetilde{s})$.

Soundness of F. The simulation of F fails when $(m_b, \widetilde{r})$ is queried and answered with some value $s \neq \widetilde{s}$ before $\widetilde{s}$ appears in $\text{HIST}\,[H]$.

Let q_F^1 denote the number of oracle queries $\mathcal{A}_1$ made to F during the find stage. Since $\widetilde{r}$ is a uniformly-distributed random variable throughout the find stage, we certainly have $\Pr[\text{F incorrect in the find stage}] \leq q_F^1/\sharp r$.

Moreover, throughout the guess stage, $\mathcal{A}_2$ cannot gain any information about $\widetilde{r}$ without knowing $H(\widetilde{s})$ because H is a random function. Hence, letting q_F^2 the number of oracle queries $\mathcal{A}_2$ made to F during the guess stage, we have $\Pr[\text{F incorrect in the guess stage}] \leq q_F^2/\sharp r$.

Consequently, the probability that an error occurs while $\mathcal{B}$ simulates the oracle F is upper-bounded by

$$\Pr[\text{F incorrect}] \leq \frac{q_F^1 + q_F^2}{\sharp r} = \frac{q_F}{\sharp r} \ .$$

Soundness of G. The simulation is perfect.

Soundness of H. The simulation is perfect.

Plaintext Extraction. The simulation of $\mathbb{D}_{sk}$ fails when $\mathbb{D}_{sk}$ returns $\perp$ although the query $c = (c_1, c_2)$ is a valid ciphertext. Let then m, r, s, t, h, w, K be the unique random variables associated to c in this case. Obviously, c was rejected through event RJ_3, because a rejection through RJ_1 or RJ_2 refutes the validity of c. Therefore, if $\mathbb{D}_{sk}$ is incorrect for c, we must have

$$\underbrace{\left((m, r) \notin \text{HIST}\,[F] \vee s \notin \text{HIST}\,[H]\right)}_{:= \neg\mathsf{E}_F \qquad\qquad\quad := \neg\mathsf{E}_H} \wedge \underbrace{\left((w, c_1) \mapsto K \notin \text{HIST}\,[G]\right)}_{:= \neg\mathsf{E}_G} \ .$$

Hence,

$$\begin{aligned}
\Pr[\mathbb{D}_{sk} \text{ incorrect for } c] &= \Pr[(\neg\mathsf{E}_F \vee \neg\mathsf{E}_H) \wedge \neg\mathsf{E}_G] \\
&= \Pr[((m,r) \neq (m_b, \widetilde{r})) \wedge (\neg\mathsf{E}_F \vee \neg\mathsf{E}_H) \wedge \neg\mathsf{E}_G] + \\
&\quad \Pr[((m,r) = (m_b, \widetilde{r})) \wedge (\neg\mathsf{E}_F \vee \neg\mathsf{E}_H) \wedge \neg\mathsf{E}_G] \\
&\leq \Pr[((m,r) \neq (m_b, \widetilde{r})) \wedge (\neg\mathsf{E}_F \vee \neg\mathsf{E}_H)] + \Pr[((m,r) = (m_b, \widetilde{r})) \wedge \neg\mathsf{E}_G] \\
&= \Pr[((m,r) \neq (m_b, \widetilde{r})) \wedge \neg\mathsf{E}_F] + \Pr[((m,r) \neq (m_b, \widetilde{r})) \wedge (\mathsf{E}_F \wedge \neg\mathsf{E}_H)] + \\
&\quad \Pr[((m,r) = (m_b, \widetilde{r})) \wedge \neg\mathsf{E}_G] \ .
\end{aligned}$$

1. ASSUME $(m,r) \neq (m_b, \widetilde{r})$ AND $(m,r) \notin \text{HIST}[F]$. Since F is a random function, $F(m,r)$ is a uniformly distributed random value unknown to $\mathcal{A}$. The fact that c is a valid ciphertext implies that $F(m,r) = s$, which happens with probability

$$\Pr[c \text{ is valid} \wedge (m,r) \notin \text{HIST}[F]] = \frac{1}{\sharp s} \, .$$

2. ASSUME $(m,r) \neq (m_b, \widetilde{r})$ AND $(m,r) \in \text{HIST}[F] \wedge s \mapsto h \notin \text{HIST}[H]$. Suppose that $s \neq \widetilde{s}$. Since H is a random function, $H(s)$ is a uniformly distributed random value unknown to $\mathcal{A}$. The validity of c implies that $(m, t \oplus H(s)) \mapsto s \in \text{HIST}[F]$, which happens with probability

$$\Pr[(m, t \oplus H(s)) \mapsto s \in \text{HIST}[F]] \leq \frac{q_F}{\sharp r} \, .$$

Now assume $s = \widetilde{s}$. In this case, we must have $(m,r) \mapsto \widetilde{s} \in F$ which occurs with probability

$$\Pr[(m,r) \mapsto \widetilde{s} \in \text{HIST}[F]] \leq \frac{q_F}{\sharp s} \, .$$

3. ASSUME $(m,r) = (m_b, \widetilde{r})$ AND $(w, c_1) \mapsto K \notin \text{HIST}[G]$. This implies $s = \widetilde{s}$, $t = \widetilde{t}$, $w = \widetilde{w}$ and $c_1 \neq y$. Hence, if c is valid, we must have $E_K(m_b) = c_2$ for a uniformly distributed K. By virtue of Lemma 1, this is bounded by

$$\Pr_K[E_K(m_b) = c_2] \leq \nu + \frac{1}{\sharp m} \, .$$

Gathering all preceding bounds, we get

$$\Pr[c \text{ is valid} \wedge \mathbb{D}_{sk} \text{ incorrect for } c]$$
$$\leq \frac{1}{\sharp s} + q_F \left(\frac{1}{\sharp r} + \frac{1}{\sharp s} \right) + \nu + \frac{1}{\sharp m},$$

which, taken over all queries of $\mathcal{A}$, leads to

$$\Pr[\mathbb{D}_{sk} \text{ incorrect}] \leq q_{\mathbb{D}_{sk}} \left(\frac{1}{\sharp s} + q_F \left(\frac{1}{\sharp r} + \frac{1}{\sharp s} \right) + \nu + \frac{1}{\sharp m} \right) \, .$$

Conclusion. We have

$$\Pr[\mathcal{B} \text{ incorrect}] = \Pr[F \text{ incorrect}] + \Pr[\mathbb{D}_{sk} \text{ incorrect}]$$
$$\leq \frac{q_F}{\sharp r} + q_{\mathbb{D}_{sk}} \left(\frac{1}{\sharp s} + q_F \left(\frac{1}{\sharp r} + \frac{1}{\sharp s} \right) + \nu + \frac{1}{\sharp m} \right) \, .$$

4.4 Reduction Cost

Success Probability. Let us suppose that $\mathcal{A}$ distinguishes $\mathbb{E}_{pk}$ within a time bound τ with advantage ε in less than q_F, q_H, q_G, $q_{\mathbb{D}_{sk}}$ oracle calls. Defining $\Pr_2[\cdot] = \Pr[\cdot \mid \neg(\mathcal{B} \text{ incorrect})]$, this means that

$$\Pr_2[\mathcal{A} \to b] \geq \frac{1 + \varepsilon}{2} \, .$$

Suppose also that E_K is (τ, ν)-indistinguishable. Assuming that the oracles are correctly simulated, if none of the events E_1, E_2 or E_3 occurs, then $\mathcal{A}$ never asked $\tilde{r}$ to the random oracle F, neither did it learn the key $\tilde{K}$ under which m_b was encrypted in $\tilde{c}_2$ (this is due to the randomness of F and G). This upper-limits the information leakage on b by ν, since $\mathcal{A}$'s running time is bounded by τ. Noting $\mathsf{E}_{win} = \mathsf{E}_1 \vee \mathsf{E}_2 \vee \mathsf{E}_3$, this implies

$$\mathrm{Pr}_2[\mathcal{A} \to b \mid \neg \mathsf{E}_{win}] \leq \frac{1+\nu}{2} \ .$$

We then get

$$\frac{1+\varepsilon}{2} \leq \mathrm{Pr}_2[\mathcal{A} \to b] \leq \mathrm{Pr}_2[\mathcal{A} \to b \mid \neg \mathsf{E}_{win}] + \mathrm{Pr}_2[\mathsf{E}_{win}]$$

$$\leq \frac{1+\nu}{2} + \mathrm{Pr}_2[\mathsf{E}_{win}] ,$$

whence $\mathrm{Pr}_2[\mathsf{E}_{win}] \geq (\varepsilon - \nu)/2$. But $\mathrm{Pr}_2[\mathcal{B} \to \tilde{w}] = \mathrm{Pr}_2[\mathsf{E}_{win}]$ and finally,

$$\mathrm{Pr}[\mathcal{B} \to \tilde{w}] \geq \mathrm{Pr}_2[\mathcal{B} \to \tilde{w}] - \mathrm{Pr}[\mathcal{B} \text{ incorrect}]$$

$$\geq \frac{\varepsilon - \nu}{2} - \frac{q_\mathsf{F}}{\sharp r} -$$

$$q_{\mathbb{D}_{sk}} \left(\frac{1}{\sharp s} + q_\mathsf{F} \left(\frac{1}{\sharp r} + \frac{1}{\sharp s} \right) + \nu + \frac{1}{\sharp m} \right) \ .$$

Hence $\mathcal{B}$ succeeds with non-negligible probability.

Total Number of Calls to $\mathcal{O}^{\text{PCA}}$. Checking that a pair of the form (w, y) satisfies $y = \mathcal{E}_{pk}(w, *)$ is done thanks to the plaintext-checking oracle $\mathcal{O}^{\text{PCA}}$. Therefore, oracle F makes at most $q_\mathsf{F} \cdot q_\mathsf{H}$ queries to $\mathcal{O}^{\text{PCA}}$, and oracle G makes at most $q_\mathsf{G} \cdot 1$ queries to $\mathcal{O}^{\text{PCA}}$. Moreover, it is easy to see that oracle $\mathcal{O}^{\mathbb{D}_{sk}}$ makes at most $q_{\mathbb{D}_{sk}}(q_\mathsf{F} + q_\mathsf{G})$ calls since in the worst case $\mathbb{D}_{sk}$ has to call $\mathcal{O}^{\text{PCA}}$ for all elements $((*, r) \mapsto s) \in \mathrm{H}\textsc{ist}\,[\mathsf{F}]$ and for all elements $((w, c_1) \mapsto K) \in \mathrm{H}\textsc{ist}\,[\mathsf{G}]$. In conclusion, the total number of calls actually needed by $\mathcal{B}$ is upper-bounded by

$$q_{\text{PCA}} \leq q_\mathsf{F}\,q_\mathsf{H} + q_\mathsf{G} + q_{\mathbb{D}_{sk}}(q_\mathsf{F} + q_\mathsf{G}) \ .$$

Total Running Time. The reduction algorithm runs in time bounded by

$$\tau_\mathcal{B} = \tau + (q_\mathsf{F}\,q_\mathsf{H} + q_\mathsf{G} + q_{\mathbb{D}_{sk}}(q_\mathsf{F} + q_\mathsf{G}))\,(\tau_{\text{PCA}} + O(1)) \ .$$

This completes the proof of Theorem 1.

5 Concluding Remarks

A very popular way to (symmetrically) encrypt a plaintext is to use a *stream cipher*. The simplest example is the Vernam cipher where a plaintext m is processed bit-by-bit to form the ciphertext c under the secret key K. With this cipher, each plaintext bit m_i is XOR-ed with the key bit K_i to produce the ciphertext bit $c_i = m_i \oplus K_i$. If K is truly random and changes for each plaintext m being encrypted, then the system is unconditionally secure. This ideal situation is, however, impractical for a real-world implementation. To resolve the key management problem, a stream cipher is usually combined with a public-key cryptosystem. Contrary to the obvious solution consisting in encrypting a random session key with an asymmetric cryptosystem and then using that key with a stream cipher, our hybrid scheme achieves provable security. Compared to a purely asymmetric solution, our scheme presents the advantage to encrypt *long* messages *orders of magnitude faster* thanks to the use of a symmetric cryptosystem.

Another merit of our scheme resides in its generic nature. The set of possible applications of the new conversion scheme is similar to that of REACT: it concerns *any* asymmetric function that is OW-PCA under a conjectured intractability assumption. A specificity of REACT is that it can operate "on the fly". The session key does not depend on the plaintext to be encrypted and can thus be computed in advance. This property is particularly advantageous when the asymmetric encryption is expensive, as is the case for discrete-log based cryptosystems. Remark that our scheme does not allow "on-the-fly" encryption, that was the price to pay for shortening the ciphertext.

In other cases such as for RSA (with a low encryption exponent, *e.g.*, 3 or $2^{16}+1$) or Rabin cryptosystem, on-the-fly encryption is not an issue and our scheme may be preferred because the resulting ciphertext is shorter. Furthermore, it is worth noting that for a deterministic asymmetric encryption scheme $\mathcal{E}_{pk}$, the notions of OW-PCA and OW-CPA are identical: the validity of a pair (m, c) of plaintext/ciphertext can be publicly checked as $c = \mathcal{E}_{pk}(m)$. So, our method allows one to construct, for example, an efficient IND-CCA2 hybrid encryption scheme whose security relies on the hardness of inverting the RSA function or factoring large numbers.

In conclusion, our generic conversion may be seen as the best alternative to REACT when the underlying asymmetric encryption is relatively fast or when memory/bandwidth savings are a priority.

References

1. Mihir Bellare, Anand Desai, David Pointcheval, and Phillip Rogaway. Relations among notions of security for public-key encryption schemes. Full paper (30 pages), February 1999. An extended abstract appears in H. Krawczyk, ed., Advances in Cryptology – CRYPTO '98, volume 1462 of Lecture Notes in Computer Science, pages 26–45, Springer-Verlag, 1998.
2. Mihir Bellare and Phillip Rogaway. Random oracles are practical: A paradigm for designing efficient protocols. In *First ACM Conference on Computer and Communications Security*, pages 62–73. ACM Press, 1993.
3. Mihir Bellare and Phillip Rogaway. Optimal asymmetric encryption. In A. De Santis, editor, *Advances in Cryptology – EUROCRYPT '94*, volume 950 of *Lecture Notes in Computer Science*, pages 92–111. Springer-Verlag, 1995.

4. Victor Boyko. On the security properties of OAEP as an all-or-nothing transform. Full paper (28 pages), August 1999. An extended abstract appears in M. Wiener, ed., Advances in Cryptology – CRYPTO '99, volume 1666 of Lecture Notes in Computer Science, pages 503–518, Springer-Verlag, 1999.

5. Eiichiro Fujisaki and Tatsuaki Okamoto. How to enhance the security of public-key encryption at minimum cost. *IEICE Transaction on of Fundamentals of Electronic Communications and Computer Science* **E83-A**(1): 24–32, January 2000.

6. Eiichiro Fujisaki and Tatsuaki Okamoto. Secure integration of asymmetric and symmetric encryption schemes. In M. Wiener, editor, *Advances in Cryptology – CRYPTO '99*, volume 1666 of *Lecture Notes in Computer Science*, pages 537–554. Springer-Verlag, 1999.

7. Eiichiro Fujisaki, Tatsuaki Okamoto, David Pointcheval, and Jacques Stern. RSA-OAEP is secure under the RSA assumption. In J. Kilian, editor, *Advances in Cryptology – CRYPTO 2001*, volume 2139 of *Lecture Notes in Computer Science*, pages 260–274. Springer-Verlag, 2001.

8. Oded Goldreich. On the foundations of modern cryptography. In B. Kaliski, editor, *Advances in Cryptology – CRYPTO '97*, volume 1294 of *Lecture Notes in Computer Science*, pages 46–74. Springer-Verlag, 1997.

9. Oded Goldreich. *Modern cryptography, probabilistic proofs and pseudo-randomness*, volume 17 of *Algorithms and Combinatorics*. Springer-Verlag, 1999.

10. Shafi Goldwasser and Silvio Micali. Probabilistic encryption. *Journal of Computer and System Sciences*, 28:270–299, 1984.

11. Moni Naor and Moti Yung. Public-key cryptosystems provably secure against chosen ciphertext attacks. In *22nd ACM Annual Symposium on the Theory of Computing (STOC '90)*, pages 427–437. ACM Press, 1990.

12. Tatsuaki Okamoto and David Pointcheval. REACT: Rapid Enhanced-security Asymmetric Cryptosystem Transform. In D. Naccache, editor, *Topics in Cryptology – CT-RSA 2001*, volume 2020 of *Lecture Notes in Computer Science*, pages 159–175. Springer-Verlag, 2001.

13. David Pointcheval. Chosen-ciphertext security for any one-way cryptosystem. In H. Imai and Y. Zheng, editors, *Public Key Cryptography*, volume 1751 of *Lecture Notes in Computer Science*, pages 129–146. Springer-Verlag, 2000.

14. Charles Rackoff and Daniel R. Simon. Non-interactive zero-knowledge proof of knowledge and chosen ciphertext attack. In J. Feigenbaum, editor, *Advances in Cryptology – CRYPTO '91*, volume 576 of *Lecture Notes in Computer Science*, pages 433–444. Springer-Verlag, 1992.

15. Ronald L. Rivest. All-or-nothing encryption and the package transform. In E. Biham, editor, *Fast Software Encryption*, volume 1267 of *Lecture Notes in Computer Science*, pages 210–218. Springer-Verlag, 1997.

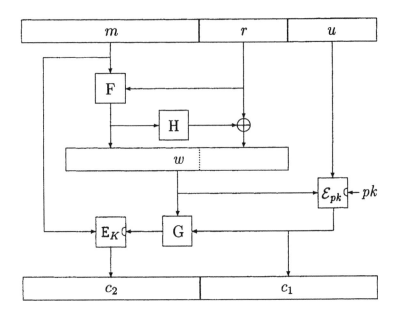

Fig. 1. Description of GEM in encryption mode.

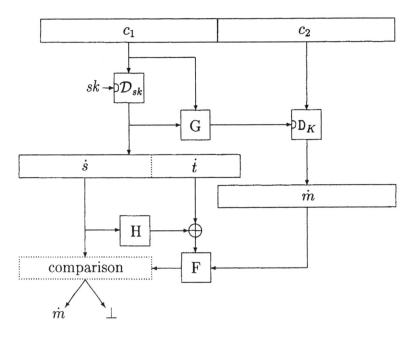

Fig. 2. Description of GEM in decryption mode.

Securing "Encryption + Proof of Knowledge" in the Random Oracle Model

Masayuki Abe

NTT Information Sharing Platform Laboratories
Nippon Telegraph and Telephone Corporation
1-1 Hikari-no-oka, Yokosuka-shi, Kanagawa-ken, 239-0847 Japan
abe@isl.ntt.co.jp

Abstract. To create encryption schemes that offer security against adaptive chosen ciphertext attacks, this paper shows how to securely combine a simple encryption scheme with a proof of knowledge made non-interactive with a hash function. A typical example would be combining the ElGamal encryption scheme with the Schnorr signature scheme. While the straightforward combination will fail to provide security in the random oracle model, we present a class of encryption schemes that uses a proof of knowledge where the security can be proven based on the random oracle assumption and the number theoretic assumptions. The resulting schemes are useful as any casual party can be assured of the (in)validity of the ciphertexts.

1 Introduction

Submitting ciphertexts to validity testing is crucial to achieve security against adaptive chosen message attacks. So far, many validity testing methods have been suggested in a variety of settings to strengthen weak encryption schemes against such strong attacks, e.g., [22,3,5,9,4,13]. Their validity tests, however, can be done only if the verifier knows the decryption key; hence they are *privately checkable*. This type of validity testing is, in general, quite efficient. For instance, just compute a hash function and compare the result with a known string. A practical drawback, however, is that when a ciphertext is rejected, no one but the owner of the decryption key can be convinced. Suppose that Alice sent a ciphertext to Bob and it was rejected by private validity testing. Alice may claim that the ciphertext is valid, while Bob claims the opposite. Alice does not want to reveal the message to support herself and Bob does not want to reveal the private key.

There are some encryption schemes where anyone can check the validity of ciphertexts; hence *publicly checkable* schemes, e.g., [12,21,20,19]. In these schemes, the validity test verifies a non-interactive zero-knowledge proof of knowledge/membership. So the overhead for encryption/checking is relatively large. The schemes in [20], which are secure in the random oracle model [2], cost three or four exponentiations for creating a zero-knowledge proof and two or three exponentiations for checking it, though they have extra security for threshold

B. Preneel (Ed.): CT-RSA 2002, LNCS 2271, pp. 277–289, 2002.
© Springer-Verlag Berlin Heidelberg 2002

decryption. The scheme in [5] is not publicly checkable but it can provide a proof of legitimate rejection according to the technique of [4], though it is still costly.

The most efficient publicly checkable scheme is the combination of ElGamal encryption [6] and Schnorr signature [18] formally discussed in [21,19]. This scheme is at least twice as fast as other publicly checkable ones but its security is proven only in the stronger model. In [21], an on-line knowledge extractor that does not rewind the prover is assumed, and in [19], the attacker is restricted to yield group elements via the group operation only and not to use the group elements to compute any non-group value except for hashing them. The difficulty posed by such schemes in proving the security, in which a proof of knowledge is used rather than membership, is that the simulator has to rewind the prover to extract the knowledge to answer to decryption queries, and this rewinding simulation will not terminate in polynomial time as this rewinding causes another rewinding which results in an avalanche.

This paper provides a way of securely combining a simple encryption scheme and a proof of knowledge so that the resulting scheme features a publicly checkable validity test and security against adaptive chosen message attacks can be proven based on the underlying number theoretic assumptions and the random oracle assumption. Among the many combinations possible, we show two concrete schemes: ElGamal(Diffie-Hellman) + Schnorr and RSA [17] + Guillou-Quisquater [11]. The latter is based on the higher order residuosity problem and the former is based on the Gap-Diffie-Hellman problem [14]. Although all examples in this paper use proof of knowledge made non-interactive by the Fiat-Shamir technique [7], our generic scheme shows that the proof-part is not necessarily a proof of knowledge but a signature scheme. This is a natural consequence as our security proof does not attempt to extract the knowledge from the proof part. In some cases, we may even be able to use somewhat weaker signature schemes than the strongest ones.

The rest of this paper is organized as follows. Section 2 presents some concrete examples that belong to the class for which we prove security. A generic description of the class and the security proof is given in Section 3. Section 4 concludes this paper with some discussion on the advantages and drawbacks of the encryption schemes addressed here. It also states some open problems.

2   Concrete Examples

This section presents two concrete examples that belongs to the class of provably secure publicly checkable secure encryption schemes we describe in Section 3. Although we only present simple schemes, it is possible to derive variations or adaptations of symmetric encryption as shown in the generic description.

2.1   ElGamal Encryption + Schnorr Signature

This scheme is secure assuming the intractability of the Gap-Diffie-Hellman problem (Gap-DH) [14]. Informally, Gap-DH is to compute g^{ab} from g^a and g^b having access to the decision Diffie-Hellman oracle that decides if given (g^a, g^b, g^c) satisfies $ab = c$. Let $\mathcal{G}$ be a probabilistic polynomial-time algorithm that generates a description of a group, say (p, q, g), where Gap-DH is intractable. Here, g is an element of the group defined by p and the order of g is q.

Key Generation: $(p, q, g) \leftarrow \mathcal{G}(1^\kappa)$. Select private key $x \in_U \mathbb{Z}_q^*$. Public key is $y(= g^x \bmod p), p, q, g$ and hash functions, $H : \langle g \rangle^2 \to \{0, 1\}^{\ell_m}, G : \{0, 1\}^* \to \mathbb{Z}_q$ where ℓ_m is the message length.

Encryption: Ciphertext C of message $m \in \{0, 1\}^{\ell_m}$ is $C = (h, c, u, s)$ such that

$$h = g^r \bmod p, \quad c = m \oplus H(h, y^r \bmod p), \quad u = G(c, g^w), \quad s = w - ur \bmod q,$$

where $r, w \in_U \mathbb{Z}_q$ and $\oplus$ denotes bitwise XOR.

Validity Test: Given a ciphertext, check if $u \stackrel{?}{=} G(c, g^s h^u \bmod p)$.

Decryption: For a valid ciphertext that passes the above test, output $c \oplus H(h, h^x \bmod p)$ as a plaintext.

2.2   RSA + GQ Signature

Since RSA is the most common encryption scheme in the current public-key infrastructure it useful to provide an instantiation that suits RSA public-key systems, though the resulting cost is higher than may be expected from the previous section. The security of the following scheme is proven under the RSA assumption.

Key Generation: Choose RSA public key (n, e) and private key d according to security parameter k. Length of e must also be a polynomial in κ. Select a hash function $H : \mathbb{Z}_n^* \times \mathbb{Z}_n^* \to \{0, 1\}^{\ell_m}, G : \{0, 1\}^* \to \{0, 1\}^\ell$ where ℓ is also a security parameter.

Encryption: Ciphertext C of message $m \in \{0, 1\}^{\ell_m}$ is $C = (h, c, u, s)$ such that

$$h = r^e \bmod n, \quad c = m \oplus H(h, r), \quad u = G(c, w^e \bmod n), \quad s = wr^u \bmod n,$$

where $r, w \in_U \mathbb{Z}_n^*$.

Validity Test: Given a ciphertext, check if $u \stackrel{?}{=} G(c, s^e h^{-u} \bmod n)$.

Decryption: For a valid ciphertext that passes the above test, output $c \oplus H(h, h^d \bmod n)$ as a plaintext.

2.3   Other Combinations

We can also create combinations of Paillier encryption [16] and a proof of knowledge from the Schnorr and Guillou-Quisquater schemes, and the Oakmoto-Uchiyama encryption [15] and the statistical zero-knowledge proof by Fujisaki [8].

3   Generic Description and Proof of Security

3.1   Notations

Here we describe, in general, public-key encryption, symmetric-key encryption, and signature schemes. These follow standard definitions. We then define "bridge functions" which is necessary to join the encryption and signature schemes. All algorithms below are polynomial-time with regard to the security parameters.

Public-key encryption scheme $\mathbf{E}^{\mathsf{pub}} = (\mathcal{G}^{\mathsf{pub}}, \mathcal{E}^{\mathsf{pub}}, \mathcal{D}^{\mathsf{pub}})$:

$\mathcal{G}^{\mathsf{pub}}(1^\kappa) \to (ek, dk)$ A probabilistic key-generation algorithm that takes security parameter κ and outputs private decryption key dk and public encryption key ek. Public-key ek determines message space $\mathcal{M}^{\mathsf{pub}}$, randomness space $\mathcal{R}^{\mathsf{pub}}$ and ciphertext space $\mathcal{C}^{\mathsf{pub}}$.

$\mathcal{E}^{\mathsf{pub}}_{ek}(m; r) \to h$ A (probabilistic) encryption algorithm that encrypts message $m \in \mathcal{M}^{\mathsf{pub}}$ into ciphertext h with regard to public-key ek and randomness $r \in \mathcal{R}^{\mathsf{pub}}$.

$\mathcal{D}^{\mathsf{pub}}_{dk}(h) \to m$ A decryption algorithm that satisfies $\mathcal{D}^{\mathsf{pub}}_{dk}(\mathcal{E}^{\mathsf{pub}}_{ek}(m; r)) = m$ for any $(dk, ek) \in \mathcal{G}^{\mathsf{pub}}(1^\kappa)$, $r \in \mathcal{R}^{\mathsf{pub}}$, and $m \in \mathcal{M}^{\mathsf{pub}}$.

To treat asymmetric session-key delivery schemes as encryption schemes, we may allow $\mathcal{D}^{\mathsf{pub}}_{dk}(\mathcal{E}^{\mathsf{pub}}_{ek}(m; r)) = m'$ when a function, $K(m) \to m'$, is available. For instance, in the Diffie-Hellman scheme, $m' = g^i$ and $m = i$. The security notions for encryption schemes also apply to such cases.

Symmetric-key encryption scheme $\mathbf{E}^{\mathsf{sym}} = (\mathcal{K}^{\mathsf{sym}}, \mathcal{M}^{\mathsf{sym}}, \mathcal{E}^{\mathsf{sym}}, \mathcal{D}^{\mathsf{sym}})$:

$\mathcal{K}^{\mathsf{sym}}$ Key space

$\mathcal{M}^{\mathsf{sym}}$ Message space

$\mathcal{E}^{\mathsf{sym}}_k(m) \to c$ An encryption algorithm for message $m \in \mathcal{M}^{\mathsf{sym}}$ and common key $k \in \mathcal{K}^{\mathsf{sym}}$.

$\mathcal{D}^{\mathsf{sym}}_k(c) \to m$ A decryption algorithm such that $\mathcal{D}^{\mathsf{sym}}_k(\mathcal{E}^{\mathsf{sym}}_k(m)) = m$ for any $k \in \mathcal{K}^{\mathsf{sym}}$ and $m \in \mathcal{M}^{\mathsf{sym}}$.

Signature scheme $\mathbf{S} = (\mathcal{G}^{\mathsf{sig}}, \mathcal{S}^{\mathsf{sig}}, \mathcal{V}^{\mathsf{sig}})$:

$\mathcal{G}^{\mathsf{sig}}(1^\kappa) \to (sk, vk)$ A probabilistic key-generation algorithm that outputs private signing key sk and public verification key vk.

$\mathcal{S}^{\mathsf{sig}}_{sk}(m) \to \sigma$ A (probabilistic) signature-generation algorithm that outputs signature σ for message m with signing key sk.

$\mathcal{V}^{\mathsf{sig}}_{vk}(m, \sigma) \to 1/0$ A signature verification algorithm that outputs 1 if σ is in the domain of $\mathcal{S}^{\mathsf{sig}}_{sk}(m)$, or outputs 0 otherwise.

To seamlessly joins the schemes, the following polynomial-time computable functions must exist.

$J_s(m, r, ek) \rightarrow sk$ A function that generates a private signing key of signature scheme **S** from m, r, ek used in $\mathcal{E}_{ek}^{\text{pub}}(m; r)$. It is also necessary that output sk have the same distribution as the signing keys generated by $\mathcal{G}^{\text{sig}}$ when m, r and ek distribute uniformly in their respective spaces.

$J_p(h, ek) \rightarrow vk$ A function that, from the ciphertext h and the public key ek where $h = \mathcal{E}_{ek}^{\text{pub}}(m; r)$, generates the public key of the signature scheme that corresponds to sk generated by $J_s(m, r, ek)$.

$J_d(vk, dk) \rightarrow h$ A probabilistic function that generates h from signature verification key vk and decryption key sk such that $vk = J_p(h, ek)$. (Decryption key dk is used only for generality. It can be encryption key ek.)

Intuitively, J_s and J_d together work as $\mathcal{G}_s$ to generate a temporary key-pair for the signature scheme. J_d can be understood as the inversion function of J_p. It is needed only for the security proof. These functions would be as simple as outputting one of the input variables.

3.2 Generic Description

A publicly checkable encryption scheme; $\mathbf{E}^{\text{pc}} = (\mathcal{G}, \mathcal{E}, \mathcal{V}, \mathcal{D})$;

$\mathcal{G}(1^\kappa) \rightarrow (x, y)$ Execute $\mathcal{G}_p(1^\kappa) \rightarrow (dk, ek)$ and select a hash function, $H : \mathcal{C}^{\text{pub}} \times \mathcal{M}^{\text{pub}} \rightarrow \mathcal{K}^{\text{sym}}$. Output private decryption key $x = (dk, H)$ and public encryption key $y = (ek, H)$.

$\mathcal{E}_y(m; r) \rightarrow C$ Divide r into two parts $(z, w) \in \mathcal{M}^{\text{pub}} \times \mathcal{R}^{\text{pub}}$, and compute $h = \mathcal{E}_{ek}^{\text{pub}}(z; w)$. Compute session key $k = H(h, z)$ and encrypt m as $c = \mathcal{E}_k^{\text{sym}}(m)$. Generate signing key $s = J_s(z, w, ek)$ and sign c with s as $\sigma = \mathcal{S}_{sk}^{\text{sig}}(c)$. Output ciphertext $C = (h, c, \sigma)$.

$\mathcal{V}_y(C) \rightarrow b$ Parse C into (h, c, σ). Generate signature verification key $v = J_p(h, e)$ and output the result of signature verification, $b := \mathcal{V}_{vk}^{\text{sig}}(c, \sigma)$.

$\mathcal{D}_x(C) \rightarrow m$ Parse C into (h, c, σ). Decrypt h as $\mathcal{D}_{dk}^{\text{pub}}(h) \rightarrow z$ with decryption key dk embedded in x. Compute session key $k = H(h, z)$ and retrieve message $m = \mathcal{D}_k^{\text{sym}}(c)$ if $\mathcal{V}_y(C) = 1$, otherwise $m = \perp$.

3.3 Security Definitions

This section briefly reviews the security definitions and intractability assumptions. We use well-known notions and popular notations for encryption schemes such as OW(one-way), IND(indistinguishable), CPA(chosen plaintext attack), CCA2(adaptive chosen ciphertext attack) without definition by referring to [1].

The following set of problems is defined with regard to the encryption scheme.

Definition 1. *Let $(dk, ek) \leftarrow \mathcal{G}^{pub}(1^\kappa)$ and $h = \mathcal{E}^{pub}_{ek}(m; r)$ for $m \in_U \mathcal{M}^{pub}$, $r \in_U \mathcal{R}^{pub}$. Let $\mathcal{T}$ be a set of all possible (m, h, ek) that satisfies $m = \mathcal{D}^{pub}_{dk}(h)$. Let $\mathcal{F}$ be a set of all random (m, h, ek) taken from each domain.*

- *The computational problem (CP) is, given (h, ek), to output m such that $(m, h, ek) \in \mathcal{T}$.*
- *The decisional problem (DP) is, given (m, h, ek) taken randomly from $\mathcal{T}$ or $\mathcal{F}$ with equal probability, to decide which set, $\mathcal{T}$ or $\mathcal{F}$, the given instance belongs to.*
- *The gap-problem (GP) is to solve CP having access to the decision oracle for DP, which returns yes if the query is in $\mathcal{T}$, otherwise no.*

The gap-problem is a rather new class of problems and was introduced in [13]. From the definition, the following relations among those problems are derived easily.

$$\text{DP easy} \Leftrightarrow \text{GP} = \text{CP}$$
$$\text{GP easy} \Leftrightarrow \text{DP} = \text{CP}$$

It follows that a gap-problem is hard if the corresponding computational problem is hard and strictly harder than the decisional problem. For instance, the computational Diffie-Hellman problem seems to be hard and harder than the decision Diffie-Hellman problem. It is clear that if $\mathcal{E}^{pub}$ is deterministic, the decision problem is easy (anyone can encrypt the message and see if it equals the ciphertext) and the gap-problem is equivalent to the computational problem.

The gap-problems are related to the following attack model.

Definition 2. *(PCA: plaintext checking attack) The adversary is given access to a plaintext checking oracle that takes plaintext m, ciphertext h and public-key ek and outputs 1 if m is produced by decrypting h with decryption key dk corresponding to ek.*

OW-PCA security is related to the gap-problems. For instance, the ElGamal encryption scheme is OW-PCA under the Gap Diffie-Hellman assumption [14]. The RSA cryptosystem is also OW-PCA under the Gap-RSA assumption, which is actually equivalent to the RSA assumption as RSA encryption is deterministic.

For signature schemes, we require EUF(existentially unforgeable) against CMA(chosen message attacks). More precisely, we require the signature scheme be EUF-CMA1 as defined below. This is a slightly weaker requirement than the original EUF-CMA defined in [10], which allows polynomially many queries to the signing oracle.

Definition 3. *(EUF-CMA1) Consider that adversary $(\mathcal{A}_1, \mathcal{A}_2)$ plays the following game;*

1. *$(sk, vk) \leftarrow \mathcal{G}^{sig}(1^\kappa)$*
2. *$(m, \omega) \leftarrow \mathcal{A}_1(vk)$*
3. *$\sigma \leftarrow \mathcal{S}^{sig}_{sk}(m)$*
4. *$(\tilde{m}, \tilde{\sigma}) \leftarrow \mathcal{A}_2(\sigma, \omega)$*

A signature scheme is EUF-CMA1 if, for any poly-time algorithm $(\mathcal{A}_1, \mathcal{A}_2)$, $(m, \sigma) \neq (\tilde{m}, \tilde{\sigma})$ and $\mathcal{V}_{vk}^{sig}(\tilde{m}, \tilde{\sigma}) = 1$ hold only with negligible probability.

Note that $m = \tilde{m}$ is allowed in the above definition as is true in the original definition of EUF-CMA in [10]. A similar notion EUF-RMA1 (existentially unforgeable against randomly chosen message attacks with 1 query) can be defined by choosing m uniformly from the message space.

Furthermore, we need **S** to be simulatable in the random oracle model. That is, **S** involves a hash function, say G, and there exists a polynomial-time algorithm such that, given public verification key v and a polynomially bounded list of input-output pairs for G, say L_G , output message-signature pair (m, σ) and a pair of input-output values, say io, which is consistent with L_G and makes the resulting signature valid. By $(\sigma, \mathsf{io}) = \mathcal{S}_v^{sim}(m, \mathsf{L}_G)$ we denote such a simulation. We say that **S** provides the property of SIM^R if such a simulation is successful with overwhelming probability. SIM^R is featured in many practical EUF-CMA signature schemes, especially in the ones made from the honest verifier zero-knowledge proofs focused on in this paper.

3.4   Security Proof

Theorem 1. *If $\mathbf{E}^{pub}$ is OW-PCA, $\mathbf{E}^{sym}$ is IND-CPA, and $\mathbf{S}$ is EUF-CMA1 and SIM^R, then $\mathbf{E}^{pc}$ is IND-CCA2 in the random oracle model.*

Proof. We assume that $\mathbf{E}^{pub}$ is not IND-CCA2. That is, there exists polynomial-time adversary $\mathcal{A}dv$ that succeeds in breaking the indistinguishability of $\mathbf{E}^{pub}$ with adaptive chosen ciphertext attacks. We show that if such $\mathcal{A}dv$ exists we can construct a polynomial-time simulator that breaks either $\mathbf{E}^{pub}$ or $\mathbf{E}^{sym}$ or $\mathbf{S}$ using $\mathcal{A}dv$ as a black-box. In the adaptive chosen ciphertext attacks, $\mathcal{A}dv$ accesses the encryption oracle $(\mathcal{EO})$, the random oracle (H) and the decryption oracle $(\mathcal{DO})$. $\mathcal{EO}$ is given two equal length messages m_0 and m_1 chosen from $\mathcal{M}^{sym}$, and returns challenge ciphertext $C = \mathcal{E}_y(m_b; r)$ for $b \in_U \{0, 1\}$ and random r. H is given query $z_i \in \mathcal{M}^{pub}$ and returns $k_i \in_U \mathcal{K}^{sym}$. $\mathcal{DO}$ is given ciphertext C_i and returns $\mathcal{D}_x(C_i)$. $\mathcal{A}dv$ also accesses random oracle G embedded in the signature scheme, which does not explicitly appear in the following high level description.

We first construct $\mathcal{S}_1$ that breaks the OW-PCA property of $\mathbf{E}^{pub}$ assuming that $\mathbf{E}^{sym}$ and **S** is secure in the above sense. Let $(\mathbf{h}, \mathbf{ek}, \mathcal{CO})$ be a randomly chosen OW-PCA instance where $\mathcal{CO}$ is a plaintext checking oracle for $\mathbf{E}^{pub}$.

Construction of $\mathcal{S}_1$:
Simulator $\mathcal{S}_1$ maintains a list, say L_H , whose entry consists of four-tuples, $(\mathsf{msg}, \mathsf{enc}, \mathsf{key}, \mathsf{ans})$, where msg stores messages in $\mathcal{M}^{pub}$, enc stores encryption of msg with $\mathcal{E}^{pub}$, key stores $H(\mathsf{msg})$, and ans stores yes/no returned from checking oracle $\mathcal{CO}$ in response to query $(\mathsf{msg}, \mathsf{enc}, \mathbf{ek})$. L_H is empty at the beginning. $\mathcal{S}_1$ also maintains list L_G for defining hash function G in **S** . L_G is empty at the beginning, too. By convention, "$*$" matches any string and "$-$" denotes an empty string in the lists. At initialization $\mathcal{S}_1$ runs $\mathcal{G}(1^\kappa) \rightarrow (x, y)$ and replace ek in y with $\mathbf{ek}$ (accordingly, x is rendered useless). $\mathcal{S}_1$ then runs $\mathcal{A}dv$ by giving

y and answers the queries from Adv in the following way. When Adv stops, it outputs $b' \in \{0,1\}$.

[$\mathcal{S}_1$: Simulation of Oracles]

$\mathcal{EO}(m_0, m_1)$: Return challenge ciphertext $C = (h, c, \sigma)$ where $h = \mathbf{h}$, $c = \mathcal{E}_k^{sym}(m_b)$ where $k \in_U \mathcal{K}^{sym}$ and $b \in_U \{0,1\}$, and $(\sigma, \text{io}) = \mathcal{S}_{vk}^{sim}(c, \mathsf{L}_G)$ where $vk = J_p(h, ek)$. $\mathcal{S}^{sim}$ fails if io is inconsistent with L_G, or $(*, h, *, *)$ is already in L_H. Otherwise, append io to L_G and $(-, h, k, \text{yes})$ to L_H.

$H(h_i, z_i)$: a1. If $h_i = h$ and $\mathcal{CO}(z_i, h, ek) = \text{yes}$, then output z_i and stop. ($\mathcal{S}_1$ wins.)

 a2. If $h_i \neq h$, $\mathcal{CO}(z_i, h_i, ek) = \text{yes}$, and $(-, h_i, (k_i), \text{yes})$ exists in L_H, then return the k_i in the entry and replace '$-$' with z_i.

 a3. Return $k_i \in_U \mathcal{K}^{sym}$ and append $(z_i, h_i, k_i, \text{ans})$ to L_H where ans is the answer of $\mathcal{CO}(z_i, h_i, ek)$.

$\mathcal{DO}(C_i)$: Parse C_i into (h_i, c_i, σ_i).

 b1. If $C_i = C$ or $\mathcal{V}_y(C_i) = 0$, return $\bot$. (If G is not yet defined with regard to the query needed for verification, define it with a random value.)

 b2. If $\mathcal{V}_y(C_i) = 1$ and $h_i = h$, simulation fails.

 b3. Otherwise, search L_H for $(*, h_i, (k_i), \text{yes})$.

 c1. If it exists, return $m_i = \mathcal{D}_{k_i}^{sym}(c_i)$ by using k_i in the entry.

 c2. Otherwise, choose $k_i \in_U \mathcal{K}^{sym}$, append $(-, h_i, k_i, \text{yes})$ to L_H, and return $m_i = \mathcal{D}_{k_i}^{sym}(c_i)$.

Queries for hash function G are simulated with random values chosen from the appropriate domain defined by $J_p(h, ek)$. List L_G maintains the input-output correspondence of G.

Evaluation of $\mathcal{S}_1$: We first claim that the simulation of $\mathcal{EO}$ works unless $\mathcal{S}^{sim}$ fails or there is $(*, h, *, *)$ in L_H before the query is made to $\mathcal{EO}$. The first case occurs with negligible probability as we assume that $\mathbf{S}$ is SIM^R. Since $h = \mathbf{h}$ and $\mathbf{h}$ is in the randomly chosen instance, such a failure happens with negligible probability if the size of h is polynomial against the security parameter. As $\mathbf{E}^{pub}$ is assumed to be one-way, the size of its ciphertext space must be at least $|\mathcal{M}^{pub}| + |\mathcal{R}^{pub}|$, which is polynomial in κ. Thus, $\mathcal{EO}$ is simulated correctly with overwhelming probability.

We next claim that event b2 happens only with negligible probability. Assume that Adv sends valid ciphertext $C_i = (h_i, c_i, \sigma_i)$ where $h_i = h$. Since any query to $\mathcal{DO}$ must not be identical to challenge ciphertext C, the remaining part of the query must satisfy $(c_i, \sigma_i) \neq (c, \sigma)$. So b2 means that Adv outputs a forged Schnorr signature with regard to public key $vk = J_p(h, ek)$ by a chosen message attack[1]. Since **S** is assumed to be EUF-CMA1 and only one signature is issued with regard to public-key vk, b2 happens only with negligible probability.

From the above observation we see that $h_i \neq h$ for all valid ciphertext asked to $\mathcal{DO}$ with overwhelming probability. Now we claim that $\mathcal{DO}$ is simulated successfully. Observe that when ciphertext C_i is asked to $\mathcal{DO}$, there are two cases; $\mathcal{A}dv$ has already asked $H(h_i, z_i)$ with correct z_i, i.e., the one such that $\mathcal{CO}(z_i, h_i, ek) = \mathsf{yes}$, or such query has not asked yet. If such a query has been already made, L_H has an entry that contains correct k_i that corresponds to z_i. Thus, m_i returned from case c1 is the same as the one returned from the real $\mathcal{DO}$. (Although $\mathcal{A}dv$ may have created c_i independently from such correct k_i, it does not matter since even the real $\mathcal{DO}$ decrypts c_i with correct k_i.) If such query has not been asked, we can temporary choose k_i as done in case c2 and later detect the corresponding query to H and backpatch as in case a2 so that the answer looks consistent. Thus, m_i from case c2 is also the same as the one from the real $\mathcal{DO}$.

Unless a1 happens, the simulation of H is perfect since all return values, k_i, are selected randomly from $\mathcal{K}^{\mathsf{sym}}$ in a3 and c2.

What remains to prove is that $b' = b$ happens with negligible advantage unless a1 happens. As we have already seen, $h_i \neq h$ for all valid ciphertexts asked to $\mathcal{DO}$ with overwhelming probability. Hence all k_i are independent of k due to the randomness of H, and so are all m_i and b. (This is the reason why h_i must be input to H. Otherwise, m_i could be related to b.) Thus, interacting with $\mathcal{DO}$ does not help $\mathcal{A}dv$ in guessing b. Next, assume that a1 does not happen. One can then see that h is independent of c due to the ideal randomness of H. Furthermore, σ is related to c only in a trivial way. Therefore, only c is relative to b. However, since $\mathbf{E}^{\mathsf{sym}}$ is indistinguishable, b can be guessed solely from c only with negligible advantage. Thus, if $b = b'$ happens with probability $1/2 + \epsilon$, a4 happens and $\mathcal{S}_1$ wins with probability close to ϵ.

Construction of $\mathcal{S}_2$: We next construct $\mathcal{S}_2$ that breaks the IND-CPA property of $\mathbf{E}^{\mathsf{sym}}$ assuming that $\mathbf{E}^{\mathsf{pub}}$ and **S** are secure. $\mathcal{S}_2$ sends (m_0, m_1) to the external symmetric encryption oracle $\mathcal{E}^{\mathsf{sym}}$ that has a randomly chosen secret key k. $\mathcal{S}_2$ finally decides which message corresponds to the ciphertext c returned from $\mathcal{E}^{\mathsf{sym}}$.

At initialization, $\mathcal{S}_2$ runs $G(1^\kappa) \to (x, y)$. This time, x is used as it is. Queries from $\mathcal{A}dv$ are answered as follows.

[$\mathcal{S}_2$: Simulation of Oracles]

$\mathcal{EO}(m_0, m_1)$:   Pass (m_0, m_1) to the external symmetric encryption oracle $\mathcal{E}^{\mathsf{sym}}$. The oracle selects $b \in_U \{0, 1\}$ and creates challenge ciphertext $\mathbf{c} = \mathcal{E}_k^{\mathsf{sym}}(m_b)$ with $k \in_U \mathcal{K}^{\mathsf{sym}}$. It

---

[1] When $\mathcal{E}^{\mathsf{sym}}$ is a random permutation with regard to fixed length messages we allow **S** to be EUF-RMA1 where the message is chosen uniformly from a fixed domain. In general, however, EUF-CMA1 is needed as we do not see how c distributes over the message space of **S** when the corresponding plaintext is chosen by the adversary.

then returns $C = (h, c, \sigma)$ where $c = \mathbf{c}$, $h = \mathcal{E}_{ek}^{\mathsf{pub}}(z; w)$
for $z \in_U \mathcal{M}^{\mathsf{pub}}$, $w \in_U \mathcal{R}^{\mathsf{pub}}$, and $\sigma = \mathcal{S}_{sk}^{\mathsf{sig}}(c)$ where
$sk = J_s(z, w, ek)$. If (h, z) has been already asked to H,
simulation fails.

$H(h_i, z_i)$:   a0. If H is already defined for this input, return the
assigned value.
a1. If $h_i = h$ and $z_i = z$, simulation fails.
a2. If $h_i \neq h$ or $z_i \neq z$, return $k_i \in_U \mathcal{K}^{\mathsf{sym}}$.

$\mathcal{DO}(C_i)$:   Parse C_i into (h_i, c_i, σ_i).

b1. If $C = C_i$ or $\mathcal{V}_y(C_i) = 0$, return $\perp$.
b2. If $\mathcal{V}_y(C_i) = 1$ and $h_i = h$, simulation fails.
b3. Otherwise, return $m_i = \mathcal{D}_{k_i}^{\mathsf{sym}}(c_i)$ where $k_i = H(h_i, z_i)$, $z_i = \mathcal{D}_{dk}^{\mathsf{pub}}(h_i)$. (This is possible as $\mathcal{S}_2$ has dk in x. Define H for input (h_i, z_i) if it has not been asked yet.)

Simulation of G in $\mathbf{S}$ is done with using only random values from the appropriate domain. When Adv outputs b', $\mathcal{S}_2$ outputs it.

Evaluation of $\mathcal{S}_2$: The simulation of $\mathcal{EO}$ fails if (h, z) has been asked to H before $\mathcal{EO}$ is invoked. As h is generated with random z and w, such a failure happens with negligible probability.

We claim that case a1 happens with negligible probability if $\mathbf{E}^{\mathsf{pub}}$ is OW-PCA. It suffices to show that the simulation of H and $\mathcal{DO}$ works just as the plaintext checking oracle in PCA attacks and not more than that. Suppose that Adv is given m_i for c_i in b3. Since k_i in b3 is independent of z_i due to the randomness of H, Adv can get no information about z_i from m_i except for the fact that H is defined as k_i for input (h_i, z_i). Adv can now exploit this fact to check whether some z_i' is the correct decryption of h_i or not by examining if decrypting c_i with k_i' obtained by asking (h_i, z_i') to H results in m_i or not. This argument is irrelevant to whether valid σ_i can be made without knowing z_i or not.

Next, we claim that case b2 does not happen since $C_i \neq C$ and $h_i = h$ implies $(c_i, \sigma_i) \neq (c, \sigma)$ which contradicts the assumption that $\mathbf{S}$ is EUF-CMA1. Note that each h_i is used only once with overwhelming probability. Independence of all returned m_i in b3 from b determined by c can be argued in the same way as done in the evaluation of $\mathcal{S}_1$ based on the fact that $h_i \neq h$ for all valid queries to $\mathcal{DO}$. Thus, the simulation works with overwhelming probability and all values viewed by Adv are independent of b except for c. Therefore $\mathcal{S}_2$ wins if Adv wins.

Construction of $\mathcal{S}_3$:
Finally, we construct $\mathcal{S}_3$ that breaks $\mathbf{S}$ assuming that $\mathbf{E}^{\mathsf{pub}}$ and $\mathbf{E}^{\mathsf{sym}}$ are secure. Let $\mathbf{vk}$ be a given randomly selected signature verification key for which $\mathcal{S}_3$ forges a signature by accessing the external signing oracle that has $\mathbf{sk}$ corresponding

to **vk**. At initialization, $\mathcal{S}_3$ runs $G(1^\kappa) \to (x, y)$. Queries from $\mathcal{A}dv$ are answered as follows.

[$\mathcal{S}_3$: Simulation of Oracles]

$\mathcal{E}\mathcal{O}(m_0, m_1)$:   Select $b \in_U \{0, 1\}$, and compute $h = J_d(\mathbf{vk}, dk)$ by using dk in x. Then compute $z = \mathcal{D}_{dk}^{\mathsf{pub}}(h)$, $k = H(h, z)$, $c = \mathcal{E}_k^{\mathsf{sym}}(m_b)$. ($H$ is defined with k for input (h, z). If H has been already defined at this point, simulation fails.) Then ask the external signing oracle to sign c with key **sk** and get signature σ. Return $C = (h, c, \sigma)$.

$H(h_i, z_i)$:   a0. If H is already defined for this input, return the assigned value.
a1. If $h_i = h$ and $z_i = z$, simulation fails.
a2. If $h_i \neq h$ or $z_i \neq z$, return $k_i \in_U \mathcal{K}^{\mathsf{sym}}$.

$\mathcal{D}\mathcal{O}(C_i)$:   Parse C_i into (h_i, c_i, σ_i).

b1. If $C = C_i$ or $\mathcal{V}_y(C_i) = 0$, return $\bot$.
b2. Else if $h_i = h$, output (c_i, σ_i) and stop.
b3. Otherwise, return $m_i = \mathcal{D}_{k_i}^{\mathsf{sym}}(c_i)$ where $k_i = H(h_i, z_i)$, $z_i = \mathcal{D}_{dk}^{\mathsf{pub}}(h_i)$, and dk in y.

Queries to G in **S** are passed to the corresponding external oracle. When $\mathcal{A}dv$ outputs b', $\mathcal{S}_3$ outputs it.

Evaluation of $\mathcal{S}_3$: The simulation of $\mathcal{E}\mathcal{O}$ works if H has not been defined at (h, z) before. Clearly, such a failure happens only with negligible probability.

We can claim that case a1 happens with negligible probability if $\mathbf{E}^{\mathsf{pub}}$ is OW-PCA in the same way as was done for $\mathcal{S}_2$.

We next claim that if case b2 and a1 do not happen and $\mathbf{E}^{\mathsf{sym}}$ is IND-CPA, $\mathcal{A}dv$ can win only with negligible advantage over $1/2$. The case for a1 has already been made. It suffices to show that case b2 should happen if $\mathcal{A}dv$ is to win with meaningful advantage. Observe that k_i in case b3 is independent of k due to the randomness of H. Hence, m_i does not have more information about b than c has regardless of the relation between c and c_i. Therefore, if b2 does not happen, issuing queries to $\mathcal{D}\mathcal{O}$ does not help in guessing b. Thus, $\mathcal{A}dv$ can win only with the same probability as breaking the IND-CPA property of $\mathbf{E}^{\mathsf{sym}}$. Since this is assumed to be infeasible, there is a contradiction. Thus, b2 must happen and $\mathcal{S}_3$ gets a forged signature-message pair (c_i, σ_i) with regard to $J_p(h, ek) = \mathbf{vk}$.

4   Conclusion

This paper studied encryption schemes that use proof of knowledge but whose security can be proven in the random oracle model under the standard intractability assumptions of number theoretic problems.

A weakness of such schemes is the lack of homomorphism in the encryption part. This is inevitable as it is essential for the security proofs that the message be encrypted with a session key generated by a hash function. On the other hand, such a hybrid structure allows long messages to be encrypted efficiently.

It should also be noted that public verifiability of ciphertexts does not imply security in terms of threshold setting. The main reason is that, in threshold setting, computation of H must be done in public (shown to all decryptors) so the simulator has to extract correct z_i given to H but the schemes addressed here do not provide such a back door, unlike the schemes that combine signature schemes derived from proofs of language [20].

References

1. M. Bellare, A. Desai, D. Pointcheval, and P. Rogaway. Relations among notions of security for public-key encryption schemes. In H. Krawczyk, editor, *Advances in Cryptology – CRYPTO '98*, volume 1462 of *Lecture Notes in Computer Science*, pages 26–45. Springer-Verlag, 1998.

2. M. Bellare and P. Rogaway. Random oracles are practical: a paradigm for designing efficient protocols. In *First ACM Conference on Computer and Communication Security*, pages 62–73. Association for Computing Machinery, 1993.

3. M. Bellare and P. Rogaway. Optimal asymmetric encryption. In Alfredo De Santis, editor, *Advances in Cryptology – EUROCRYPT '94*, volume 950 of *Lecture Notes in Computer Science*, pages 92–111. Springer-Verlag, 1995.

4. R. Canetti and S. Goldwasser. An efficient threshold public key cryptosystem secure against adaptive chosen ciphertext attack. In Jacques Stern, editor, *Advances in Cryptology – EUROCRYPT '99*, volume 1592 of *Lecture Notes in Computer Science*, pages 90–106. Springer-Verlag, 1999.

5. R. Cramer and V. Shoup. A practical public key cryptosystem provably secure against adaptive chosen ciphertext attack. In H. Krawczyk, editor, *Advances in Cryptology – CRYPTO '98*, volume 1462 of *Lecture Notes in Computer Science*, pages 13–25. Springer-Verlag, 1998.

6. T. ElGamal. A public key cryptosystem and a signature scheme based on discrete logarithms. In G. R. Blakley and D. Chaum, editors, *Advances in Cryptology – CRYPTO '84*, volume 196 of *Lecture Notes in Computer Science*, pages 10–18. Springer-Verlag, 1985.

7. U. Feige, A. Fiat, and A. Shamir. Zero-knowledge proofs of identity. *Journal of Cryptology*, 1:77–94, 1988.

8. E. Fujisaki. A simple approach to secretly sharing a factoring witness in publicly-verifiable manner. (unpublished manuscript), 2001.

9. E. Fujisaki and T. Okamoto. Secure integration of asymmetric and symmetric encryption schemes. In M. Wiener, editor, *Advances in Cryptology – CRYPTO '99*, volume 1666 of *Lecture Notes in Computer Science*, pages 537–554. Springer-Verlag, 1999.

10. S. Goldwasser, S. Micali, and R. Rivest. A digital signature scheme secure against adaptive chosen-message attacks. *SIAM Journal of Computing*, 17(2):281–308, April 1988.

11. L. C. Guillou and J.-J. Quisquater. A practical zero-knowledge protocol fitted to security microprocessor minimizing both transmission and memory. In C. G. Günther, editor, *Advances in Cryptology – EUROCRYPT '88*, volume 330 of *Lecture Notes in Computer Science*, pages 123–128. Springer-Verlag, 1988.

12. M. Naor and M. Yung. Public-key cryptosystems provably secure against chosen ciphertext attacks. In *Proceedings of the 22st annual ACM Symposium on the Theory of Computing*, pages 427–437, 1990.

13. T. Okamoto and D. Pointscheval. The gap-problems: a new class of problems for the security of cryptographic schemes. In *PKC 2001*, Lecture Notes in Computer Science. Springer-Verlag, 2001.

14. T. Okamoto and D. Pointscheval. REACT: Rapid enhanced-security asymmetric cryptosystem transform. In *RSA '2001*, Lecture Notes in Computer Science. Springer-Verlag, 2001.

15. T. Okamoto and S. Uchiyama. A new public-key cryptosystem as secure as factoring. In K. Nyberg, editor, *Advances in Cryptology – EUROCRYPT '98*, volume 1403 of *Lecture Notes in Computer Science*, pages 308–318. Springer-Verlag, 1998.

16. P. Paillier. Public-key cryptosystems based on composite degree residuosity classes. In Jacques Stern, editor, *Advances in Cryptology – EUROCRYPT '99*, volume 1592 of *Lecture Notes in Computer Science*, pages 223–238. Springer-Verlag, 1999.

17. R. L. Rivest, A. Shamir, and L. M. Adleman. A method for obtaining digital signatures and public-key cryptosystems. *Communications of the ACM*, 21(2):120–126, 1978.

18. C. P. Schnorr. Efficient signature generation for smart cards. *Journal of Cryptology*, 4(3):239–252, 1991.

19. C.P. Schnorr and M. Jakobsson. Security of signed elgamal encryption. In T. Okamoto, editor, *Advances in Cryptology – ASIACRYPT 2000*, volume 1976 of *Lecture Notes in Computer Science*, pages 73–89. Springer-Verlag, 2000.

20. V. Shoup and R. Gennaro. Securing threshold cryptosystems against chosen ciphertext attack. In K. Nyberg, editor, *Advances in Cryptology – EUROCRYPT '98*, volume 1403 of *Lecture Notes in Computer Science*, pages 1–16. Springer-Verlag, 1998.

21. Y. Tsiounis and M. Yung. On the security of El Gamal based encryption. In H. Imai and Y. Zheng, editors, *First International Workshop on Practice and Theory in Public Key Cryptography – PKC '98*, volume 1431 of *Lecture Notes in Computer Science*, pages 117–134. Springer-Verlag, 1998.

22. Y. Zheng and J. Seberry. Practical approaches to attaining security against adaptively chosen ciphertext attacks. In E. F. Brickell, editor, *Advances in Cryptology – CRYPTO '92*, volume 740 of *Lecture Notes in Computer Science*, pages 292–304. Springer-Verlag, 1993.

Nonuniform Polynomial Time Algorithm to Solve Decisional Diffie-Hellman Problem in Finite Fields under Conjecture

Qi Cheng[1] and Shigenori Uchiyama[2,*]

[1] School of Computer Science
University of Oklahoma, Norman, OK 73019, USA
[2] NTT Laboratories
1-1 Hikarinooka, Yokosuka-shi, 239-0847 Japan

Abstract. In this paper, we show that curves which are defined over a number field of small degree but have a large torsion group over the number field have considerable cryptographic significance. If those curves exist and the heights of torsions are small, they can serve as a bridge for prime shifting, which results a nonuniform polynomial time algorithm to solve DDH on finite fields and a nonuniform subexpontial time algorithm to solve elliptic curve discrete logarithm problem. At this time we are unable to prove the existence of those curves. To the best of our knowledge, this is the first attempt to apply the ideas related to the Uniform Boundedness Theorem(UBT), formerly known as Uniform Boundedness Conjecture, in cryptography.

1   Introduction

Since the proposal of the concept of public-key cryptography, certain kinds of computational problems such as the integer factoring, the discrete logarithm problem, the Diffie-Hellman problem [3] and the RSA problem [15] have been much researched for the last two decades. So far, these computational problems are widely believed to be intractable. One of the most important topics in cryptography is to propose a practical and provably-secure cryptographic scheme under such a reasonable computational assumption. Here, we usually say a cryptographic scheme is provably secure if it is proven to be as secure as such an intractable computational problem. Besides, in order to prove the security of a cryptographic scheme, we sometimes use some other kinds of computational problems what we call decisional problems such as the Decisional Diffie-Hellman problem, the DDH problem for short. This kind of decision problem was firstly introduced in [6] to prove the semantical security of a public-key encryption scheme from a cryptographic point of view. Since then, such a decisional problem has been typically employed to prove the semantical security of public-key

---

* Part of the research was done while the first author was a student in the University of Southern California and the second author was visiting there. The first author was partially support by NSF grant CCR-9820778.

B. Preneel (Ed.): CT-RSA 2002, LNCS 2271, pp. 290–299, 2002.
© Springer-Verlag Berlin Heidelberg 2002

encryption schemes such as the ElGamal and Cramer-Shoup encryption schemes [4,17,2]. More precisely, the Cramer-Shoup encryption scheme is secure against adaptive chosen ciphertext attack under the DDH and the universal one-way hash assumptions. The DDH problem is especially useful and has a lot of applications, so it has been very attractive to cryptography. With respect to applications and a survey of the DDH problem, see the following excellent papers [13,1].

Here, we briefly review the definitions, basic properties of the DDH and related problems. In the following, G denotes a multiplicative finite cyclic group generated by an element g from G, and let l be the order of G. From a cryptographic point of view, we may assume that l is prime.

- **The Discrete Logarithm problem**: Given two elements x and y, to find an integer m so that $y = x^m$.
- **The Diffie-Hellman problem**: Given two elements g^x and g^y, to find g^{xy}.
- **The Decisional Diffie-Hellman problem**: Given two distributions (g^x, g^y, g^{xy}) and (g^x, g^y, g^z), where x, y, z are randomly chosen from $\mathbf{Z}/l\mathbf{Z}$, to distinguish between these two distributions. In other words, given three elements g^x, g^y and g^z, where x, y, z are chosen at random from $\mathbf{Z}/l\mathbf{Z}$, to decide whether $xy \equiv z \pmod{l}$ or not.

Here, it is well-known that the Diffie-Hellman problem can be efficiently reduced to the Discrete Logarithm problem and the DDH problem can be efficiently reduced to the Diffie-Hellman problem. When it comes to relationships between these problems, Maurer and Wolf [10] showed that the Discrete Logarithm problem can be reduced to the Diffie-Hellman problem, if there exists some auxiliary group defined over $\mathbf{F}_l$ and it has some nice properties. More precisely, the Maurer and Wolf's idea is given by the following. An auxiliary group can be taken as the rational points on an elliptic curve defined over $\mathbf{F}_l$ whose order is sufficiently smooth, then we can easily solve the Discrete Logarithm problem over this elliptic curve by using the Pohlig-Hellman algorithm. Furthermore, since we can reduce the Discrete Logarithm problem over G to that over this elliptic curve by employing the Diffie-Hellman oracle, the Discrete Logarithm problem over G can be reduced to the Diffie-Hellman problem. Namely, in this case, we can say that the Diffie-Hellman problem is as hard as the Discrete Logarithm problem. So far, the best known algorithm for these problems over a general group, is a generic algorithm such as the Baby-Step Giant-Step (BSGS) and Pholig-Hellman. Their run time are given by $O(\sqrt{l})$, where l is the order of the base group G. Besides, Shoup [16] showed that the lower bound on computation of the DDH problem is the same as that of the Discrete Logarithm problem under the generic model, i.e., the lower bound is given by $c\sqrt{l}$, where c is some constant, for the DDH problem as well as the Discrete Logarithm problem. More precisely, Shoup showed that an algorithm such as the BSGS is the best possible generic algorithm for the DDH, Diffie-Hellman and Discrete Logarithm problems.

On the other hand, very recently, Joux and Nguyen [8] presented very interesting examples such that the DDH problem is easy while the Diffie-Hellman problem is as intractable as the Discrete Logarithm problem over certain groups of the rational points on elliptic curves defined over finite fields. It is obvious that the DDH problem over an elliptic curve defined over a finite field is very easy if we can compute a pairing such as the Weil and Tate pairing. Actually, we assume that $\langle\,,\,\rangle_l$ is the l-th Tate pairing, where l is prime and also the DDH problem over the group generated by a point P whose order is prime l, then we have $\langle xP, yP\rangle_l = \langle P, P\rangle_l^{xy}$ and $\langle zP, P\rangle_l = \langle P, P\rangle_l^z$. So, in this case, deciding whether $xy \equiv z \pmod{l}$ or not is very easy unless $\langle P, P\rangle_l = 1$. Anyhow, in such a case, the DDH problem can be solved in polynomial time on the size of the input. Here we note that we are not able to evaluate the Tate pairing for all elliptic curves but special classes of curves such as supersingular and trace 2 elliptic curves. Besides, as mentioned above, according to the result by Maurer and Wolf, if we can generate some auxiliary group for these elliptic curves which satisfy certain good properties, the Diffie-Hellman problem is as hard as the Discrete Logarithm problem. That is, Joux and Nguyen presented supersingular and trace 2 elliptic curves with such good auxiliary groups (see for details in [8]).

Here, this observation raises the following question: Is there an efficient reduction from the DDH problem in a finite field to the DDH problem over some special elliptic curve? This paper will explore the possibility.

More precisely, this paper proposes an attack against the DDH problem in finite fields under some reasonable assumption. Suppose that our target field is $\mathbf{F}_p$. l is the largest prime factor of $p - 1$. The correctness of our algorithm relies on a conjecture, which concerns about the number of K-rational torsion points on an elliptic curve over number field K. Let $[K : \mathbf{Q}] = d$ and $\mathcal{E}$ an elliptic curve defined over K. The celebrated Uniform Boundedness Theorem asserts that the number of torsion points of E rational over K is bounded by a constant B_d depending only on d. Effective version of UBT shows that the bound B_d depends exponentially on d. We conjecture that for prime l, there exist a number field K and an elliptic curve $\mathcal{E}/K$ such that $[K : \mathbf{Q}] \leq \log^{O(1)} l$ and $\mathcal{E}$ has non-trivial K-rational l-torsion points. In addition, we assume (1) $\mathcal{E}$ has multiplicative reduction E' at a place above p; (2) All the K-rational l-torsions reduce to non-singular points on E'; (3) all the torsion points has reasonable size. We can efficently map the elements in the l-part of the $\mathbf{F}_p^*$ to the points on E'. Suppose that p-adic representation of $\mathcal{E}$ of certain precision is given, we then find the p-adic representations of the coresponding points on $\mathcal{E}$ upto to the precision. Since the degree of K is low, we employ **LLL**-algorithm to get the minimum polynomials of the coordinates of the points. There certainly exists another prime r, such that $l|r-1$ and $\mathcal{E}$ has good reduction E'' at a place above r. The coordinates of the coresponding points on E'' will be computed. The Tate-pairing on the l-part of $E'(\mathbf{F}_r)$ is non-trivial and is efficently computable. Hence we have the reduction of DDH problem in finite fields to DDH problem over special elliptic curves.

This paper is organized as follows. First, we introduce the reduction from the multiplicative group of a finite field to the additive group on a singular cubic curve. In the section 4, we formulate the conjecture. In the section 5, we describe an algorithm to lift the points over finite field to torsions over number field. In the section 6, we discuss the idea of shifting prime and prove the main theorem. In the section 7, we apply the idea to elliptic curve discrete logarithm problem.

2   Non-uniform algorithm

In this paper, when we talk about non-uniform polynomial time algorithm to solve DDH problem, we mean that given a cyclic group G with generator g, there exists a circuit C_G, depending on G, such that if the input of the circuit is g^x, g^y, g^z, output 1 iff $z \equiv xy \pmod{|G|}$, and the size of the circuit C_G grows polynomially with $\log |G|$. If we know how to construct C_G efficiently, then we have a uniform polynomial time algorithm to solve DDH problem.

Perhaps the best known non-uniform algorithm in cryptography is the reduction from DH problem to DL problem, proposed by Maurer and Wolf[10]. Given an arbitrary finite field $\mathbf{F}_p$, it is not known how to construct an elliptic curve over $\mathbf{F}_p$ with sufficiently smooth order. Sometimes, even the existence of such a curve is in question.

3   Reduction from Finite Field to Singular Cubic Curve

We first fix some notations. Suppose we are given a prime power q, $g \in \mathbf{F}_q^*$ generate a subgroup S. Assume that $l = |S|$ is a prime. $a, b, c \in S$. Certainly there exist three integers x, y and z such that $a = g^x, b = g^y, c = g^z$. We want to determine whether $xy \equiv z \bmod l$. This is called the Decision Diffie-Hellman(DDH) problem in a finite field.

There is an analogy problem in elliptic curve cryptography. Given a curve E defined over $\mathbf{F}_q$. W.l.o.g., assume that the order of $E(\mathbf{F}_q)$ is a prime l. Let G be a generator of $E(\mathbf{F}_q)$. $A, B, C \in E(\mathbf{F}_q)$. There exist three integers x, y and z such that $A = xG$, $B = yG$ and $C = zG$. The DDH problem is to determine whether $xy \equiv z \bmod l$.

It is believed that in general the DDH problem over an elliptic curve is harder than the DDH problem in a finite field if the groups have the same size. However, in some special case, most notably when the Tate-pairing is non-trivial and is easy to compute, DDH problem over the elliptic curve admits polynomial time algorithm.

In this paper, we study how to reduce the DDH problem in a finite field to the DDH problem over some special elliptic curve. It turns out that the isomorphism from the multiplicative group of a finite field to the additive group of a cubic curve is well-known, and is efficiently computable, as shown in the following proposition. It is not however scientifically interesting because the cubic curve is singular, hence the Tate-pairing is trivial.

Proposition 1. *Let K be a finite field and E/K be a curve given by*

$$y^2 + a_1 xy + a_3 y = x^3 + a_2 x^2 + a_4 + a_6$$

with discriminant $\Delta = 0$. Suppose E has a node S, and let

$$y = \alpha_1 x + \beta_1 \quad \text{and} \quad y = \alpha_2 x + \beta_2$$

be the two distinct tangent lines to E at S. The the map ϕ

$$E_{ns} \to \bar{K}^*$$
$$(x, y) \to \frac{y - \alpha_1 x - \beta_1}{y - \alpha_2 x - \beta_2}$$

is an isomorphism of abelian groups, which can be computed efficiently. The reverse map can also be computed efficiently.

4   Conjecture

We can imagine that there is a global elliptic curve, which has a multiplicative reduction at p or a place above p. The Tate pairing on this global curve is non-trivial. By lifting points from the singular cubic curve to the global curve, we will reduce the DDH problem in finite field to the DDH problem over the global elliptic curve.

The approach sounds appealing. But we are required to lift points over finite field to torsion points over number field. The group order is usually very large in cryptography application. One of the obstacles is to describe the global curve, since the curve, as well as the torsions, may be defined over a number field of very high degree. Fortunately, the current research seems to indicate that the maximum possible number of torsions over a number field grows exponentially with the degree of the number field.

Even if the curve and its torsions are defined over a number field with low degree, the size of the curve is too large to even be written down. We can, however, write down the p-adic version of this curve up to the certain precision.

The biggest obstacle is that some of its l-torsions may have huge size too. Notice that if E is represented by Weiestrass equation $y^2 = x^3 + ax + b$, $a, b \in \mathbf{Z}_K$, then the l-torsions have very large coordinates. But at least all the y-coordinates share a lot of common factors. We could hope that the Weiestrass equation will be birationally equivalent to an equation with all torsions in reasonable size.

Our expectation of the curve are summarized in the conjecture 1.

Definition 1. *The height of an integer n is defined to be $\log |n|$. The height of a rational number $\frac{n}{m}$, $n, m \in \mathbf{Z}$, $(n, m) = 1$, is defined to be $\log |n| + \log |m|$.*

Conjecture 1. There is a constant c, such that for any given prime p, there exists an elliptic curves $\mathcal{E}/K$, $[K : Q] \leq (\log p)^c$, $K = \mathbf{Q}[x]/(k(x)) = \mathbf{Q}(\alpha)$, and

1. The heights of all the coefficients of $k(x)$ are bounded by $(\log p)^c$. $k(x) \in \mathbf{Z}[x]$ is separable over $\mathbf{F}_p$ and it splits completely over $\mathbf{F}_p$.

2. There are l-torsion points $P_1, P_2, ..., P_l = 0 \in \mathcal{E}(K)$, where l is the largest prime divisor of $p - 1$ (we assume that $l^2 \ /\!\!\!| \ p - 1$). They form a group. $\mathcal{E}$ has a multiplicative reduction at place v above p. all the P_i's reduce to non-singular points.
3. $\mathcal{E}$ can be represented by an equation

$$f(x, y) = 0$$

such that for any $1 \leq i \leq l$, if $P_i = (x_i, y_i)$, the minimum polynomials in $\mathbf{Q}[x]$ for x_i and y_i are $g_i(x)$ and $h_i(x)$, then the height of all the coefficients of g_i and h_i are bounded by $(\log p)^c$.

We would like to mention several very deep results proved by Mazur[11], Kamienny[9], Merel[12], Parent[14], Hindry and Silverman[7] respectively about the torsions on the elliptic curves. It is these results that motivate us to formulate the conjecture.

Proposition 2. *Let $\mathcal{E}$ be an elliptic curve over a number field K. We denote the order of the torsion subgroup of $\mathcal{E}(K)$ by N. Let $d = [K : \mathbf{Q}]$. Suppose that p is a prime divisor of N, and p^n the largest power of p dividing N*

1. *If $d = 1$, the torsion subgroup of $\mathcal{E}(\mathbf{Q})$, is isomorphic to one of the following groups: $\mathbf{Z}/m\mathbf{Z}$ for $m \leq 10$, or $\mathbf{Z}/2\mathbf{Z} \times \mathbf{Z}/2m\mathbf{Z}$ for $m \leq 4$.*
2. *If $d = 2$, there is a positive integer B such that $N \leq B$. Moreover, $p \in \{2, 3, 5, 7, 11, 13\}$.*
3. *We have*

$$p \leq (1 + 3^{d/2})^2$$
$$p^n \leq 65(3^d - 1)(2d)^6.$$

4. *If the curve $\mathcal{E}$ has good reduction everywhere, then the number of K-rational torsion points of $\mathcal{E}$ is bounded by $1\,977\,408d \log d$*

5   Torsion-to-Torsion Lift

In this section, we describe an algorithm to lift the points over finite field (which are certainly torsions) to torsion points over number field. We use the same notation as in the conjecture section.

First we fix one embedding:

$$K \to \mathbf{Q}_p,$$

by mapping α to one of p-adic roots of $k(x)$. If a p-adic number a is an image of $x \in K$ by this embedding, then the minimum polynomial over $\mathbf{Q}$ of x can be computed in polynomial time on $\log p$ and the size of the minimum polynomial. In the computing, we only need know the first m digits of a, where m is also bounded by polynomial on $\log p$ and the size of the minimum polynomial.

Let
$$\tilde{f}(x,y) = 0$$
be p-adic representation of $f(x,y) = 0$ upto m digits, where m will be set later. Let $F(x,y)$ be the curve on $\tilde{f}$ mod p. $F(x,y) = 0$ is a singular cubic curve with a node. Suppose we want to solve the DDH problem in subgroup S of $\mathbf{F}_p^*$ with order l, where $l|p-1$ is a prime. For any point $Q = (a,b) = \phi^{-1}(x)$ on F,
$$lQ = 0 \bmod p.$$

There is a l-torsion point $P = (x,y)$ on f which will reduce to Q. We first find out the p-adic representation of P.

Let $P = (a + a_1 p + a_2 p^2 + \cdots, b + b_1 p + b_2 p^2 + \cdots)$. From
$$l(a + a_1 p, b + b_1 p) = 0 \bmod p^2,$$
by using squaring technique, combined with the curve equation,
$$\tilde{f}(a + a_1 p, b + b_1 p) = 0 \pmod{p^2},$$
we can solve a_1 and b_1 efficiently. Note that a_1 or b_1 only occur in linear terms. Similarly, from
$$\begin{cases} l(a + a_1 p + a_2 p^2, b + b_1 p + b_2 p^2) = 0 \pmod{p^3} \\ \tilde{f}(a + a_1 p + a_2 p^2, b + b_1 p + b_2 p^2) = 0 \pmod{p^3} \end{cases}$$
we can get a_2 and b_2. Generalize this idea, we will obtain the p-adic representation up to m digital in time $(m \log p)^{O(1)}$.

On the other hand, we know that the minimum polynomials of x and y have coefficients whose sizes are bounded by $(\log p)^c$. Those polynomial, denoted by g and h respectively, can be computed using **LLL**-algorithm when m is big enough. By factoring g and h over K, we get the representation of x and y as elements in K.

6   Shifting Prime

There certainly exists a prime r, satisfying

1. $l|r-1$ and $l^2 \nmid r-1$.
2. There is a place u over r with degree 1 in K.
3. The reduction of f at place u is non-singular, hence it is an elliptic curve.

Fix an embedding:
$$K \to \mathbf{Q}_r.$$

Let $t(x,y) = 0$ be the reduction of f at place u. All the l-torsion points will reduce to a $\mathbf{F}_r$-rational points on g. They form a subgroup in $t(\mathbf{F}_r)$. The key observation here is that the calculate of Tate-pairing on this subgroup is non-trivial and efficient. This concludes the whole reduction, which can be illustrated by following picture.

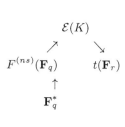

Theorem 1. *Adopt the notation in the conjecture 1. Assume that the conjecture is true. Given the p-adic representation of f upto to m digits and the reduction of f modulo the place u, there is a polynomial time algorithm to solve DDH problem in the l-subgroup of $\mathbf{F}_p^*$.*

The algorithm is summarized in the following:

Non-uniform information: p; The p-adic representation of $\mathcal{E}$ upto to m digits; Prime r; The reduction of f at place u.

Input: $a, b, c \in \mathbf{F}_p$.

1. From p-adic representation of $\mathcal{E}$, compute $E' = \mathcal{E} \bmod p$. We get a singular cubic curve with a node.
2. Compute $A' = \phi^{-1}(a)$, $B' = \phi^{-1}(b)$, $C' = \phi^{-1}(c)$ on E'.
3. Fix $K \to \mathbf{Q}_p$.
4. Lift A', B' and C' to A, B and C on $\mathcal{E}$. We compute the p-adic representations of A, B and C first. Then we use **LLL**-algorithm to compute $A, B, C \in K^2$.
5. Fix $K \to \mathbf{Q}_r$. Compute the reduction of A, B and C on E''. Denote them by A'', B'' and C''.
6. Solve DDH problem of A'', B'' and C''. Output the answer.

Here we note that, since the prime l satisfies that $l^2 \nmid r - 1$, so we can evaluate the l-th Tate-pairing $\langle A'', A'' \rangle_l \neq 1$ (See [8,5]). That is, we can solve the DDH problem A'', B'' and C''.

The algorithm is non-uniform, because for every p, we need the p-adic representation of f, the special prime r and the reduction of f modulo the place u. We don't know how to compute these parameters at this time.

7 Application to the Elliptic Curve Discrete Logarithm

Let $\mathcal{E}$ be a curve defined over a number field K. $\mathcal{E}_{\mathrm{tor}}(K)$ has order l. For simplicity, we assume that l is a prime. Assume that $\mathcal{E}$ has good reduction $E'/\mathbf{F}_q$ at place v. W.l.o.g, we assume that $|E'(\mathbf{F}_q)| = l$.

Corollary 1. *The discrete logarithm on $E'(\mathbf{F}_q)$ has nonuniform subexpontial algorithm if there is a constant c such that,*

1. *$K = \mathbf{Q}[x]/k(x) = \mathbf{Q}[\alpha]$, $[K : Q] \leq (\log p)^c$. The heights of all the coefficients of $k(x)$ are bounded by $(\log p)^c$. $k(x) \in \mathbf{Z}[x]$ is separable and it splits completely over $\mathbf{F}_p$.*

2. *There are l-torsion points $P_1, P_2, ..., P_l = 0 \in \mathcal{E}(K)$. They form a group.*
3. *$\mathcal{E}$ can be represented by an equation*

$$f(x, y) = 0$$

such that for any $1 \leq i \leq l$, if $P_i = (x_i, y_i)$, the minimum polynomials in $\mathbf{Q}[x]$ for x_i and y_1 are $g_i(x) =$ and $h_i(x) = 0$, then the height of all the coefficients of g_i and h_1 are bounded by $(\log p)^c$.

Sketch of the proof: If the conditions in the corollary hold, then there must exist a prime p such that $r|p - 1$ and the reduction of $\mathcal{E}$, denoted by $E''/\mathbf{F}_p$, at place u over p is non-singular. Because $\mathcal{E}$ has a torsion group over K, $E''(\mathbf{F}_p)$ has l-part subgroup and the Tate-pairing of elements in the subgroup can be efficiently computed. We can apply FR-algorithm here [5]. Hence the discrete logarithm over the r-part of $E'(\mathbf{F}_q)$ has non-uniform subexpontial time solution.

8   Discussion

In conjecture 1, we don't require that the size of curve $\mathcal{E}$ is small. In fact, it is impossible. Since curve $\mathcal{E}$ has a large torsion group of prime order, it must have bad reductions at a lot of small primes. The j-invariant of $\mathcal{E}$ will have very large height, hence the plain representation of $\mathcal{E}$ will take huge amount of space.

The p-adic representation of $\mathcal{E}$ is not complete, in the sense that we only have the representation up to the precision of polynomial bound $(\log p)^{O(1)}$. However, since the size of torsions are assumed to be small, we will get complete information of torsions.

Acknowledgements

We thank Dr. Ming-Deh Huang, Dr. Sheldon Kamienny and Dr. Len Adleman for helpful discussions.

References

1. D. Boneh, "The Decisional Diffie-Hellman Problem," *Proc. of ANTS-IV*, LNCS 1423, Springer-Verlag, pp.48–63, 1998.
2. R. Cramer and V. Shoup, "A Practical Public Key Cryptosystem Provably Secure against Adaptive Chosen Ciphertext Attack," *Proc. of Crypto'98*, LNCS 1462, Springer-Verlag, pp.13–25, 1998.
3. W. Diffie and M.E. Hellman, "New Directions in Cryptography," *IEEE Transactions on Information Theory*, IT-22(6), pp.644–654, 1976.
4. T. ElGamal, "A public key cryptosystem and a signature scheme based on discrete logarithms," *IEEE Transactions on Information Theory*, 33, pp.469–472, 1985.
5. G. Frey, H.G. Ruck, A remark concerning m-divisibility and the discrete logarithm in the divisor class group of curves. *Mathematics of Computation*, Volume 62, Issue 206, 1994, 865-874.

6. S. Goldwasser and S. Micali, "Probabilistic Encryption," *Journal of Computer and System Sciences*, 28, pp.270–299, 1984.

7. M. Hindry, J. Silverman, Sur le nombre de points de torsion rationnels sur une courbe elliptique. (French. English, French summary) [On the number of rational torsion points on an elliptic curve] C. R. Acad. Sci. Paris Ser. I Math. 329 (1999), no. 2, 97–100.

8. A. Joux and K. Nguyen, "Separating Decisional Diffie-Hellman from Diffie-Hellman in cryptographic groups," *Cryptology ePrint Archive: Report 2001/003*, http://eprint.org/2001/003

9. S. Kamienny, Torsion points on elliptic curves, *Bull. Amer. Math. Soc. (N.S.)* 23 (1990), no. 2, 371–373.

10. U. Maurer and S. Wolf, "The relationship between breaking the Diffie-Hellman protocol and computing discrete logarithms," *SIAM J. Comput.*, 28(5), pp.1689–1731, 1999.

11. B. Mazur, Rational points on modular curves, In: Modular Functions of One Variable, V, *Lecture Notes in Mathematics,* Vol. 601. New York: Spring-Verlag, 1976.

12. L. Merel, Bornes pour la torsion des courbes elliptiques sur les corps de nombres. (French) [Bounds for the torsion of elliptic curves over number fields] *Invent. Math.* 124 (1996), no. 1-3, 437–449.

13. M. Naor and O. Reingold, "Number theoretic constructions of efficient pseudo random functions," *Proc. FOCS'97*, pp.458–467, 1997.

14. P. Parent, Bornes effectives pour la torsion des courbes elliptiques sur les corps de nombres. (French. French summary) [Effective bounds for the torsion of elliptic curves over number fields] J. Reine Angew. Math. 506 (1999), 85–116.

15. R. Rivest, A. Shamir and L.M. Adleman, "A Method for Obtaining Digital Signatures and Public Key Cryptosystems," *Communications of the ACM*, 21(2), pp.120–126, 1978.

16. V. Shoup, "Lower bounds for discrete logarithms and related problems," *Proc. of Eurocrypto'97*, LNCS 1233, Springer-Verlag, pp.256–266, 1997.

17. Y. Tsiounis and M. Yung, "On the security of ElGamal based encryption," *Proc. of PKC'98*, LNCS 1431, Springer-Verlag, pp.117–134, 1998.

Secure Key-Evolving Protocols for Discrete Logarithm Schemes

Cheng-Fen Lu and ShiuhPyng Winston Shieh[*]

Computer Science and Information Engineering Department
National Chiao Tung University, Taiwan 30050
{cflu,ssp}@csie.nctu.edu.tw

Abstract. This paper addresses the security and efficiency of key-evolving protocols. We identify forward-secrecy and backward-secrecy as the security goals for key-evolving and present two protocols to achieve these goals. The first protocol is operated in Z_p^* and is efficient for the secret-key holder; the second one is operated in Z_n^*, and is efficient for the public-key holder. For both protocols, we provide proofs and analysis for correctness, security and efficiency.

Keywords: Key Management, Key-Evolving, Forward-Secrecy, Backward-Secrecy.

1 Introduction

Over the past 20 years, public key cryptography has provided many signature schemes and public key cryptosystems. However, if a signing key or decryption key is compromised, it is regarded a total break of the system. To avoid this undesirable situation, one common practice is assigning each key a certain usage period and updating the key when entering a new period. Therefore, the security and efficiency of key updating or key-evolving becomes an important topic for key management [11].

Key-evolving is commonly employed for symmetric cryptosystems, where the sender and the receiver of a message share a common master key. Instead of using this shared key directly, they use it to derive a set of sub-keys and each sub-key is valid for a certain period or for a specific application [1].

In this paper, we focus on the key-evolving of asymmetric key systems, where the communication parties no longer possess the same master key. Instead, the secret key holder (the decryptor or the signer) generates the public and secret key base and the public key holder (the encryptor or the verifier) retrieves this public key base. In the beginning, the public key base is first signed by a Certificate Authority (CA). And then it is retrieved and verified by the public key holder.

Without a key-evolving protocol, the certificate retrieval and verification operations must be performed periodically. With a key-evolving protocol, the certificate retrieval and verification operation is performed only once for the public

---

[*] This work is supported in part by Ministry of Education, National Science Council of Taiwan, and Lee & MTI Center, National Chiao Tung University.

B. Preneel (Ed.): CT-RSA 2002, LNCS 2271, pp. 300–309, 2002.
© Springer-Verlag Berlin Heidelberg 2002

key base. Thereafter, both parties update their corresponding public or secret keys for the future periods.

The goal of secure key-evolving is to reduce the damage in case of a secret key compromise. The corresponding security notions have been investigated in papers of key management and key agreement [12,8,14]. There the properties such as forward-secrecy, backward-secrecy, key-independence were defined. Informally, forward-secrecy refers to that the compromise of one or several secret keys does not compromise previous secret keys. Likewise, backward-secrecy refers to that the compromise of one or several secret keys does not compromise future secret keys. Key-independence means the secret keys used in different periods are basically independent. Thus, even if the attacker finds out the secret key of a certain period, it gives him little advantage in finding the secret keys of other periods.

One related concept of the forward-secrecy is the *perfect forward-secrecy*, which assures that the "compromise of long-term keys does not compromise past session keys" [7,11]. Also in recent papers of *forward-secure* signature, signatures schemes with forward-secrecy properties were proposed [4,9]. The related concept of the backward-secrecy is the resistance to the *known-key attack*, which assures the compromise of past sessions keys will allow neither a passive adversary to compromise future session keys nor an active adversary to impersonate in the future [5,17,16,11]. Recently, Tzeng proposed two public key encryption schemes which also enjoy the property of key-independence but they require the help of the extra trusted agent (TA) in initial key distribution and at each key-evolving [15].

The security notions for the key-evolving protocol have been little investigated. Thus, we redefined the above security notions in the key-evolving setting. Besides, we presented and analyzed two key-evolving protocols for discrete logarithm schemes. The advantage of these key-evolving protocols is that no trusted agent is needed and they are applicable for both public key encryption schemes and signature schemes.

This paper is organized as follows. Section 2 describes the model and definitions. Section 3 presents the protocol based on the difficulty of computing discrete logarithm in Z_p^*. Section 4 presents the protocol based on the difficulty of factoring a large composite n. Section 5 discusses the possible extensions of the protocols. Section 6 concludes this paper.

2   Model and Definitions

Let (PKB, SKB) denote the pair of the public/secret key base, which is used to derive the key-pair in $i - th$ period, (PK_i, SK_i).

Definition 1 *A key-evolving protocol consists of three algorithms (KG, g, f):*

1. *Public/Secret Key Base Generation Algorithm $KG(k)$*
   Given a security parameter k, the secret key holder generates the public/ secret key base (PKB, SKB). SKB is kept secret at the secret key holder and PKB is then distributed to the users in the form of public certificate.

2. *Public Key-Evolving Algorithm $g()$*
   Given the period i and PKB, the public key holder computes $PK_i = g(PKB, i)$.
3. *Secret Key-Evolving Algorithm $f()$*
   Given the period i and SKB, the secret key holder computes $SK_i = f(SK_{i-1})$ or $f(SKB, i)$.

Next, some desirable properties of the key-evolving protocols are presented as follows:

Definition 2 *A key-evolving protocol is forward-secret if the compromise of SK_i will not compromise SK_j for all $j < i$.*

Definition 3 *A key-evolving protocol is backward-secret if the compromise of SK_i will not compromise SK_j for all $j > i$.*

Definition 4 *A key-evolving protocol is key-independent if it is forward-secret and backward-secret.*

Definition 5 *A key-evolving protocol is t-bounded key-independent if*

1. *The compromise of a set of secret keys C and $|C| > t$ will compromise all the secret keys.*
2. *The compromise of a set of secret keys C and $|C| \leq t$ does not help to compromise additional secret keys.*

3   Protocol 1

The first protocol is based on the difficulty of computing discrete logarithm in Z_p^*. It applies a technique of Feldman for the construction of a non-interactive verifiable secret sharing scheme [6]. This technique uses the secret sharing scheme of Shamir [13], where a polynomial $f(x)$ of degree t is employed. Polynomials have the following properties:

I. Given $t+1$ distinct points on the polynomial $f(x)$ degree t, namely (x_0, y_0), $(x_1, y_1), \cdots, (x_t, y_t)$ and $y_i = f(x_i)$, all the $t+1$ coefficients can be determined. In other words, the polynomial can be uniquely determined.
II. Given t distinct points on the polynomial $f(x)$ degree t, namely (x_1, y_1), $(x_2, y_2), \cdots, (x_t, y_t)$ and $y_i = f(x_i)$, the $t+1$ coefficients cannot be uniquely determined.

The Protocol is presented in Figure 1, consisting of three algorithms: First, the secret key holder executes the key base generation algorithm and publishes the public key base $PKB = (P_0, P_1, \cdots P_t) = (g^{a_0}, g^{a_1}, \cdots, g^{a_t})$, where $a_0, a_1, \cdots, a_t$ are random. In contrast, $SKB = (a_0, a_1, \cdots a_t)$, the set of coefficients of $f(x)$, is kept secret. Also he publishes a hash function h that works as a simple randomizer and takes the index of the period to a number in Z_q, i.e. $h(i) \in Z_q$. Depending on the application, this $h(\cdot)$ can be required to be collision-free or be a permutation in Z_q.

1. Public/Secret Key Base Generation Algorithm $KG(1^n, t)$

   (1) The secret key holder chooses a $n-$bit prime $p = 2q + 1$ with that q is a prime of at least 160-bits long, i.e. $q > 2^{160}$. Let G_q denote the subgroup of the quadratic residues modulo p and g the generator of G_q.

   (2) The secret key holder chooses a t-degree polynomial $f(x)$ with its coefficients randomly chosen from Z_q and $f(x) = \sum_{j=0}^{t}(a_j x^j) \pmod{q}$. $SKB = (a_0, a_1, \cdots, a_t)$, the coefficients of $f(x)$, will be kept secret and used to derive the secret keys for later periods.

   (3) The secret key holder publishes the public key base information and a hash function, including:

$$PKB = (P_0, P_1, \cdots P_t) = (g^{a_0}, g^{a_0}, \cdots, g^{a_t}),$$

$$\text{and } h : N \to Z_q.$$

2. Public Key-Evolving Algorithm

   Given the index i, the public key holder will update its public key of period i as follows:

$$PK_i = \prod_{j=0}^{t}(P_j)^{h(i)^j} \pmod{p}.$$

3. Secret Key-Evolving Algorithm

   Given the index i, the secret key holder will update the key of period i as:

$$SK_i = f(h(i)) \pmod{q}.$$

Fig. 1. Key-evolving protocol using Feldman's technique

The public key holder retrieves and verifies PKB and $h(\cdot)$. Suppose that he needs the public key of the period i. He would then compute $PK_i = \prod_{j=0}^{t}(P_j)^{h(i)^j}$ (mod p).

SKB is the long-term system secret, protected separately from SK_i at the secret-key holder. Its protection follows the paper of Shamir secret-sharing scheme [13]. At the beginning of the period i, the secret key holder evaluates the secret key as $SK_i = f(h(i)) \pmod{q}$. The analysis of this protocol is as follows:

Correctness. The relation of $PK_i = g^{SK_i}$ is established because

$$PK_i = \prod_{j=0}^{t}(P_j)^{h(i)^j} \pmod{p} = \prod_{j=0}^{t}(g^{a_j})^{h(i)^j} \pmod{p}$$

$$= g^{\sum_{j=0}^{t} a_j \cdot h(i)^j} \pmod{q} = g^{f(h(i))} \pmod{q} = g^{SK_i}.$$

Security. The following lemmas summarize the security property.

Lemma 1. *The ability of the attacker, who is able to compute the corresponding SK of a given PK, is equivalent to the solving of the Discrete Logarithm problem in Z_p^*.*

Proof: ($\Leftarrow$) This is trivial because $SK = \log_g PK$.
($\Rightarrow$) We can build a DL-oracle using this attacker. Suppose we want to find $x = \log_g y$. We choose $PK = yg^a$ and give it the attacker. If the attacker outputs the secret key SK corresponding to PK, we can compute $x = \log_g y = \log_g (PK/g^a) = \log_g PK - a = SK - a$. Q.E.D.

Conjecture 1 *The protocol in Figure 1 is t-bounded key-independent.*

Remark:
Given $t+1$ sets of keys of $(SK_{i_0}, i_0), (SK_{i_1}, i_1), \cdots, (SK_{i_t}, i_t)$, the attacker can determine the coefficients of $f(x)$ and thus the secret keys of all periods.

If less than $t+1$ secrets keys are compromised, then the polynomial for the secret key-evolving can not be unique determined, i.e. it has at least $q \geq 2^{160}$ free choices for the coefficients. Therefore, the only way to compute one extra secret key corresponding to a public key seems to compute the discrete logarithm in Z_p^*, which is intractable.

The interesting problems include (1) building a DL oracle using the successful attacker of breaking t-bounded key-independence property and (2) using the successful attacker to construct the oracle for Diffie-Hellman (DH) or Decisional Diffie-Hellman (DDH) problem.

Efficiency. For the secret key holder, this protocol is very efficient, requiring only the evaluation of a t-degree polynomial over Z_q. For the public key holder, this protocol requires t modular exponentiations and t modular multiplications. However, to reduce the on-line computation, the t modular exponentiations can be pre-computed.

4   Protocol 2

This protocol is based on the difficulty of computing discrete logarithms in Z_n^*, where n is the product of several large primes. It uses the same technique of Maurer-Yacobi in the design of non-interactive public-key distribution system [10]. Recently, this technique was also suggested by Anderson to build a forward-secret signature scheme [2]. All operations are performed in Z_n^*, where factoring n is hard and g is the element of the maximal order in Z_n^*. A lemma from Maurer and Yacobi showed that every square modulo n has a discrete logarithm to the base of g. And the knowledge of the factorization of n is the trapdoor for solving the discrete logarithm problem; namely,

1. For the secret key holder, with the trapdoor knowledge of the factors of n, he can compute $SK_i = log_g(PK_i)$.
2. For someone without this trapdoor knowledge, given PK_i, the computation of SK_i is intractable.

This protocol is presented in Figure 2, consisting of three algorithms.
In the beginning, the secret key holder first executes the key base generation

1. Public/Secret Key Base Generation Algorithm

   The secret key holder chooses a k-bit composite $n = p_1 p_2 \cdots p_r$ where the factorization of n is intractable and $(p_1 - 1)/2, (p_2 - 1)/2, \cdots, (p_r - 1)/2$ are pairwise relatively prime. Next, he chooses an element g of the maximal order in Z_n^*.

   Then he broadcasts the public key base $PKB = (n, g, H, PK_0)$, where the following requirements are met.

   (a) $SKB = (p_1, p_2, \cdots, p_r)$, the set of factors of n, is kept secret to the secret key holder,

   (b) $H(\cdot) : \{0,1\}^* \to Z_n^*$ is a cryptographic hash function,

   (c) PK_0 is a random number.

2. Public Key-Evolving Algorithm

   Given the period i, the public key holder evaluates for $j = 1$ to i,

   $$PK_j = (H(PK_{j-1}))^2 \pmod{n} \in \mathrm{QR}_n.$$

3. Secret Key-Evolving Algorithm

   In the beginning of period i, the secret key holder first compute the public key by one hash-and-square

   $$PK_i = (H(PK_{i-1}))^2 \pmod{n} \in \mathrm{QR}_n.$$

   With the knowledge of the factors of n, the secret key holder computes

   $$SK_i = log_g(PK_i).$$

Fig. 2. Key-Evolving based on Maurer-Yacobi Scheme

algorithm and publishes the public key base $PKB = (n, g, H, PK_0)$, where n and g are defined as above, H is a cryptographic hash function, and PK_0 is a random number. The secret key base SKB, consisting of the factors of n, is kept secret.

The public key holder retrieves and verifies the public key base PKB. Suppose that he needs the public key of the period i. He would then perform a series of hash-and-square operations to get $PK_1 = (H(PK_0))^2 \pmod{n}$, $PK_2 = (H(PK_1))^2 \pmod{n}, \cdots, PK_i = (H(PK_{i-1}))^2 \pmod{n}$. If he could store some of these intermediate values, he can reduce the hash-and-square computations.

SKB is the long-term system secret, protected separately from SK_i at the secret holder. Its protection follows the paper of Maurer and Yacobi [10]. At the beginning of the period i, the secret key holder performs the hash-and-square $PK_i = (H(PK_{i-1}))^2 \pmod{n}$ and then calculates the corresponding secret key $SK_i = log_g(PK_i)$ with his trapdoor knowledge. The analysis of this protocol is as follows:

Correctness. $PK_i = g^{SK_i}$ is established because SK_i is computed by the secret key holder using the factors of n (trapdoor information).

Security. We will show that the security of this protocol depends on the following relations between the computing discrete logarithm in Z_n^* and factoring n [3,10,11].

I. If the discrete logarithm problem in Z_n^* can be solved efficiently, then n can be factored efficiently.

II. If n can be factored efficiently and if the discrete logarithm in every $Z_p^*, p|n$ can be solved efficiently, then the discrete logarithm problem in Z_n^* can be solved efficiently.

The following lemmas summarize the security property. First, we show that successful attacker can be used to construct a DL oracle in Z_n^*. Next, we state a lemma from Maurer and Yacobi about the factorization of n using a DL-oracle [10]. For detailed algorithm, please refer to the original paper. Finally, we prove the security in the random oracle model.

Lemma 2. *The ability of the attacker, who is able to compute the corresponding SK of a given PK, is equivalent to the solving of the Discrete Logarithm problem over Z_n^*.*

Proof: ($\Leftarrow$) This is trivial because $SK = \log_g PK$.
($\Rightarrow$) We can build a DL-oracle using this attacker. Suppose we want to find $x = \log_g y$. We choose $PK = yg^a$ and give it the attacker. If the attacker outputs the secret key SK corresponding to PK, we can compute $x = \log_g y = \log_g (PK/g^a) = \log_g PK - a = SK - a$. Q.E.D.

Lemma 3. *[10] Let n be the product of distinct odd primes $p_1, \cdots, p_r$ and let g be primitive in each of the prime fields $GF(p_i)$ for $1 \le i \le r$. Then computing discrete logarithms modulo n to the base of g is at least as difficult as factoring n completely.*

Theorem 1. *In the random oracle model, the key-evolving protocol in Figure 2 is key-independent if factoring n is hard.*

Proof: We will show how to factor n if the successful attacker breaks the key-independent property. First, we will build a DL oracle using the successful attacker. Next, we follow the algorithm of Maurer-Yacobi to factor n using this DL oracle.

Given y in the maximal cyclic group of Z_n^*, we will use the successful attacker to compute $x = \log_g y$. First, we will generate a set of random numbers $\{r_i | i = 1, 2, \cdots, T+1\}$. And we control the random oracle to output $\{r_1, r_2, \cdots, r_{k-1}\}$ at the first $k-1$ queries. Thus, the following relations hold.

$$r_1 = H(PK_0), \qquad PK_1 = r_1^2$$
$$r_2 = H(PK_1), \qquad PK_2 = r_2^2$$
$$\vdots \qquad\qquad \vdots$$
$$r_{k-1} = H(PK_{k-2}), \qquad PK_{k-1} = r_{k-1}^2.$$

At the beginning of the k periods, we force the random oracle to output yg^a, i.e.

$$yg^a = H(PK_{k-1}), PK_k = y^2 g^{2a}.$$

For periods from $k + 1$ to $T + 1$, the random oracle proceeds as usual.

Second, we give (PK_i, SK_i) and the above public key set to the attacker A. If he successes to break forward-secrecy and forward-secrecy, A would return (PK_j, SK_j) for some $j \neq i$.

The probability that $j = k$ is $\frac{1}{T}$. If this happens, we have $SK_j = SK_k = \log_g PK_k = \log_g y^2 g^{2a} = 2\log_g y + 2a = 2x + 2a$. Because $PK_j = PK_k$ is a square in Z_n^*, SK_j will be even. Though the order $\lambda(n)$ is unknown, the discrete logarithm of y can be computed as follows:

$$x = \frac{1}{2} SK_j - a.$$

Given this DL oracle, we follow Maurer-Yacobi's method to factor n. Therefore, if factoring n is hard, this key-evolving protocol is key-independent in the random oracle model. Q.E.D.

Efficiency. This protocol is especially efficient for the public key holder because only i hash-and-square operations are needed and this can be pre-computed. Also if the public key holder stores some intermediate values, the load of the public key-evolving is further reduced. Thus it is suitable for public key holders with low computation power such as mobile devices or smart cards. The load of secret key holder is mainly the computation of discrete logarithm of $Z_p^*, p|n$. The computation of discrete logarithm relies on the choices of p [10]; for example, one can employ Pohlig-Hellman algorithm when $p|n, (p-1)/2$ is chosen to be the product of primes of moderate size.

5   Possible Extensions

The possible extensions of the protocols include: First, the immediate problem is to prove the conjecture 1 using stronger assumption such as DH or DDH assumption or using DL assumption in a restricted model.

Second, can we provide a public-key holder efficient protocol in Z_p^* instead of in Z_n^*? And is it possible to use this key-evolving model for specific algebraic structure or RSA related schemes?

Third, since each DL key pair is associated with a period, it will be desirable to provide period-stamping or time-stamping service of the public key. Furthermore, applying these two party protocols to multi-party computation would be an interesting task.

6   Conclusion

We first discussed about the importance of key-evolving and then identified several security notions, namely forward-secrecy, backward-secrecy, and key independence. Next, two concrete constructions were proposed. One protocol is based on a technique of Feldman, and it is efficient for the secret-key holder; the other one is a variant of the Maurer-Yacobi protocol, and it is efficient for the public-key holder. Besides, we provide proofs and analysis for correctness, security and efficiency. Finally, we pointed out the possible extensions for future work.

References

1. M. Abdalla and M. Bellare. Increasing the lifetime of a key: a comparative analysis of the security of re-keying techniques. In T. Okamoto, editor, *Advances in Cryptology – ASIACRYPT ' 2000*, Kyoto, Japan, 2000.
2. R.J. Anderson. Two remarks on public key cryptology. In *Rump Session Eurocrypt'97*.
3. E. Bach. Discrete logarithm and factoring. *Report no. UCB/CSD 84/186, Comp. Sc. Division (EECS), University of California, Berkeley*, June 1984.
4. M. Bellare and S. K. Miner. A forward-secure digital signature scheme. In *Proc. 19th International Advances in Cryptology Conference – CRYPTO '99*, pages 431–448, 1999.
5. D. E. Denning and M. S. Sacco. Timestamps in key distribution protocols. *Communications of the ACM*, 24(7):533–536, 1981.
6. P. Feldman. A practical scheme for non-interactive verifiable secret sharing. In *28th Symposium on Foundations of Computer Science (FOCS)*, pages 427–437. IEEE Computer Society Press, 1987.
7. C. G. Guenther. An identity-based key-exchange protocol. In Jean-Jacques Quisquater and Joos Vandewalle, editors, *Advances in Cryptology - EuroCrypt '89*, pages 29–37, Berlin, 1989. Springer-Verlag. Lecture Notes in Computer Science Volume 434.
8. Y. Kim, A. Perrig, and G. Tsudik. Simple and fault-tolerant key agreement for dynamic collaborative groups. In *Proceedings of the 7th ACM conference on Computer and communications security (CCS-00)*, pages 235–244. ACM Press, 2000.
9. H. Krawczyk. Simple forward-secure signatures from any signature scheme. In Sushil Jajodia and Pierangela Samarati, editors, *Proceedings of the 7th ACM Conference on Computer and Communications Security (CCS-00)*, pages 108–115. ACM Press, 2000.
10. U. M. Maurer and Y. Yacobi. A non-interactive public-key distribution system. *Designs, Codes and Cryptography*, vol. 9, no. 3:305–316, 1996.
11. A. J. Menezes, P. C. van Oorschot, and S. A. Vanstone. *Handbook of Applied Cryptography*. Boca Raton, 1997.
12. A. Perrig. Efficient collaborative key management protocols for secure autonomous group communication. In *International Workshop on Cryptographic Techniques and E-Commerce (CrypTEC '99)*, July 1999.
13. A. Shamir. How to share a secret. *Communication of ACM*, pages 612–613, (Nov. 1979).
14. M. Steiner, G. Tsudik, and M. Waidner. Key agreement in dynamic peer groups. *IEEE Transactions on Parallel and Distributed Systems*, 11(8):769–780, August 2000.
15. W. Tzeng and Z. Tzeng. Robust key-evolving public key encryption schemes. Record 2001/009, Cryptology ePrint Archive, 2001.

16. Y. Yacobi. A key distribution s "paradox". In Alfred J. Menezes and Scott A. Vanstone, editors, *Advances in Cryptology - Crypto '90*, pages 268–273, Berlin, 1990. Springer-Verlag. Lecture Notes in Computer Science Volume 537.
17. Y. Yacobi and Z. Shmuely. On key distribution systems. In Gilles Brassard, editor, *Advances in Cryptology - Crypto '89*, pages 344–355, Berlin, 1989. Springer-Verlag. Lecture Notes in Computer Science Volume 435.

Author Index

Lecture Notes in Computer Science

For information about Vols. 1–2190
please contact your bookseller or Springer-Verlag